Lecture Notes in Computer Science　　10054

Commenced Publication in 1973
Founding and Former Series Editors:
Gerhard Goos, Juris Hartmanis, and Jan van Leeuwen

FoLLI Publications on Logic, Language and Information

Subline of Lectures Notes in Computer Science

More information about this series at http://www.springer.com/series/7407

Maxime Amblard · Philippe de Groote
Sylvain Pogodalla · Christian Retoré (Eds.)

Logical Aspects of Computational Linguistics

Celebrating 20 Years of LACL (1996–2016)

9th International Conference, LACL 2016
Nancy, France, December 5–7, 2016
Proceedings

 Springer

Editors
Maxime Amblard
Campus Scientifique
LORIA (UMR 7503) – Sémagramme
  Campus Scientifique
Vandœuvre-lès-Nancy
France

Philippe de Groote
Inria Nancy
Villers-lès-Nancy
France

Sylvain Pogodalla
Inria Nancy
Villers-lès-Nancy
France

Christian Retoré
LIRMM, Université de Montpellier
Montpellier
France

ISSN 0302-9743   ISSN 1611-3349 (electronic)
Lecture Notes in Computer Science
ISBN 978-3-662-53825-8   ISBN 978-3-662-53826-5   (eBook)
DOI 10.1007/978-3-662-53826-5

Library of Congress Control Number: 2016956486

LNCS Sublibrary: SL1 – Theoretical Computer Science and General Issues

Printed on acid-free paper

This Springer imprint is published by Springer Nature
The registered company is Springer-Verlag GmbH Germany
The registered company address is: Heidelberger Platz 3, 14197 Berlin, Germany

This 20th anniversary LACL volume is dedicated to the memory of
Alexander Dikovsky, chair of LACL 2012,
and of
Joachim Lambek, invited speaker at LACL 1997,
who both passed away in 2014.

Foreword

The ninth edition of LACL (Logical Aspect of Computational Linguistics) that took place in 2016 in Nancy, marked the 20th anniversary of the conference.

The first edition of the conference that I launched in 1996 was also held in Nancy, and then as now, I had the pleasure of coorganizing it with my friend Philippe de Groote, who at that time headed our Calligramme research group.

What a pleasure to see that, 20 years later, there is still a need for such a conference on the relation between formal logic and computational linguistics, or perhaps we should say between formal linguistics and computational logic—since nowadays "computational linguistics" hardly evokes "logic". Not only did the LACL field keep on developing since 1996, but today it encompasses new triggering questions between logic, linguistics and computer science that we had no idea in 1996. This anniversary took place at LORIA, which was the first venue not only for that reason: LORIA includes an important department of computational linguistics where the two other editors of this volume, Maxime Amblard and Sylvain Pogodalla, former PhD students of mine, hold a position in Philippe de Groote's Sémagramme research group, the continuation of Calligramme.

I would like to thank Philippe, Sylvain, and Maxime, in the order with which they started making contributions to the Logical Aspects of Computational Linguistics, for organising in Nancy this beautiful LACL anniversary.

September 2016

Christian Retoré
Program Chair of LACL 1996 and 2016
(first and ninth editions)

Preface

We are pleased to provide the proceedings of the 9th International Conference on Logical Aspects of Computational Linguistics, LACL 2016, which was held in Nancy, France, during December 5–7, 2016. LACL aims to be a forum for the exchange of ideas involving all aspects of formal logic within computational linguistics, from syntactic parsing to formal semantics and discourse interpretation.

Previous LACL conferences where held in Nancy (1996, 1997), Grenoble (1998), Le Croisic (2001), Bordeaux (2005), Montpellier (2011), Nantes (2012), and Toulouse (2014).

The proceedings of this ninth edition comprise four invited contributions, by Maria Aloni (Universiteit van Amsterdam, The Netherlands), Johan Bos (Rijksuniversiteit Groningen, The Netherlands), Shalom Lappin (Göteborgs Universitet, Sweden), and Louise McNally (Universita Pompeu Frabra, Barcelona, Spain), 19 contributed papers, and six short abstracts selected from 39 submissions. Each paper received three reviews, and sometimes more, provided by the Program Committee and additional reviewers, listed herein.

We would like to thank all those who submitted papers for consideration at LACL, the four invited speakers, and all conference participants. We want to thank our international team of reviewers, who often gave extensive comments to authors. We very much hope that these comments will be of use to those who submitted papers for their future research.

We are also grateful to our institutional sponsors and supporters: the Association for Logic, Language and Information (FoLLI), the computer science laboratory in Nancy (LORIA), the French National Institute for Computer Science and Applied Mathematics (Inria), the National Center for Scientific Research (CNRS), the University of Lorraine, the Région Lorraine, and the Communauté Urbaine du Grand Nancy. We would also like to express our gratitude to the Organizing Committee and all the people whose efforts made this meeting possible.

September 2016

Maxime Amblard
Philippe de Groote
Sylvain Pogodalla
Christian Retoré

Organization

Organizing Committee

Maxime Amblard   LORIA, Université de Lorraine, Nancy, France
Anne-Lise Charbonnier   Inria Nancy, France
Philippe de Groote   LORIA, Inria Nancy, France
Sylvain Pogodalla   LORIA, Inria Nancy, France

Program Committee

Maxime Amblard   LORIA, Université de Lorraine, Nancy, France
Nicholas Asher   IRIT, CNRS and Université Paul Sabatier, Toulouse, France
Denis Béchet   LINA, University of Nantes, France
Daisuke Bekki   Ochanomizu University, Japan
Raffaella Bernardi   University of Trento, Italy
Gemma Boleda   Universitat Pompeu Fabra, Barcelona, Spain
Heather Burnett   CNRS, Université de Paris 7, France
Wojciech Buszkowski   Adam Mickiewicz University, Poznań, Poland
Stergios Chatzikyriakidis University of Gothenburg, Sweden
Robin Cooper   University of Gothenburg, Sweden
Philippe de Groote   LORIA, Inria Nancy, France
Valeria De Paiva   Nuance Communications, Cupertino, USA
Markus Egg   Humboldt-Universität Berlin, Germany
Annie Foret   IRISA, University of Rennes 1, France
Nissim Francez   Technion, Haifa, Israel
Makoto Kanazawa   National Institute of Informatics, Japan
Greg Kobele   University of Chicago, USA
Marcus Kracht   Universität Bielefeld, Germany
Hans Leiß   Universität München, Germany
Robert Levine   The Ohio State University, USA
Zhaohui Luo   Royal Holloway College, University of London, UK
Alda Mari   IJN, CNRS, Paris and University of Chicago, USA
Michael Moortgat   Utrecht Institute of Linguistics - OTS, The Netherlands
Richard Moot   CNRS (LaBRI) and Bordeaux University, France
Glyn Morrill   Universitat Politècnica de Catalunya, Spain
Larry Moss   University of Indiana, USA
Sylvain Pogodalla   LORIA, Inria Lorraine, France
Carl Pollard   The Ohio State University, USA
Jean-Philippe Prost   LIRMM, Université Montpellier 2, France

Myriam Quatrini   Institut de Mathématiques de Luminy, Aix-Marseille
                                   Université, France
Christian Retoré   LIRMM, Université de Montpellier, France
Mehrnoosh Sadrzadeh   Queen Mary College, University of London, UK
Serguei Soloviev   IRIT, Université de Toulouse-3, France
Stephanie Solt   Zentrum für Allgemeine Sprachwissenschaft, Berlin,
                                   Germany
Edward Stabler   UCLA and Nuance Communications, Cupertino, USA
Mark Steedman   University of Edinburgh, UK
Jakub Szymanik   University of Amsterdam, The Netherlands
Isabelle Tellier   Lattice, Université Paris 3, France
Laure Vieu   IRIT, CNRS, Toulouse, France
Marek Zawadowski   University of Warsaw, Poland

Additional Reviewers

Bonfante, Guillaume   LORIA, Université de Lorraine, Nancy, France
Grudzinska, Justyna   Institute of Philosophy, University of Warsaw, Poland
Perrier, Guy   LORIA, Université de Lorraine, Nancy, France
Valentin, Oriol   Universitat Politècnica de Catalunya, Spain

Abstracts of Invited Talks

FC Disjunction in State-Based Semantics

Maria Aloni

Institute for Logic, Language, and Computation, University of Amsterdam,
Amsterdam, The Netherlands

In a state-based semantics sentences are interpreted with respect to states (defined as sets of possible worlds) rather than single possible worlds. This feature makes state-based semantics particularly suitable to capture the inherent epistemic and/or alternative-inducing nature of disjunctive words in natural language. In the first part of the talk, I will discuss three notions of disjunction that have been proposed in a state-based semantics with emphasis on their potential to account for Free Choice (FC) inferences when combined with a possibility modal:

(1) FC inferences

    a. Wide scope: $\Diamond a \vee \Diamond b \models \Diamond a \wedge \Diamond b$

    b. Narrow scope: $\Diamond (a \vee b) \models \Diamond a \wedge \Diamond b$

The first notion $\vee_1$ corresponds to disjunction in classical logic; the second notion $\vee_2$ has been independently proposed by Yang and Väänänen [5] and Hawke and Steinert-Threlkeld [3]; the third notion $\vee_3$ corresponds to inquisitive disjunction as in Ciardelli and Roelofsen [2] (see also Kit Fine's truthmaker semantics). Team/assertion logic $\vee_2$ in combination with a context-sensitive notion of modality à la Veltman [4] derives wide scope FC inference (as discussed in [3]). Inquisitive/truthmaker $\vee_3$ combined with Aloni's [1] alternative-sensitive notion of modality derives narrow scope FC inference. Neither combinations however can account for both wide scope and narrow scope FC. Furthermore, when free choice inducing sentences occur embedded under negation, both systems predict weaker readings than attested in ordinary language use. In the second part of the talk, I will present a third state-based system, adopting $\vee_2$, which derives both wide scope and narrow scope FC while solving the negation problem. Merits and shortcomings of this novel system will be discussed as well as its potential to be extended to account for free choice indefinites.

References

1. Aloni, M.: Free choice, modals and imperatives. Nat. Lang. Sem. **15**, 65–94 (2007)
2. Ciardelli, I., Roelofsen, F.: Inquisitive logic. J. Philos. Logic **40**(1), 55–94 (2011)
3. Hawke, P., Steinert-Threlkeld, S.: Informational dynamics of 'Might' assertions. In: van der Hoek, W., Holliday, W., Wang, W. (eds.) Proceedings of Logic, Rationality, and Interaction (LORI-V), pp. 143–155 (2015)
4. Veltman, F.: Defaults in update semantics. J. Philos. Logic **25**, 221–261 (1996)
5. Yang, F., Väänänen, J.: Propositional team logics (2016, submitted)

The Parallel Meaning Bank: A Large Corpus of Translated Texts Annotated with Formal Meaning Representations

Johan Bos

University of Groningen, Groningen, The Netherlands

Several large corpora annotated with meaning representations are nowadays available such as the Groningen Meaning Bank [4], the AMR Corpus [1], or Treebank Semantics [5]. These are usually resources for a single language. In this paper I present a project with the aim to develop a meaning bank for translations of texts — in other words, a parallel meaning bank. The languages involved are English, Dutch, German and Italian. The idea is to use language technology developed for English and project the outcome of the analyses to the other languages. There are five steps of processing:

- Tokenisation: segmentation of words, multi-word expressions and sentences, using Elephant, a statistical tokenizer [7];
- Semantic Tagging: mapping word tokens to semantic tags (abstracting over traditional part-of-speech tags and named entities and a bit more);
- Symbolisation: assigning appropriate non-logical symbols to word tokens (combining lemmatization and normalisation);
- Syntactic Parsing: based on Combinatorial Categorial Grammar [6, 9];
- Semantic Parsing: based on Discourse Representation Theory, using the semantic parser Boxer [3];

The first aim of the project is to provide appropriate compositional semantic analyses for the aforementioned language taking advantage of the translations. The second aim is to study the role of meaning in translations: even though you would expect that meaning is preserved in translations, human translators often perform little tricks involving meaning shifts and changes to arrive at better translations [2, 8].

References

1. Banarescu, L., Bonial, C., Cai, S., Georgescu, M., Grifftt, K., Hermjakob, U., Knight, K., Koehn, P., Palmer, M., Schneider, N.: Abstract meaning representation for sembanking. In: Proceedings of the 7th Linguistic Annotation Workshop and Interoperability with Discourse, pp. 178–186 (2013)
2. Bos, J.: Semantic annotation issues in parallel meaning banking. In: Proceedings of the Tenth Joint ACL-ISO Workshop on Interoperable Semantic Annotation (ISA-10), Reykjavik, Iceland, pp. 17–20 (2014)

3. Bos, J.: Open-domain semantic parsing with boxer. In: Megyesi, B. (ed.) Proceedings of the 20th Nordic Conference of Computational Linguistics (NODALIDA 2015), pp. 301–304 (2015)
4. Bos, J., Basile, V., Evang, K., Venhuizen, N., Bjerva, J.: The Groningen meaning bank. In: Ide, N., Pustejovsky, J. (eds.) The Handbook of Linguistic Annotation. Springer, Berlin (2017)
5. Butler, A.: The Semantics of Grammatical Dependencies, vol. 23. Emerald Group Publishing Limited (2010)
6. Curran, J.R., Clark, S., Bos, J.: Linguistically motivated large-scale NLP with c&c and boxer. In: Proceedings of the ACL 2007 Demo and Poster Sessions, Prague, Czech Republic, pp. 33–36 (2007)
7. Evang, K., Basile, V., Chrupala, G., Bos, J.: Elephant: sequence labeling for word and sentence segmentation. In: Proceedings of the 2013 Conference on Empirical Methods in Natural Language Processing (EMNLP 2013), Seattle, Washington, pp. 1422–1426 (2013)
8. Langeveld, A.: Vertalen wat er staat. Synthese, De arbeiderspers (1986)
9. Lewis, M., Steedman, M.: A* ccg parsing with a supertag-factored model. In: Proceedings of the 2014 Conference on Empirical Methods in Natural Language Processing (EMNLP), pp. 990–1000. Association for Computational Linguistics, Doha, Qatar, October 2014

Bayesian Inference in a Probabilistic Type Theory

Shalom Lappin[1,2,3]

[1] Department of Philosophy, Linguistics, and Theory of Science,
University of Gothenburg, Gothenburg, Sweden
[2] Department of Philosophy, King's College, London, UK
[3] School of Electronic Engineering and Computer Science,
Queen Mary University of London, London, UK
shalom.lappin@gu.se

Classical semantic theories [8], as well as dynamic [7] and underspecified frameworks [5] use categorical type systems. A type T identifies a set of possible denotations for expressions in T. The theory specifies combinatorial operations for deriving the denotation of an expression from the values of its constituents. These theories cannot represent the gradience of semantic properties that is pervasive in speakers' judgements concerning truth, predication, and meaning relations.

There is a fair amount of evidence indicating that language acquisition in general crucially relies on probabilistic learning [2]. It is not clear how a reasonable account of semantic learning could be constructed on the basis of the categorical type systems that either classical or revised semantic theories assume. Such systems do not appear to be efficiently learnable from the primary linguistic data (with weak learning biases). There is little (or no) psychological data to suggest that classical categorical type systems provide biologically determined constraints on semantic learning.

A semantic theory that assigns probability rather than truth conditions to sentences is in a better position to deal with gradience and learning. Gradience is intrinsic to the theory by virtue of the fact that values are assigned to sentences in the continuum of real numbers [0, 1], rather than Boolean values in {0, 1}. A probabilistic account of semantic learning is facilitated if the target of learning is a probabilistic representation of meaning. Both semantic interpretation and semantic learning are characterised as reasoning under uncertainty.

[4] propose a probabilistic re-formulation of [3]'s Type Theory with Records (TTR). They specify a rich type theory, ProbTTR, in which probability is distributed over situation types [1]. An Austinian proposition is a judgement that a situation is of a particular type, and we treat it as probabilistic. It expresses a subjective probability in that it encodes the belief of an agent concerning the likelihood that a situation is of that type. The core of an Austinian proposition is a type judgement of the form $s : T$, which is expressed probabilistically as $p(s : T) = r$, where $r \in [0, 1]$. ProbTTR provides the basis for a compositional probabilistic semantics of natural language.

Joint work with Robin Cooper, Simon Dobnik, and Staffan Larsson, University of Gothenburg.

We specify a Bayesian learning and inference component for ProbTTR. We formulate Bayes' theorem in type theoretic terms and use it to develop a Naive Bayesian Classifier for learning basic predicate types from observation and mentor led instruction. We extend this component to a type theoretic version of Bayesian Networks [6, 9, 10], which we propose as a framework for semantic learning.

The basic types and type judgements at the foundation of our probabilistic type system correspond to perceptual judgements concerning objects and events in the world, rather than to entities in a model, and set theoretic constructions defined on them. We incorporate a theory of learning and inference with Bayesian Networks into ProbTTR. Our account grounds meaning in learning how to make observational judgements concerning the likelihood of situations obtaining in the world.

References

1. Barwise, J., Perry, J.: Situations and Attitudes. Bradford Books, MIT Press, Cambridge (1983)
2. Clark, A., Lappin, S.: Linguistic Nativism and the Poverty of the Stimulus. Wiley-Blackwell, Chichester, West Sussex, Malden (2011)
3. Cooper, R.: Type theory and semantics in flux. In: Kempson, R., Asher, N., Fernando, T. (eds.) Handbook of the Philosophy of Science. Philosophy of Linguistics, vol. 14. Elsevier (2012). General editors: Dov M. Gabbay, Paul Thagard and John Woods
4. Cooper, R., Dobnik, S., Lappin, S., Larsson, S.: Probabilistic type theory and natural language semantics. Linguist. Issues Lang. Technol. **10**, 1–43 (2015)
5. Fox, C., Lappin, S.: Expressiveness and complexity in underspecified semantics. Linguist. Anal., Festschrift for Joachim Lambek **36**, 385–417 (2010)
6. Halpern, J.: Reasoning About Uncertainty. MIT Press, Cambridge (2003)
7. Kamp, H., Reyle, U.: From Discourse to Logic: Introduction to Modeltheoretic Semantics of Natural Language, Formal Logic and Discourse Representation Theory. Kluwer, Dordrecht (1993)
8. Montague, R.: Formal Philosophy: Selected Papers of Richard Montague. Yale University Press, New Haven (1974). Ed. and with an introduction by Richmond H. Thomason
9. Murphy, K.: A brief introduction to graphical models and bayesian networks, University of British Columbia (2001). http://www.cs.ubc.ca/~murphyk/Bayes/bnintro.html
10. Pearl, J.: Bayesian decision methods. In: Shafer, G., Pearl, J. (eds.) Readings in Uncertain Reasoning, pp. 345–352. Morgan Kaufmann (1990)

Combining Formal and Distributional Semantics: An Argument from the Syntax and Semantics of Modification

Louise McNally

Universitat Pompeu Fabra, Barcelona, Spain

The lexical semantics of content words has historically generated comparatively little interest among formal semanticists, except when a class of content words proves to be sensitive to some sort of logical or grammatical phenomenon, as has happened, for example, with verbal aspect and with gradability. In contrast, the "distributional turn" in computational semantics has focused primarily on the content word lexicon and has had a varied and arguably difficult relationship with logical approaches to meaning, including the semantics of function words like *and* or *the* (see [1] for a recent overview). How the insights and benefits of these two approaches can be combined, if at all, is currently a matter of active research (see e.g. [4] and research reported on there).

In this talk, I reflect on my experience when I turned to distributional models in an effort to better address the interaction of lexical and compositional semantics. After considering the possibility of a full-blown distributional semantics for both content and function words, I have opted to explore a mixed model based on Discourse Representation Theory (DRT, [11]), perhaps most similar in spirit to the recent mixes of formal and distributional semantics found in [10, 12, 13] (though the latter do not use DRT), but motivated and implemented a bit differently.

Specifically, I argue that certain kinds of modification constructions, including (broadly) noun incorporation, point to a well-established distinction in natural language between productive compositional operations that combine descriptive contents unmediated by reference, resulting in complex kind- or type-level descriptions, and composition operations that are crucially mediated by tokenlevel reference (see [6] on the notion of kind; see e.g. [5, 15, 18], for the relevance of the type/token distinction in nominal modification). I propose, building on the results of a distributional study in [3], that compositional distributional methods can provide interesting complex type-level descriptions, and show, following [14, 16], how these can be represented and put to use in a semantics based on DRT.

I gratefully acknowledge my co-authors on the papers that have most directly inspired this work: Gemma Boleda, Berit Gehrke, Marco Baroni, Alexandra Spalek, Scott Grimm, and Nghia Pham. I also thank Carla Umbach and the participants in the 2016 ESSLLI Workshop 'Referential Semantics One Step Further: Incorporating Insights from Conceptual and Distributional Approaches to Meaning (RefSemPlus)' and the FloSS and Meaning in Context groups for discussion of the larger issues. This research was supported by Spanish MINECO grants FFI2010-09464-E and FFI2013-41301-P, AGAUR grant 2014SGR698, and an ICREA Academia award.

I close by arguing that this mixed approach sheds light on the often-reported but never explained observation that incorporation constructions, such as the Hindi example in (1) from [8], often carry steoreotypicality implications that non-incorporated counterparts do not (see [7] on the relevance of the type/token distinction for this phenomenon; see [9] for discussion and analysis of related examples such as the contrast in (2) and (3), from Catalan).

(1) anu sirf **puraanii kitaab** becegii
    Anu only old book sell-FUT
    'Anu will only sell old books.'
(2) Té **una parella**.
    has a partner
    'S/he has a partner.'
(3) Té **parella**.
    has partner
    'S/he has a partner.' (so, s/he's married/can now dance/...)

I also suggest more speculatively that the mixed distributional/DRT approach promotes rethinking the analysis of certain function words as "instructions" in the spirit of the Procedural/Conceptual Meaning distinction in Relevance Theory (see e.g. [2]) and ideas that have long informed the literature on information structure and dialog (see e.g. [17]).

References

1. Baroni, M., Bernardi, R., Zamparelli, R.: Frege in space. Linguist. Issues Lang. Technol. **9**(6), 5–110 (2014)
2. Blakemore, D.: Semantic Constraints on Relevance. Blackwell, Oxford (1987)
3. Boleda, G., Baroni, M., Pham, N.T., McNally, L.: Intensionality was only alleged: On adjective-noun composition in distributional semantics. In: Proceedings of IWCS 2013. Potsdam (2013)
4. Boleda, G., Herbelot, A.: Formal distributional semantics: Introduction to the introduction to a proposed special issue of Computational Linguistics on Formal Distributional Semantics (2016, special issue)
5. Bouchard, D.: Sèriation des adjectifs dans le sn et formation de concepts. Recherches Linguistiques de Vincennes **34**, 125–142 (2005)
6. Carlson, G.N.: Reference to kinds in English. Ph.D. thesis, University of Massachusetts at Amherst (1977)
7. Carlson, G.N.: Weak indefinites. In: Coene, M., D'Hulst, Y. (eds.) From NP to DP: On the Syntax and Pragma-Semantics of Noun Phrases, vol. 1, Benjamins, pp. 195–210 (2003)
8. Dayal, V.: Hindi pseudo-incorporation. Nat. Lang. Linguist. Theory **29**, 123–167 (2011)
9. Espinal, M.T., McNally, L.: Bare singular nominals and incorporating verbs in Spanish and Catalan. J. Linguist. **47**, 87–128 (2011)
10. Garrette, D., Erk, K., Mooney, R.: Integrating logical representations with probabilistic information using Markov logic. In: Proceedings of IWCS 2011 (2011)

11. Kamp, H.: A theory of truth and semantic representation. In: Groenendijk, J., Janssen, T., Stokhof, M. (eds.) Formal Methods in the Study of Language, vol. 1, pp. 277–322. Mathematisch Centrum, Amsterdam (1981)

12. Lenci, A.: "going dynamic" in distributional semantics. Presented at the ESSLLI 2016 Workshop "Referential Semantics One Step Further: Incorporating Insights from Conceptual and Distributional Approaches to Meaning" (2016)

13. Lewis, M., Steedman, M.: Combined distributional and logical semantics. Trans. Assoc. Comput. Linguist. 1(179–192) (2013)

14. McNally, L.: Kinds, descriptions of kinds, concepts, and distributions (2015), ms., Universitat Pompeu Fabra

15. McNally, L., Boleda, G.: Relational adjectives as properties of kinds. In: Bonami, O., Cabredo Hofherr, P. (eds.) Empirical Issues in Syntax and Semantics, vol. 5, pp. 179–196 (2004). http://www.cssp.cnrs.fr/eiss5/mcnally-boleda/mcnally-boleda-eiss5.pdf

16. McNally, L., Boleda, G.: Conceptual vs. referential affordance in concept composition. In: Winter, Y., Hampton, J. (eds.) Concept Composition and Experimental Semantics/Pragmatics. Springer, Berlin (to appear)

17. Vallduví, E.: The informational component. Garland Press, New York (1992)

18. Zamparelli, R.: Layers in the determiner phrase. Ph.D. thesis, U. Rochester (1995)

Contents

Abstracts of Short Talks

Language Games

Nicholas Asher and Soumya Paul[⊠]

Institut de Recherche en Informatique de Toulouse, Toulouse, France
nicholas.asher@irit.fr, soumya.paul@gmail.com

Abstract. In this paper we summarize concepts from earlier work and demonstrate how infinite sequential games can be used to model strategic conversations. Such a model allows one to reason about the structure and complexity of various kinds of winning goals that conversationalists might have. We show how to use tools from topology, set-theory and logic to express such goals. Our contribution in this paper is to offer a detailed examination of an example in which a player 'defeats himself' by going inconsistent, and to introduce a simple yet revealing way of talking about unawareness. We then demonstrate how we can use ideas from epistemic game theory to define various solution concepts and justify rationality assumptions underlying a conversation.

Keywords: Strategic reasoning · Conversations · Dialogues · Infinite games · Epistemic game theory

1 Introduction

A strategic conversation involves (at least) two people (agents) who have opposing interests concerning the outcome of the conversation. A debate between two political candidates is an instance. Each candidate has a certain number of points she wants to convey to the audience, and each wants to promote her own position and damage that of her opponent or opponents. In other words, each candidate wants to win. To achieve these goals each participant needs to plan for anticipated responses from the other. Debates are thus a sequence of exchange of messages at the end of which an agent may win, lose or draw. Similar strategic reasoning about what one says is a staple of board room or faculty meetings, bargaining sessions, etc.

It is therefore natural to model such conversations as games. Attempts to this end have been made in the past with the most notable of them being the use of signaling games [24] and the closely related persuasion games [15]. In a signaling game one player with a knowledge of the actual state sends a signal and the other player who has no knowledge of the state chooses an action, usually upon an interpretation of the received signal. The standard setup supposes that both players have common knowledge of each other's preference profiles as well as their own over a set of commonly known set of possible states, actions and signals.

The authors thank ERC grant 269427 for supporting this research.

M. Amblard et al. (Eds.): LACL 2016, LNCS 10054, pp. 1–17, 2016.
DOI: 10.1007/978-3-662-53826-5_1

However for modeling non-cooperative strategic contexts of sequential dynamic games, signaling games suffer from many drawbacks. Some of them can be summarised as follows (see [4] for a more comprehensive discussion):

- A game that models a non-cooperative setting, that is a setting where the preferences of the players are opposed, must be zero-sum. However, it has been shown [11] that under the zero-sum criterion, in equilibrium, the sending and receiving of any message has no effect on the receiver decision.
- In order to use games as part of a general theory of meaning, one has to make clear how to construct the game-context, which includes providing an interpretation of the game's ingredients (types, messages, actions). Franke [13] extended the setting of signaling games to that of *interpretation games* to address this issue. Such games encode a 'canonical context' for an utterance, in which relevant conversational implicatures may be drawn. The game structure is determined by the set of 'sender types'. Interpretation games model the interpretation of the messages and actions of a signaling game in a co-operative context for 'Gricean agents' quite well. But in the non-cooperative setting, things are much more intricate and problems remain (again see [4]).
- Signaling games are one-shot and fail to capture the dynamic nature of a strategic conversation. One can attempt to encode a sequence of moves of a particular player as a single message m sent by that player but then one runs into the problem of assigning correct utilities for m because such utilities depend again on the possible set of continuations of m.
- Finally, there is an inherent asymmetry associated with the setting of a signaling game - one player is informed of the state of the world but the other is not; one player sends a message but the other does not. Conversations (like debates), on the other hand, are symmetric - all participants should (and usually do) get equal opportunities to get their messages across.

Strategic conversations are thus special and have characteristics unique to them which have not been captured by previous game-theoretic models. Some of these important characteristics are as follows.

- Conversations are sequential and dynamic and inherently involve a 'turn-structure' which is important in determining the merit of a conversation to the participants. In other words, it is important to keep track of "who said what".
- A 'move' by a player in a linguistic game typically carries more semantic content than usually assumed in game theory. What a player says may have a set of 'implicatures', may be 'ambiguous', may be 'coherent/incoherent' or 'consistent/inconsistent' to what she had said earlier in the conversation. She may also 'acknowledge' other people's contributions or 'retract' her previous assertions. These features too have important consequences on the existence and complexity of winning strategies.
- Conversations typically have a 'Jury' who evaluates the conversation after it has ended and determines if one or more of the players have reached their goals – determines the winner. Players will spin the description of the game to

their advantage and so may not present an accurate view of what happened. The Jury can be a concrete or even a hypothetical entity who acts as a 'passive player' in the game. For example, in a courtroom situation there is a physical Jury who gives the verdict, whereas in a political debate the Jury is the audience or the citizenry in general. This means that the winning conditions of the players are affected by the Jury in that, they depend on what they believe that the Jury expects them to achieve.

- Epistemic elements thus naturally creep into such games. In particular, the players and the Jury have 'types'. In addition the players also have 'beliefs' about the types of the other players and that of the Jury. They strategize based on their beliefs and also update their beliefs after each turn.
- Lastly but most importantly, conversations do not have a 'set end'. When two people or a group of people engage in a conversation they do not know at the outset how many turns it will last or how many chances each player will get to speak (if at all). Sure in a 'conducted' conversation such as a political debate or a courtroom debate, there is usually a moderator whose job is to ensure that each player receives his or her fair chance to put their points across but even such a moderator does not know at the outset how the conversation will unfold and how many turns each player will receive. Players thus cannot strategize for a set horizon while starting a conversation. This rules out backward induction reasoning for both the players and we analysts.

With the above aspects in mind, [4] model conversations as infinite games over a countable 'vocabulary' V which they call Message Exchange games (ME games). In this paper, we first summarize the main results of [4] in a compact fashion but also add some new remarks concerning first order definability of conversational goals. We then add a more nuanced analysis of a particular dialogue excerpt (our example 4) and prove a theorem beyond the scope of [4], showing how unexpected moves can complicate the search for winning strategies. Finally in Sect. 3, we break new ground and add an epistemic layer to ME games.

Let's now turn to the basics of ME games. The intuitive idea behind an ME game is that a conversation proceeds in turns where in each turn one of the players 'speaks' or plays a string of letters from her own vocabulary. However, the player does not play just any sequence of arbitrary strings but sentences or sets of sentences that 'make sense'. To ensure this, the vocabulary V should have an exogenous semantics built-in. In order to achieve this, we exploit a semantic theory for discourse, SDRT [1]. SDRT develops a rich language to characterize the semantics and pragmatics of moves in dialogue. This means that we can exploit the notion of entailment associated with the language of SDRSs to track commitments of each player in an ME game. In particular, the language of SDRT features variables for dialogue moves that are characterized by contents that the move commits its speaker to. Crucially, some of this content involves predicates that denote rhetorical relations between moves—like the relation of *question answer pair* (qap), in which one move answers a prior move characterized by a question. The vocabulary V of an ME game thus contains a countable set of discourse constituent labels $\mathsf{DU} = \{\pi, \pi_1, \pi_2, \ldots\}$, and a finite set of discourse relation symbols $\mathcal{R} = \{\mathsf{R}, \mathsf{R}_1, \ldots \mathsf{R}_n\}$, and formulas $\phi, \phi_1, \ldots$ from some fixed language for describing elementary discourse move contents. V consists of formulas of the form $\pi \colon \phi$, where ϕ is a description of the content of the discourse unit labelled by π in a logical language like the language of higher order logic used, e.g., in Montague Grammar, and $\mathsf{R}(\pi, \pi_1)$, which says that π_1 stands in relation R to π. One such relation R is qap. Thus, each discourse relation symbolized in V comes with constraints as to when it can be coherently used in context and when it cannot.

2   Message Exchange Games

In this section we formally define Message Exchange games and state some of their properties and their use in modeling strategic conversations as explored at length in [4]. For simplicity, we shall develop the theory for the case of conversations that involve two participants, which we shall denote by Player 0 and Player 1. It will be straightforward to generalize it to the case where there are more than two players. Thus, in what follows, we shall let i range over the set of players $\{0, 1\}$. Furthermore, Player $-i$ will always denote Player $(1 - i)$, the opponent of Player i.

We first define the notion of a 'Jury'. As noted in Sect. 1, a Jury is any entity or a group of entities that evaluates a conversation and decides the winner. A Jury thus 'groups' instances of conversations as being winning for Player 0 or Player 1 or both.

For any set A let A^* be the set of all finite sequences over A and let A^ω be the set of all countably infinite sequences over A. Let $A^\infty = A^* \cup A^\omega$ and $A^+ = A^* \setminus \{\epsilon\}$. Now, let V be a vocabulary as defined at the end of Sect. 1 and let $V_i = V \times \{i\}$. This is to make explicit the 'turn-structure' of a conversation as alluded to in the introduction.

Definition 1. *A* Jury *$\mathcal{J}$ over $(V_0 \cup V_1)^\omega$ is a tuple $\mathcal{J} = (\mathsf{win}_1, \mathsf{win}_2)$ where* $\mathsf{win}_i \subseteq (V_0 \cup V_1)^\omega$ *is the* winning condition *or* winning set *for Player i.*

Given the definition of a Jury over $(V_0 \cup V_1)^\omega$ we define a Message Exchange game as:

Definition 2. *A Message Exchange game (ME game) $\mathcal{G}$ over $(V_0 \cup V_1)^\omega$ is a tuple $\mathcal{G} = ((V_0 \cup V_1)^\omega, \mathcal{J})$ where $\mathcal{J}$ is a Jury over $(V_0 \cup V_1)^\omega$.*

Formally the ME game $\mathcal{G}$ is played as follows. Player 0 starts the game by playing a non-empty sequence in V_0^+. The turn then moves to Player 1 who plays a non-empty sequence from V_1^+. The turn then goes back to Player 0 and so on. The game generates a play ρ_n after n (≥ 0) turns, where by convention, $\rho_0 = \epsilon$ (the empty move). A play can potentially go on forever generating an infinite play ρ_ω, or more simply ρ. Player i wins the play ρ iff $\rho \in \mathsf{win}_i$. $\mathcal{G}$ is zero-sum if $\mathsf{win}_i = (V_0 \cup V_1)^\omega \setminus \mathsf{win}_{-i}$ and is non zero-sum otherwise. Note that both player or neither player might win a non zero-sum ME game $\mathcal{G}$. The Jury of a zero-sum

ME game can be denoted simply as win where by convention win $=$ win$_0$ and win$_1 = (V_0 \cup V_1)^\omega \setminus$ win.

Plays are segmented into *rounds*—a move by Player 0 followed by a move by Player 1. A finite play of an ME game is (also) called a history, and is denoted by ρ. Let Z be the set of all such histories, $Z \subseteq (V_0 \cup V_1)^*$, where $\epsilon \in Z$ is the empty history and where a history of the form $(V_0 \cup V_1)^+ V_0^+$ is a 0-history and one of the form $(V_0 \cup V_1)^+ V_1^+$ is a 1-history. We denote the set of i-histories by Z_i. By convention $\epsilon \in Z_1$. Thus $Z = Z_0 \cup Z_1$. For $\rho \in Z$, turns(ρ) denotes the total number of turns (by either player) in ρ. A **strategy** σ_i of Player i is thus a function from the set of $-i$-histories to V_i^+. That is, $\sigma_i : Z_{-i} \to V_i^+$. A play $\rho = x_0 x_1 \ldots$ of an ME game $\mathcal{G}$ is said to **conform** to a strategy σ_i of Player i if for every prefix ρ_j of ρ, $j = i(\mod 2)$ implies $\rho_{j+1} = \rho_j \sigma_i(\rho_j)$. A strategy σ_i is called **winning** for Player i if $\rho \in$ win$_i$ for every play ρ that conforms to σ_i.

Given how we have characterized the vocabulary $(V_0 \cup V_1)$, we can assumed a fixed meaning assignment function from EDUs to formulas the describe their contents. Then, a sequence of conversational moves can be represented as a graph (DU, E, ℓ), where DU is the set of vertices each representing a discourse unit, $E \subseteq \mathsf{DU} \times \mathsf{DU}$ a set of edges representing links between discourse units that are labeled by $\ell : E \to \mathcal{R}$ with discourse relations.[1]

Example 1. To illustrate this structure of conversations, consider the following example taken from [2] from a courtroom proceedings where a prosecutor is querying the defendant. We shall return to this example later on for a strategic analysis.

a. **Prosecutor**: *Do you have any bank accounts in Swiss banks, Mr. Bronston?*
b. **Bronston**: *No, sir.*
c. **Prosecutor**: *Have you ever?*
d. **Bronston**: *The company had an account there for about six months, in Zurich.*
e. **Prosecutor**: *Thank you Mr. Bronston.*

Example 2. We can view the conversation in Example 1 as an ME game as in Fig. 1. The figure shows a weakly connected graph, which represents a fully coherent conversation, with a set of discourse constituent labels $\mathsf{DU} = \{\pi_{\mathsf{bank}}, \pi_{\neg\mathsf{bank}}, \pi_{\mathsf{bank-elab}}, \pi_{\mathsf{company}}, \pi_{\mathsf{ack}}, \ldots\}$ and a set of relations $\mathcal{R} = \{\mathsf{qap}, \mathsf{q - followup}, \mathsf{ack}, \ldots\}$. The arrows depict the individual relation instances between the DUs.

ME game messages come with a conventionally associated meaning in virtue of the constraints enforced by the Jury; an agent who asserts a content of a message commits to that content, and it is in virtue of such commitments that other agents respond in kind. While SDRT has a rich language for describing

---

[1] We note that this is a simplification of SDRT which also countenances complex discourse units (CDUs) and another set of edges in the graph representation, linking CDUs to their simpler constituents. These edges represent parthood, not rhetorical relations. We will not, however, appeal to CDUs here.

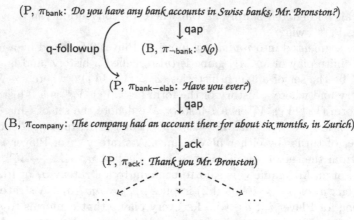

Fig. 1. An example ME game

dialogue moves, it is not explicit about how dialogue moves explicitly affect the commitments of the agents who make the moves or those who observe the moves. [25,26] link the semantics of the SDRT language with commitments explicitly (in two different ways). They augment the SDRT language with formulas that describe the commitments of dialogue participants, using a simple propositional modal syntax. Thus for any formula ϕ in the language of Montague Grammar that describes the content of a label $\pi \in \mathsf{DU}$, they add: $\neg\phi \mid \phi_1 \vee \phi_2 \mid C_i\phi$, $i \in \{0,1\} \mid C^*\phi$, with the derived operators $\wedge$, $\implies$, $\top, \bot$ are defined as usual, providing a propositional logic of commitments over the formulas that describe labels. Of particular interest are the commitment operators C_i and C^*. If ϕ is a formula for describing a content, $C_i\phi$ is a formula that says that Player i commits to ϕ and $C^*\phi$ denotes 'common commitment' of ϕ. Commitment is modeled as a Kripke modal operator via an alternativeness relation in a pointed model with a distinguished (actual) world w_0. This allows them to provide a semantics for discourse moves that links the making of a discourse move by an agent to her commitments: i's assertion of a discourse move ϕ, for instance, we will assume, entails a common commitment that i commits to ϕ, written $C^*C_i\phi$. They show how each discourse move ϕ defines an action, a change or update on the model's commitment structure; in the style of public announcement logic viz. [6,7]. For instance, if agent i asserts ϕ, then the commitment structure for the conversational participants is updated such so as to reflect the fact that $C^*C_i\phi$. Finally, they define an entailment relation $\models$ that ensures that $\phi \models C^*C_i\phi$. This semantics is useful because it allows us to move from sequences of discourse moves to sequences of updates on any model for the discourse language. See [25,26] for a detailed development and discussion.

ME games resemble infinite games that have been used in topology, set theory [19] and computer science [16] to study the descriptive complexity of different infinite sets. We can leverage some of the results from these areas to talk about the general 'shape' of conversations or to analyse the complexity of the winning conditions of the players. This has been extensively explored in [4]. We give a flavor of some of the applications here.

To do that we first need to define an appropriate topology on $(V_0 \cup V_1)^\omega$ which will allow us to characterize the descriptive complexity of the winning sets win_0 and win_1. We proceed as follows. We define the topology on $(V_0 \cup V_1)^\omega$ by defining the **open sets** to be sets of the form $A(V_0 \cup V_1)^\omega$ where $A \subseteq (V_0 \cup V_1)^*$. Such an open set will be often denoted as $\mathcal{O}(A)$. When A is a singleton set $\{x\}$ (say), we abuse notation and write $\mathcal{O}(\{x\})$ as $\mathcal{O}(x)$. The **Borel sets** are defined as the sigma-algebra generated by the open sets of this topology. The Borel sets can be arranged in a natural hierarchy called the Borel hierarchy which is defined as follows. Let Σ_1^0 be the set of all open sets. $\Pi_1^0 = \overline{\Sigma_1^0}$, the complement of the set of Σ_1^0 sets, is the set of all closed sets. Then for any $\alpha > 1$ where α is a successor ordinal, define Σ_α^0 to be the countable union of all $\Pi_{\alpha-1}^0$ sets and define Π_α^0 to be the complement of Σ_α^0. $\Delta_\alpha^0 = \Sigma_\alpha^0 \cap \Pi_\alpha^0$.

Definition 3 [19]. *A set A is called* **complete** *for a class Σ_α^0 (resp. Π_α^0) if $A \in \Sigma_\alpha^0 \setminus \Pi_\alpha^0$ (resp. $\Pi_\alpha^0 \setminus \Sigma_\alpha^0$) and $A \notin (\Sigma_\beta^0 \cup \Pi_\beta^0)$ for any $\beta < \alpha$.*

The Borel hierarchy represents the descriptive or structural complexity of the Borel sets. A set higher up in the hierarchy is structurally more complex than one that is lower down. Complete sets for a particular class of the hierarchy represent the structurally most complex sets of that class. We can use the Borel hierarchy and the notion of completeness to capture the complexity of winning conditions in conversations. For example, two typical sets in the fist level of the Borel hierarchy are defined as follows. Let $A \subseteq (V_0 \cup V_1)^+$, then

$$\text{reach}(A) = \{\rho \in (V_0 \cup V_1)^\omega \mid \rho = xy\rho', y \in A\}, \ \text{safe}(A) = (V_0 \cup V_1)^\omega \setminus \text{reach}(A)$$

A little thought convinces us that $\text{reach}(A) \in \Sigma_1^0$ and $\text{safe}(A) \in \Pi_1^0$. Let reachability be the class of sets of the form $\text{reach}(A)$ and safety be the class of sets of the form $\text{safe}(A)$.

Example 3. Returning to our example of Bronston and the Prosecutor, let us consider what goals the Jury expects each of them to achieve. The Jury will award its verdict in favor of the Prosecutor: (i) if he can eventually get Bronston to admit that (a) he had an account in Swiss banks, or (b) he never had an account in Swiss banks, or (ii) if Bronston avoids answering the Prosecutor forever. In the case of (i)a, Bronston is incriminated, (i)b, he is charged with perjury and (ii), he is charged with contempt of court. Bronston's goal is the complement of the above, that is to avoid either of the situations (i)a, (i)b and (ii). We thus see that the Jury winning condition for the Prosecutor is a Boolean combination of a reachability condition and the complement of a safety condition, which is in the first level of the Borel hierarchy.

Conversations, to be meaningful, must also satisfy certain natural constraints which the Jury might impose throughout the course of a play. Below we define some of these constraints and then go on to study the complexity of the sets satisfying them.

Let $\rho = x_0 x_1 x_2 \ldots$ be a play of an ME game $\mathcal{G}$ where $x_0 = \epsilon$ and $x_j \in V^+_{((j-1) \bmod 2)}$ is the sequence played by Player $((j-1) \bmod 2)$ in turn j. For every i define the function $\mathsf{du}_i : V^+_i \to \wp(\mathsf{DU})$ such that $\mathsf{du}_i(x_j)$ gives the set of contributions (in terms of DUs) of Player i in the jth turn. By convention, $\mathsf{du}_i(x_j) = \emptyset$ for $x_j \in V^+_{-i}$.

Definition 4. *Let* $\mathcal{G} = ((V_0 \cup V_1)^\omega, \mathcal{J})$ *be an ME game over* $(V_0 \cup V_1)^\omega$. *Let* $\rho = x_0 x_1 x_2 \ldots$ *be a play of* $\mathcal{G}$. *Then*

Consistency: ρ *is* consistent *for Player* i *if the set* $\{\mathsf{du}_i(x_j)\}_{j>0}$ *is consistent. Let* CONS_i *denote the set of consistent plays for Player* i *in* $\mathcal{G}$.

Coherence: Player i *is* coherent *on turn* $j > 0$ *of play* ρ *if for all* $\pi \in \mathsf{du}_i(x_j)$ *there exists* $\pi' \in (\mathsf{du}_i(x_k) \cup \mathsf{du}_{-i}(x_{k-1}))$ *where* $k \leq j$ *such that there exits* $\mathsf{R} \in \mathcal{R}$ *such that* $(\pi'\mathsf{R}\pi \vee \pi\mathsf{R}\pi')$ *holds. Let* COH_i *denote the set of all coherent plays for Player* i *in* $\mathcal{G}$.

Responsiveness: Player i *is* responsive *on turn* $j > 0$ *of play* ρ *if there exists* $\pi \in \mathsf{du}_i(x_j)$ *such that there exits* $\pi' \in \mathsf{du}_{-i}(x_{j-1})$ *such that* $\pi'\mathsf{R}\pi$ *for some* $\mathsf{R} \in \mathcal{R}$. *Let* RES_i *denote the set of responsive plays for Player* i *in* $\mathcal{G}$. x_j *(or abusing notation,* π*) will be sometimes called a* response move.

Rhetorical-cooperativity: Player i *is* rhetorically-cooperative *in* ρ *if she is both coherent and responsive in every turn of hers in* ρ. ρ *is rhetorically-cooperative if both the players are rhetorically-cooperative in* ρ. *Let* RC_i *denote the set of rhetorically-cooperative plays for Player* i *in* $\mathcal{G}$ *and let* RC *be the set of all rhetorically-cooperative plays.*

To define the constraints NEC and CNEC we need first the definition of an 'attack' and a 'response'. Thus

Definition 5. *Let* $\mathcal{G} = ((V_0 \cup V_1)^\omega, \mathcal{J})$ *be an ME game over* $(V_0 \cup V_1)^\omega$. *Let* $\rho = x_0 x_1 x_2 \ldots$ *be a play of* $\mathcal{G}$. *Then*

Attack: $\mathsf{attack}(\pi', \pi)$ *on Player* $-i$ *holds at turn* j *of Player* i *just in case* $\pi \in \mathsf{du}_i(x_j)$, $\pi' \in \mathsf{du}_{-i}(x_k)$ *for some* $k \leq j$, *there is an* $\mathsf{R} \in \mathcal{R}$ *such that* $\pi'\mathsf{R}\pi$ *and: (i)* π' *entails that* $-i$ *is committed to* ϕ *for some* ϕ, *(ii)* ϕ *entails that* $\neg\phi$ *holds. In such a case, we shall often abuse notation and denote it as* $\mathsf{attack}(k, j)$. *Furthermore,* x_j *or alternatively* π *shall be called an* attack move. *An attack move is* relevant *if it is also a* response *move.* $\mathsf{attack}(k, j)$ *on* $-i$ *is* irrefutable *if there is no move* $x_\ell \in V_{-i}$ *in any turn* $\ell > j$ *such that* $\mathsf{attack}(j, \ell)$ *holds and* $x_0 x_1 \ldots x_\ell$ *is consistent for* $-i$.

Response: $\mathsf{response}(\pi', \pi)$ *on Player* $-i$ *holds at turn* j *of Player* i *if there exits* $\pi'' \in \mathsf{du}_i(x_\ell)$, $\pi' \in \mathsf{du}_{-i}(x_k)$ *and* $\pi \in \mathsf{du}_i(x_j)$ *for some* $\ell \leq k \leq j$, *such that* $\mathsf{attack}(\pi'', \pi')$ *holds at turn* k *of Player* $-i$, *there exists* $\mathsf{R} \in \mathcal{R}$ *such that* $\pi'\mathsf{R}\pi$ *and* π *implies that (i) one of* i*'s commitments* ϕ *attacked in* π' *is true or (ii) one of* $-i$*'s commitments in* π' *that entails that* i *was committed to* $\neg\phi$ *is false. We shall often denote this as* $\mathsf{response}(k, j)$.

We can now define the constraints NEC and CNEC as follows.

Definition 6. *Let* $\mathcal{G} = ((V_0 \cup V_1)^\omega, \mathcal{J})$ *be an ME game over* $(V_0 \cup V_1)^\omega$. *Let* $\rho = x_0 x_1 x_2 \ldots$ *be a play of* $\mathcal{G}$. *Then*

NEC: NEC *holds for Player i in* ρ *on turn j if for all* ℓ, k, $\ell \leq k < j$, *such that* attack(ℓ, k), *there exists* m, $k < m \leq j$, *such that* response(k, m). NEC *holds for Player i for the entire play* ρ *if it holds for her in* ρ *for infinitely many turns. Let* NEC$_i$ *denote the set of plays of* $\mathcal{G}$ *where* NEC *holds for player i.*

CNEC: CNEC *holds for Player i on turn j of* ρ *if there are fewer attacks on i with no response in* ρ_j *than for* $-i$. CNEC *holds for Player i over a* ρ *if in the limit there are more prefixes of* ρ *where* CNEC *holds for i than there are prefixes* ρ *where* CNEC *holds for* $-i$. *Let* CNEC$_i$ *be the set of all plays of* $\mathcal{G}$ *where* CNEC *holds for i.*

For a zero-sum ME game $\mathcal{G}$, the structural complexities of most of the above constraints can be derived from another constraint which we call rhetorical decomposition sensitivity (RDS) which is defined as follows.

Definition 7. *Given a zero sum ME game* $\mathcal{G} = ((V_0 \cup V_1)^\omega, \text{win})$, win *is rhetorically decomposition sensitive* (RDS) *if for all* $\rho \in$ win *and for all finite prefixes* ρ_j *of* ρ, $\rho_j \in Z_1$ *implies there exists* $x \in V_0^+$ *such that* $\mathcal{O}(\rho_j x) \cap$ win $= \emptyset$.

[4] show that if Player 0 has a winning strategy for an RDS winning condition win then win is a Π_2^0 complete set. Formally,

Proposition 1 [4]. *Let* $\mathcal{G} = ((V_0 \cup V_1)^\omega, \text{win})$ *be a zero-sum ME game such that* win *is RDS. If Player 0 has a winning strategy in* $\mathcal{G}$ *then* win *is* Π_2^0 *complete for the Borel hierarchy.*

In the zero-sum setting, CONS$_0$, RES$_0$, COH$_0$, NEC$_0$ are all RDS and it is easy to observe that Player 0 has winning strategies in all these constraints (considered individually). Hence, as an immediate corollary to Proposition 1 we have

Corollary 1. CONS$_0$, RES$_0$, COH$_0$, NEC$_0$ *are* Π_2^0 *complete for the Borel hierarchy for a zero sum ME game.*

CNEC, on the other hand, is a structurally more complex constraint. This is not surprising because CNEC can be intuitively viewed as a limiting case of NEC. Indeed, this was formally shown in [4].

Proposition 2 [4]. CNEC$_i$ *is* Π_3^0 *complete for the Borel hierarchy for a zero sum ME game.*

The above results have interesting consequences in terms of first-order definability. Note that certain infinite sequences over our vocabulary $(V_0 \cup V_1)$ can be coded up using first-order logic over discrete linear orders $(\mathbb{N}, <)$, where $\mathbb{N}$ is the set of non-negative natural numbers. Indeed, for every i and for every $a \in V_i$, let a_0^i be a predicate such that given a sequence $x = x_0 x_1 \ldots$, $x_j \in (V_0 \cup V_1)$

for all $j \geq 0$, $x \models a_0^i(j)$ iff $x_j = a$. Closing under finite Boolean operations and $\forall, \exists$, we obtain the logic $\mathsf{FO}(<)$. Now for any formula $\varphi \in \mathsf{FO}(<)$ and for any play ρ of an ME game $\mathcal{G}$, $\rho \models \varphi$ can be defined in the standard way. Thus every formula $\varphi \in \mathsf{FO}(<)$ gives a set of plays $\rho(\varphi)$ of $\mathcal{G}$ defined as: $\rho(\varphi) = \{\rho \in (V_0 \cup V_1)^\omega \mid \rho \models \varphi\}$. A set $A \subseteq (V_0 \cup V_1)^\omega$ is said to be $\mathsf{FO}(<)$ definable if there exists a $\mathsf{FO}(<)$ formula φ such that $A = \rho(\varphi)$. The following result is well-known.

Theorem 1 [21]. $A \subseteq (V_0 \cup V_1)^\omega$ *is* $\mathsf{FO}(<)$ *definable iff* $A \in (\Sigma_2^0 \cup \Pi_2^0)$.

Thus $\mathsf{FO}(<)$ cannot define sets that are higher than the second level of the Borel hierarchy in their structural complexity. Thus as a corollary of Proposition 2 and Corollary 1, we have

Corollary 2. CONS_0, RES_0, COH_0, NEC_0 *are* $\mathsf{FO}(<)$ *definable but not* CNEC_i.

This agrees with our intuition because as we observed, CNEC_i is a limit constraint and $\mathsf{FO}(<)$, being local [14], lacks the power to capture it. To define CNEC_i one has to go beyond $\mathsf{FO}(<)$ and look at more expressive logics. One such option is to augment $\mathsf{FO}(<)$ with a counting predicate cnt which ranges over $(\mathbb{N} \cup \{\infty\})$ [20]. Call this logic $\mathsf{FO}(<, \mathsf{cnt})$. One can write formulas of the type $\exists^\infty x \varphi(x)$ in $\mathsf{FO}(<, \mathsf{cnt})$ which says that "there are infinitely many x's such that $\varphi(x)$ holds." Note that it is straightforward to write a formula in $\mathsf{FO}(<, \mathsf{cnt})$ that describes CNEC_i. Another option is to consider the logic $\mathscr{L}_{\omega_1\omega}(FO, <)$ which is obtained by closing $\mathsf{FO}(<)$ under infinitary boolean connectives $\bigvee_j$ and $\bigwedge_j$. We can define a strict syntactic subclass of $\mathscr{L}_{\omega_1\omega}(FO, <)$, denoted $\mathscr{L}^*_{\omega_1\omega}(FO, <)$, where every formula is of the form $O_p O_q \ldots O_t \varphi_{pq\ldots t}$, where, for $k \in \{p, q, \ldots, t-1\}$, $O_k = \bigvee_k$ iff $O_{k+1} = \bigwedge_{k+1}$ and each $\varphi_{pq\ldots t}$ is an $(FO, <)$ formula, $p, q, \ldots, t \in \mathbb{N}$. That is, in every formula of $\mathscr{L}^*_{\omega_1\omega}(FO, <)$, the infinitary connectives are not nested and occur only in the beginning. We can then show that $\mathscr{L}^*_{\omega_1\omega}(FO, <)$ can express sets in any countable level of the Borel hierarchy. We do not go into further details here.

We now turn to strategic analyses of actual conversations. Consider this example, an excerpt from the 1988 Dan Quayle-Lloyd Bentsen Vice-Presidential debate which has exercised us now for several years, from the perspective of the theory of ME games developed above.

Example 4. Quayle (Q), a very junior and politically inexperienced Vice-Presidential candidate, was repeatedly questioned about his experience and his qualifications to be President. Till a point in the debate both of them were going neck to neck. But then to rebut doubts about his qualifications, Quayle compared his experience with that of the young John (Jack) Kennedy. To that, Bentsen (BN) made a discourse move that Quayle apparently did not anticipate. We give the relevant part of the debate below:

a. **Quayle:** *... the question you're asking is, "What kind of qualifications does Dan Quayle have to be president", [...] I have far more experience than many others that sought the office of vice president of this country. I have as much experience in the Congress as Jack Kennedy did when he sought the presidency.*

b. **Bensten**: *Senator, I served with Jack Kennedy. I knew Jack Kennedy. Jack Kennedy was a friend of mine. Senator, you're no Jack Kennedy.*

c. **Quayle**: *That was unfair, sir. Unfair.*

d. **Bensten**: *You brought up Kennedy, I didn't.*

Example (4) is an example of how a player can go inconsistent in a debate, which has disastrous consequences, if the Jury enforces consistency as a necessary component of any winning condition. But the analysis depends on the semantics of discourse relations. It would seem that Quayle was unaware that (Example 4b.) was a possible move for Bentsen in a strategy of countering his commitments (we shall talk more about unawareness shortly). However, note that Quayle's commitments in (Example 4a.) are not innocuous in the first place. He brings up as a comparison one of the most revered Presidents in contemporary American history; and while it is true that John F. Kennedy, like Quayle, was a relatively inexperienced junior senator when he ran for President in 1960, Quayle could have chosen many other figures for comparison—for instance, Richard Nixon's credentials prior to his taking the post of Vice-President in 1952 were also comparable to Quayle's. But by choosing JFK as a reference and by referring to him with his nickname 'Jack' used by his advisors and friends, Quayle made the suggestion or weak-implicature, that perhaps he would be comparable in other ways to JFK. It certainly put Quayle's experience or lack thereof in a favorable light.

Notice too that Quayle did not come out with a bald assertion of this implicature in (Example 4a.). He did not say

a'. *I have as much experience in the Congress and as much Presidential potential as Jack Kennedy did when he sought the presidency.*

He sensed this would be a dangerous move, opening him up to attack and perhaps even ridicule, either from his opponent or at least in the minds of the Jury. So instead, he couched his message in an implicit form.

Our intuition is that Quayle did not anticipate a direct attack on the implicature he was drawing out. Perhaps he was not even aware that he was making such an implicature, though our discussion of alternatives suggests that something like that implicature is there and the result of a choice of Quayle's comparison. In any case, Quayle had no real counter-move or strategy prepared, we feel.

So what happened with Quayle's response? (Example 4d.) in discourse theory terms is a 'commentary' on Bentsen's attack move. Commentaries carry with them a commitment by their speaker to the content they are commenting on. Now if the commentary's target is the *content* of what Bentsen said, then this is devastating for Quayle. By saying Quayle is no Kennedy, Bentsen is implicating something stronger, that Quayle is not of Presidential material. With commentary on the *content*, Quayle then commits to that content. In so doing he commits to his not being of Presidential stature when precisely his winning condition was to constantly come back to that commitment and reaffirm it. His commitments are now inconsistent, and inconsistency can be a game-losing property in a conversation. Moreover, this was an inconsistency involving an intrinsic property of Quayle's winning condition.

There is an alternative interpretation of the commentary move (Example 4d.) by Quayle. The commentary move is not about the content of Bensten's move but rather about the fact that Bensten made this move. This seems more plausible and it commits Quayle on the face of it only to the fact that Bentsen made a particular discourse move. But by not counter-attacking Bensten, Quayle sends a message that is terrible for him. First, he commits that the attack is coherent and responsive. Second, by not replying he concedes and commits to the proposition that the content of Bensten's move *and its implicatures* are not attackable. That is, Quayle implicates he has no means to refute the content of the attack. But this in turn implies that he implicitly must commit to their content. Hence, his non-reply makes his commitments look inconsistent.

Example 4 also lends itself to an analysis from the perspective of 'unawareness' of moves available to one player by the other player. What happens when Player 0 thinks that an ME game $\mathcal{G}$ is being played over a vocabulary $(V_0 \cup V_1)$ whereas Player 1 actually has moves available to him from a larger vocabulary $W_1 \supsetneq V_1$? That is $\mathcal{G} = ((V_0 \cup W_1)^\omega, \mathcal{J})$. To answer this question, we make use of the following result.

Proposition 3. *Let V and W be countable vocabularies such that $V \subsetneq W$. Then, a Σ_0^1 complete set in X^ω jumps to Δ_2^0 in Y^ω, and all other sets stay in the same level.*

To preserve the continuity of the text, we give the proof in the appendix. Proposition 3 thus implies that a winning set win which is Σ_1^0 in an ME game $\mathcal{G} = ((V_0 \cup V_1)^\omega, \mathcal{J})$ might be Δ_2^0 in an ME game $\mathcal{G}' = ((V_0 \cup W_1)^\omega, \mathcal{J})$ where $W_1 \supsetneq V_1$. win is hence more complex structurally in $\mathcal{G}'$. The result of this might be that even if Player 0 had a winning strategy σ_0 in $\mathcal{G}$, σ_0 might not be winning for her in $\mathcal{G}'$.

Coming back now to Example 4, Quayle believed that if he just made his comparison with John F. Kennedy, to whom he refers by his colloquial nickname used by friends and members of JFK's cabinet, no matter what the response Bentsen made, that is the responses of which he was aware in V_1 would hurt his chances. He had a simple goal, which we could characterize as a Σ_1^0 goal: mentioning this comparison. As such, he also had a simple winning strategy for achieving this goal. However, in the larger set of discourse moves, W_1 Bentsen had an attack that floored Quayle. In fact, we can easily show that Quayle had no winning strategy for keeping to his winning condition *over strings in* $(V_0 \cup V_1)^\omega$; given that his winning strategy depended on *his opponent*'s use of moves in V_1, all that Bentsen had to do to defeat Quayle was to use a coherent move in W_1 to upset Quayle's strategy. This is a simple-minded yet insightful analysis of the interesting and deep notion of unawareness which we wish to fully explore in our future work. To fully understand this phenomenon, one has to appeal to the theory of epistemic games, to which we now turn.

3 Imperfect Information and Epistemic Considerations

So far we have shown how to model strategic conversations as infinite sequential games and how to reason about the complexity of certain commonly used winning goals in such conversations in terms of both their topological and logical complexities. A couple of issues that we have not addressed are:

- Yes, a conversation at the outset can be potentially infinite. But still in real life, the Jury does end the game after a finite amount of time, after a finite number of turns. By doing so, how can it be sure that it has correctly determined the outcome of the conversation? In other words, how does the Jury, at any point in a conversation gauge how the players are faring and when does it decide to call it a day?
- How does the Jury determine the winning conditions win_0 and win_1? Surely, it does not come up with a arbitrary subset of $(V_0 \cup V_1)^\omega$ with an arbitrary Borel complexity.

To address the above questions [3] introduced the model of 'weighted ME games' or WME games. A WME game is similar to a ME game except that the Jury instead of specifying the winning sets win_i as subsets of $(V_0 \cup V_1)^\omega$, determines them on-the-fly. It does so by evaluating every move of each player by assigning a 'weight' or a 'score'. The cumulative weight of a conversation ρ is then the discounted sum of these individual weights. [3] also showed that given an $\epsilon > 0$ there exists a number n_ϵ such that the Jury can stop the game after n_ϵ turns and determine the winner, being sure that no player could have done more than ϵ better than what they had already done. We do not go into the details here but refer the interested reader to that paper.

In this section, we study the exact information structure implicit in the strategic reasoning in conversations by extending framework of ME games with epistemic notions. We use the well-established theory of type-structures, first introduced in [17] and widely studied since. We assume that each player $i \in (\{0,1\} \cup \{\mathcal{J}\})$ has a (possibly infinite) set of types T_i. With each type t_i of Player i is associated a (first-order) belief function $\beta_i(t_i)$ which assigns to t_i a probability distribution over the types of the other players. That is, $\beta_i : T_i \to \Delta(\prod_{j \neq i} T_j)$. $\beta_i(t_i)$ represents the 'beliefs' of type t_i of Player i about the types of the other players and the Jury. The higher-order beliefs can be defined in a standard way by iterating the functions β_i. We assume that each type t_i of each Player i starts the game with an initial belief $\beta_i(t_i) \in \Delta(\prod_{j \neq i} T_j)$, called the 'prior belief'. The players take turns in making their moves and after every move, all the players dynamically update their beliefs through Bayesian updates. The notions of 'optimal strategies', 'best-response', 'rationality', 'common belief in rationality' etc. can then be defined in the standard way (see [12]).

Having imposed the above epistemic structure on ME games, we can now reason about the 'rationality' of the players' strategies. In order to justify or predict the outcome of games, many different solution concepts viz., Nash equilibrium, iterated removal of dominated strategies, correlated equilibrium, rationalizability etc. have been proposed [5,10,22]. Most of them have also been characterized in terms of the exact belief structure and strategic behavior of the players (see [12] for an overview). We can borrow results from this rich literature to predict or justify outcomes in strategic conversations. The details of the above is on-going work and we leave it to an ensuing paper. However, let us apply the above concepts and analyze our original example of Bronston and the Prosecutor.

To illustrate the power of types, let us return to Example 1. One conversational goal of the Prosecutor in Example 1 is to get Bronston to commit to an answer eventually (and admit to an incriminating fact) or to continue to refuse to answer (in which case he will be charged with contempt of court). Under such a situation, the response (1d.) of Bronston is clearly a clever strategic move. Bronston's response (1d.) was a strategic move aimed to 'misdirect' the Jury $\mathcal{J}$. He believed that $\mathcal{J}$ was of a type that would be convinced by his ambiguous response and neither incriminate him nor charge him with perjury nor of contempt of court. His move was indeed rational, *given his belief* about the Jury type. It turns out that while the jury of a lower court $\mathcal{J}_1$ was not convinced of Bronston's arguments and charged him with perjury, a higher court $\mathcal{J}_2$ overturned the verdict and released him. Thus his belief agreed with $\mathcal{J}_2$ but not $\mathcal{J}_1$.

Powerful as the above techniques are, one has to exercise caution and define the moves, states and the types of the players carefully. Having too rich a type space can lead to inexistence results. For example, consider the following situation.

Example 5. Two philosophers Michael and Brian must occupy a panel discussion before an audience. They both have an extremely good opinion of themselves. Each philosopher's goal is to prove that he is better than the other by talking highly of himself. They exchange dialogues where in every turn a philosopher can boast of himself as long as he wants to but eventually has to stop and concede the turn to the other philosopher. The audience, unlike the philosophers, can become impatient and decide at any moment to stop the discussion, give its verdict and leave. It offers the win to the one who has spoken 'more' of himself.

Clearly, the above game does not have an equilibrium pair of strategies. To see this, suppose without loss of generality that Michael speaks first. He has to concede the turn to Brian after saying m_1 points in his own favour (say). Brian plays next and he says b_1 points in his own favour. Now suppose the audience decides to stop the conversation after k sentences have been uttered by both the players. We can always find a k such that neither Michael nor Brian has a winning strategy. Indeed, if $b_1 > m_1$ and $k = b_1 + m_1$ then Michael cannot win. However, if $k < 2p_1$ Brian cannot win. Thus, both Michael and Brian could have done better by having said a 'bit more' about themselves in their corresponding turns. Without equilibria, it is unclear what our speakers should do in such a situation. Such examples pose a challenge to a fundamental assumption amongst linguists and philosophers that conversation is a rational activity with optimal strategies for achieving speakers' goals.

Our example in fact follows from a general result by [18], which says that if the space of types is not a separable set then there always exists a game with no equilibrium. In the above game, associating the types of a player with possible subsets of her strategies, we see that the space of types is a set with a large cardinality $(> \aleph_1)$ and hence we lose separability.

Conversationalists are aware implicitly of the dangers of such cases and debates have exogenous means of ensuring that there are optimal strategies for the speakers to follow. For instance, in debates there is usually a 'moderator' who ensures that all the participants get a fair chance to speak. She might interrupt a speaker and pass the turn on to another speaker. Note that this variant of our example game (Example 5) restores the presence of an equilibrium: each philosopher keeps speaking about himself till he is interrupted by the moderator - that is the best he can do anyway since he does not know in advance *when* he will be interrupted. More generally, we can restore separability (and hence the existence of equilibria) by limiting the set of types. One way is to require that each type (and hence each winning condition that players might countenance) be expressible in some language with a limited complexity. As long as the language is countable, separability can be restored for type spaces, and then by [18] any such game must have an equilibrium. Another way is to simply restrict the space of types to a strict subset of the entire space [8, 9]. Thus not all possible subsets of the conversational space define rational or rationalizable conversational goals. In the case of our example (Example 5) this means that our philosophers should limit the set of types that they consider possible. For example, they might expect each turn to last for a maximum of 20 min (say) so that their belief closed set is restricted to types of players who speak for a maximum of 20 min in each turn. This ensures the presence of an equilibrium.

4   Conclusion

We believe that the work summarized and extended in this paper is the start of a novel yet powerful approach to study strategic conversations. We have but scratched the surface here and there are many directions into which we would like to delve deeper in the future. One such direction, as we already mentioned, is to work out the epistemic theory of ME games in full detail. That is our current work in progress. Another is that in the present work we have considered the Jury as a 'passive' entity - it simply evaluates the play and determines the winner. However, in real life situations, the Jury can be an 'active' member of the conversation itself. It can 'applaud' or 'criticize' moves of the players. Thus, the Jury can be seen as making these moves in the game. Based on what the players observe about the Jury, they may update or change their beliefs and vice-versa. Incorporating this into our ME games requires a modification of the current framework where the Jury is another player making moves from its own set of vocabulary. We plan to explore this in future work.

Finally, in addition to the Jury, debates usually also have a moderator whose job is to conduct the debate and assign turns to the players. The moderator may also actively 'pass comments' about the moves of the players. A fair moderator gives all the players equal opportunity to speak and put their points across. However, if the moderator is unfair, he may 'starve' a particular player by not letting her enough chance to speak, respond to attacks and so on. Exploring the effects the inclusion of a moderator in such conversations is another interesting topic which we leave for future work.

A   Appendix

To prove Proposition 3 we shall refer to a result from [23].

Proposition 4 [23]. *If V is an infinite vocabulary, the subsets of V^ω of the form AV^ω, where A is a set of words of bounded length of V^* are clopen.*

We now prove Proposition 3.

Proof. First, we show that the set V^ω is closed but not open in the space W^ω. That is, $V^\omega \in (\Pi_1^0 \setminus \Sigma_1^0)$ in W^ω. Indeed, we have

$$V^\omega = \bigcap_{n \geq 0} V^n W^\omega$$

For every $n \geq 0$ we have that V^n is a set of words of bounded length of V^* and hence by Proposition 4 we have that $V^n W^\omega$ is clopen. Thus V^ω is closed. Also, V^ω is not open by the definition of open sets.

Now let $X \subset V^\omega$ be $(\Sigma_1^0 \setminus \Pi_1^0)$ in V^ω. By definition, we know that X is of the form AV^ω where $A \subset V^*$ Thus

$$X = AV^\omega = AW^\omega \cap V^\omega$$

Then since AW^ω is open (Σ_1^0) in W^ω and V^ω, as we just showed, is closed $(\Pi_1^0 \setminus \Sigma_1^0)$ in W^ω, their intersection is a Δ_2^0 set.

Next let $Y \subset X^\omega$ be $(\Pi_1^0 \setminus \Sigma_1^0)$ in V^ω. We show that Y is also closed in W^ω. Indeed, because the complement of Y in V^ω is of the form BV^ω for some $B \subset V^*$. Hence, the complement of Y in W^ω is

$$W^\omega \setminus Y = BW^\omega \cup W^*(W \setminus V)W^\omega$$

which is open.

References

1. Asher, N., Lascarides, A.: Logics of Conversation. Cambridge University Press, Cambridge (2003)
2. Asher, N., Lascarides, A.: Strategic conversation. Semant. Pragmat. **6**(2), 1–62 (2013). http://dx.doi.org/10.3765/sp.6.2
3. Asher, N., Paul, S.: Evaluating conversational success: weighted message exchange games. In: Hunter, J., Stone, M. (eds.) 20th workshop on the semantics and Pragmatics of Dialogue (SEMDIAL), New Jersey, USA, July 2016 (To appear)
4. Asher, N., Paul, S., Venant, A.: Message exchange games in strategic conversations. J. Philos. Log. (2016, in press)

5. Aumann, R.: Subjectivity and correlation in randomized strategies. J. Math. Econ. **1**, 67–96 (1974)
6. Baltag, A., Moss, L.S.: Logics for epistemic programs. Synthese **139**(2), 165–224 (2004)
7. Baltag, A., Moss, L.S., Solecki, S.: The logic of public announcements, common knowledge and private suspicions. Technical report SEN-R9922, Centrum voor Wiskunde en Informatica (1999)
8. Battigalli, P.: Rationalizability in infinite dynamic games with incomplete information. Res. Econ. **57**(1), 1–38 (2003)
9. Battigalli, P., Siniscalchi, M.: Rationalization and incomplete information. B.E. J. Theoret. Econ. **3**, 1–46 (2003)
10. Bernheim, B.D.: Rationalizable strategic behaviour. Econometrica **52**(4), 1007–1028 (1984)
11. Crawford, V., Sobel, J.: Strategic information transmission. Econometrica **50**(6), 1431–1451 (1982)
12. Dekel, E., Siniscalchi, M.: Epistemic game theory (chapter 12). In: Aumann, R.J., Hart, S. (eds.) Handbook of Game Theory with Economic Applications, vol. 4, pp. 619–702. Elsevier Publications, Cambridge (2015)
13. Franke, M.: Semantic meaning and pragmatic inference in non-cooperative conversation. In: Icard, T., Muskens, R. (eds.) ESSLLI 2008-2009. LNCS (LNAI), vol. 6211, pp. 13–24. Springer, Heidelberg (2010). doi:10.1007/978-3-642-14729-6_2
14. Gaiffman, H.: On local and non-local properties. In: Proceedings of Herbrand Symposium, Logic Colloquium 1981. North Holland (1982)
15. Glazer, J., Rubinstein, A.: On optimal rules of persuasion. Econometrica **72**(6), 119–123 (2004)
16. Grädel, E., Thomas, W., Wilke, T. (eds.): Automata, Logics, and Infinite Games: A Guide to Current Research [Outcome of a Dagstuhl Seminar, February 2001]. LNCS, vol. 2500. Springer, Heidelberg (2002)
17. Harsanyi, J.C.: Games with incomplete information played by Bayesian players, parts i-iii. Manag. Sci. **14**, 159–182 (1967)
18. Hellman, Z., Levy, Y.: Bayesian games with a continuum of states. Technical report, Bar Ilan University (2013)
19. Kechris, A.: Classical Descriptive Set Theory. Springer, New York (1995)
20. Libkin, L.: Elements of Finite Model Theory. Springer, New York (2004)
21. McNaughton, R., Papert, S.: Counter-free automata. In: Research Monograph, vol. 65. MIT Press, Cambridge (1971)
22. Nash, J.: Non-cooperative games. Ann. Math. **54**(2), 286–295 (1951)
23. Perrin, D., Pin, J.E.: Infinite Words - Automata, Semigroups, Logic and Games. Elsevier, Cambridge (1995)
24. Spence, A.M.: Job market signaling. J. Econ. **87**(3), 355–374 (1973)
25. Venant, A., Asher, N.: Dynamics of public commitments in dialogue. In: Proceedings of 11th International Conference on Computational Semantics, pp. 272–282. Association for Computational Linguistics, London, April 2015
26. Asher, N., Venant, A.: Ok or not ok? In: Semantics and Linguistic Theory 25. Cornell University Press (2015)

Polysemy and Coercion – A Frame-Based Approach Using LTAG and Hybrid Logic

William Babonnaud[1], Laura Kallmeyer[2], and Rainer Osswald[2(✉)]

[1] ENS Cachan, Université Paris-Saclay, Cachan, France
william.babonnaud@ens-cachan.fr
[2] Heinrich-Heine-Universität, Düsseldorf, Germany
{kallmeyer,osswald}@phil.uni-duesseldorf.de

Abstract. In this article, we propose an analysis of polysemy and coercion phenomena using a syntax-semantics interface which combines Lexicalized Tree Adjoining Grammar with frame semantics and Hybrid Logic. We show that this framework allows a straightforward and explicit description of selectional mechanisms as well as coercion processes. We illustrate our approach by applying it to examples discussed in Generative Lexicon Theory [23,25]. This includes the modeling of *dot objects* and associated coercion phenomena in our framework, as well as cases of functional coercion triggered by transitive verbs and adjectives.

Keywords: Systematic polysemy · Coercion · Lexical semantics · Frame semantics · Hybrid logic · Lexicalized tree adjoining grammars · Hole semantics · Underspecification · Syntax-semantics interface · Generative lexicon theory

1 Introduction

Any compositional model of the syntax-semantics interface has to cope with polysemy and coercion phenomena. Well-known examples of *inherent systematic polysemy* are the varying sortal characteristics of physical carriers of information such as *book*: Books can be bought, read, understood, put away, and remembered, and thus can refer to physical objects or abstract, informational entities, depending on the context of use. The question is then how to represent such potential meaning shifts in the lexicon and how to integrate the respective meaning components compositionally within the given syntagmatic environment. A different but related phenomenon in *selectional polysemy* [25], where an apparent selectional mismatch is resolved by *coercion* mechanisms that go beyond referential shifts provided by lexical polysemy. Examples are given by expressions like *Mary*

This work was supported by the CRC 991 "The Structure of Representations in Language, Cognition, and Science" funded by the German Research Foundation (DFG). The first author was financially supported during his stay in Düsseldorf by ENS Cachan, Université Paris-Saclay. We would like to thank the three anonymous reviewers for their helpful comments.

M. Amblard et al. (Eds.): LACL 2016, LNCS 10054, pp. 18–33, 2016.
DOI: 10.1007/978-3-662-53826-5_2

began the book and *John left the party*, where the aspectual verb *begin* selects for an event argument (here, an activity with the book as an undergoer), and *leave* selects for an argument of type *location*.

There is a considerable body of work on the compositional treatment of polysemy and coercion. One important strand of research in this domain is the *dot type* and *qualia structure* approach as part of *Generative Lexicon Theory* developed by James Pustejovsky and his colleagues [6, 22, 23, 25, 26]. A more recent development in this direction is *Type Composition Logic* [3–5, 7], which introduces an elaborate system of complex types and rules for them. The approach presented in the following takes a model-oriented perspective in that it asks for the semantic structures in terms of semantic frames that underlie the phenomena in question. We propose a compositional framework in which syntactic operations formulated in Lexicalized Tree Adjoining Grammar drive the semantic composition. On the semantic side, we use underspecified Hybrid Logic formulas for specifying the associated semantic frames.

The rest of the paper is structured as follows: Sect. 2 introduces the general model of the syntax-semantics interface adopted in this paper. Its main components are a formal model of semantic frames, a slightly adapted version of Hybrid Logic for describing such frames, and a version of Lexicalized Tree Adjoining Grammar which combines elementary trees with underspecified Hybrid Logic formulas. Section 3 shows how this framework can be fruitfully employed for a detailed modeling of systematic polysemy and coercion phenomena. It is shown how dot objects can be represented in frame semantics and how various cases of argument selection and coercion can be formally described. Section 4 gives a brief summary and lists some topics of current and future research.

2   The Formal Framework

We follow [16] in adopting a framework for the syntax-semantics interface that pairs a Tree Adjoining Grammar (TAG) with semantic frames. More concretely, every elementary syntactic tree is paired with a frame description formulated in Hybrid Logic (HL) [2]. In the following, we briefly introduce this framework; see [14, 16] for more details.

2.1   Frames

Frames [8, 10, 18] are semantic graphs with labeled nodes and edges, as in Fig. 1, where nodes correspond to entities (individuals, events, ...) and edges to (functional or non-functional) relations between these entities. In Fig. 1 all relations except *part-of* are meant to be functional.

Frames can be formalized as extended typed feature structures [15, 21] and specified as models of a suitable logical language. In order to enable quantification over entities or events, [16] propose to use Hybrid Logic, an extension of modal logic.

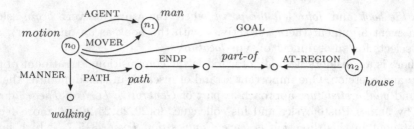

Fig. 1. Frame for the meaning of *the man walked to the house* (adapted from [15])

2.2 Hybrid Logic and Semantic Frames

Before giving the formal definition of Hybrid Logic (HL) as used in this paper, let us illustrate its use for frames with some examples. Consider the frame in Fig. 1. The types in frames are propositions holding at single nodes, the formula *motion*, for instance, is true at the node n_0 but false at all other nodes of our sample frame. Furthermore, we can talk about the existence of an attribute for a node. This corresponds to stating that there exists an outgoing edge at this node using the $\Diamond$ modality in modal logic. In frames, there may be several relations, hence several modalities, denoted by $\langle R \rangle$ with R the name of the relation. For example, $\langle \text{AGENT} \rangle man$ is true at the *motion* node n_0 in our frame because there is an AGENT edge from n_0 to some other node where *man* holds. (Note that HL does not distinguish between functional and non-functional edge labels. That is, functionality has to be enforced by additional constraints.) Finally, we can have conjunction, disjunction, and negation of these formulas. E.g., *motion* $\wedge$ $\langle \text{MANNER} \rangle walking \wedge \langle \text{PATH} \rangle \langle \text{ENDP} \rangle \top$ is also true at the *motion* node n_0.

HL extends this with the possibility to name nodes in order to refer to them, and with quantification over nodes. We use a set of nominals (unique node names), and a set of node variables. n_0 is such a nominal, the node assigned to it is the motion node in our sample frame. $x, y, \ldots$ are node variables. The truth of a formula is given with respect to a specific node w in a frame, an assignment V from nominals to nodes in the frame and an assignment g which maps variables to nodes in the frame.

There are different ways to state existential quantifications in HL, namely $\exists\phi$ and $\exists x.\phi$. $\exists\phi$ is true at w if there exists a node w' at which ϕ holds. In other words, we move to some node w' in the frame and there ϕ is true. For instance, $\exists house$ is true at any node in our sample frame. As usual, we define $\forall\phi \equiv \neg\exists(\neg\phi)$ and $\phi \rightarrow \psi \equiv \neg\phi \vee \psi$. In contrast to $\exists\phi$, $\exists x.\phi$ is true at w if there is a w' such that ϕ is true at w under an assignment of x to w'. In other words, there is a node that we name x but for the evaluation of ϕ, we do not move to that node. E.g., the formula $\exists x.\langle \text{PATH} \rangle \langle \text{ENDP} \rangle \langle part\text{-}of \rangle(x \wedge region) \wedge \exists(house \wedge \langle \text{AT-REGION} \rangle x)$ is true at the *motion* node in our sample frame.

Besides quantification, HL also allows us to use nominals or variables to refer to nodes via the @ operator: $@_n\phi$ specifies the moving to the node w denoted by n before evaluating ϕ. n can be either a nominal or a variable.

The ↓ operator allows us to assign the current node to a variable: $\downarrow x.\phi$ is true at w if ϕ is true at w under the assignment g_w^x. I.e., we call the current node x, and, under this assignment, ϕ is true at that node. E.g., $\langle\text{PATH}\rangle\langle\text{ENDP}\rangle\langle\textit{part-of}\,\rangle(\downarrow x.\textit{region} \wedge \exists(\textit{house} \wedge \langle\text{AT-REGION}\rangle x))$ is true at the *motion* node in our frame.

To summarize this, our HL formulas have the following syntax: Let Rel = Func ∪ PropRel be a set of functional and non-functional relation symbols, Type a set of type symbols, Nom a set of nominals (node names), and Nvar a set of node variables, with Node = Nom ∪ Nvar. Formulas are defined as:

(1)   Forms $::= \top \mid p \mid n \mid \neg\phi \mid \phi_1 \wedge \phi_2 \mid \langle R\rangle\phi \mid \exists\phi \mid @_n\phi \mid \downarrow x.\phi \mid \exists x.\phi$

where $p \in$ Type, $n \in$ Node, $x \in$ Nvar, $R \in$ Rel and $\phi, \phi_1, \phi_2 \in$ Forms. For more details and the formal definition of satisfiability as explained above see [14,16].

2.3   LTAG and Hybrid Logic

A *Lexicalized Tree Adjoining Grammar* (LTAG; [1,12]) consists of a finite set of *elementary trees*. Larger trees are derived via *substitution* (replacing a leaf with a tree) and *adjunction* (replacing an internal node with a tree). An adjoining tree has a unique *foot node* (marked with an asterisk), which is a non-terminal leaf labeled with the same category as the root of the tree. When adjoining such a tree to some node n of another tree, in the resulting tree, the subtree with root n from the original tree is attached at the foot node of the adjoining tree.

The non-terminal nodes in LTAG are usually enriched with feature structures [27]. More concretely, each node has a top and a bottom feature structure (except substitution nodes, which have only a top). Nodes in the same elementary tree can share features. Substitutions and adjunctions trigger unifications in the following way: In a substitution step, the top of the root of the new tree unifies with the top of the substitution node. In an adjunction step, the top of the root of the adjoining tree unifies with the top of the adjunction site and the bottom of the foot of the adjoining tree unifies with the bottom of the adjunction site. Furthermore, in the final derived tree, top and bottom must unify in all nodes.

Our framework for the syntax-semantics interface follows previous LTAG semantics approaches in pairing each elementary tree with a semantic representation that consists of a set of HL formulas, which can contain holes and which can be labeled. In other words, we apply *hole semantics* [9] to HL and link these underspecified formulas to the elementary trees. Composition is then triggered by the syntactic unifications arising from substitution and adjunction, using interface features on the syntactic trees, very similar to [11,13,17].

As a basic example consider the derivation given in Fig. 2 where the two NP trees are substituted into the two argument slots in the *ate* tree. The interface features I on the NP nodes make sure that the contributions of the two arguments feed into the AGENT and THEME nodes of the frame. Furthermore, an interface feature MINS is used for providing the label of the $\exists(\textit{eating}...)$ formula as minimal scope to a possible quantifier. The unifications lead to identities ① $= i$, ② $= x$ and

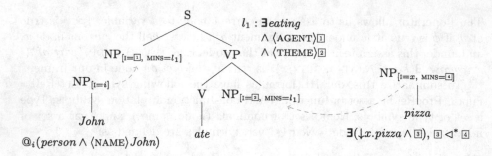

Fig. 2. Derivation of *John ate pizza*

$④ = l_1$, triggered by the feature unifications on the syntactic tree. As a result, when collecting all formulas, we obtain the underspecified representation

(2) $@_i(person \wedge \langle \text{NAME} \rangle John)$, $l_1 : \exists(eating \wedge \langle \text{AGENT} \rangle i \wedge \langle \text{THEME} \rangle x)$, $\exists(\downarrow x.pizza \wedge ③)$, $③ \lhd^* l_1$

The relation $\lhd^*$ links holes to labels: $h \lhd^* l$ signifies that the formula labeled l is a subformula of h or, to put it differently, is contained in h. In (2), the $\exists(eating...)$ formula, labeled l_1, has to be part of the nuclear scope of the quantifier, which is given by the hole $③$. Disambiguating such underspecified representations consists of "plugging" the labeled formulas into the holes while respecting the given constraints. Such a plugging amounts to finding an appropriate bijection from holes to labels. (2) has a unique disambiguation, namely $③ \rightarrow l_1$. This leads to (3), which is then interpreted conjunctively.

(3) $@_i(person \wedge \langle \text{NAME} \rangle John)$, $\exists(\downarrow x.pizza \wedge \exists(eating \wedge \langle \text{AGENT} \rangle i \wedge \langle \text{THEME} \rangle x))$

3   Application to Coercion

3.1   Dot Objects in Frames Semantics

In order to capture the full complexity of concepts while modeling them, we need a way to represent the phenomenon of *inherent polysemy*, that is, the phenomenon that certain concepts integrate two or more different and apparently contradictory senses. Consider for instance the following two sentences:

(4)   a.   The book is heavy.
         b.   The book is interesting.

Both sentences use *book* in the common way, but while in (4a) the adjective *heavy* applies to a physical object, the adjective *interesting* in (4b) requires its object to be an information. It thus appears that *book* carries two different aspects, which are arguably incompatible. However, this contradiction reveals an underlying structure in which these aspects are linked to each other. This structure appears in Pustejovsky's work [23] under the name of *dot object*. Following this approach, our frame definition of *book* encodes the lexical structure proposed by

Pustejovsky by taking two nodes with types *information* and *phys-obj* respectively, to represent both aspects, and defining an explicit relation between them, which is quite similar to what Pustejovsky calls the formal component of the concept. In the traditional definition of frames, one node should be marked as the referential, or central one, the others being connected to it by functional edges (see e.g. [21]). The necessity of fixing a referent for sense determination was also proposed in [19]. We have therefore chosen to take the physical aspect of *book* as the referential node; and since its two aspects are linked by the "has information content" relation, we define a CONTENT attribute to connect the physical object to the information it carries.[1] We thus get the following formula to express the semantics for *book*:

(5) $book \wedge \langle \text{CONTENT} \rangle information$

To ensure that the type *book* is permitted where a *phys-obj* is required, we assume general constraints which, among other things, express that books are entities of type *phys-obj*. Furthermore, we introduce a type *info-carrier* for information carrying physical objects, and therefore build our constraints in two steps:

(6)   a. $\forall (book \rightarrow info\text{-}carrier)$
        b. $\forall (info\text{-}carrier \rightarrow phys\text{-}obj \wedge \langle \text{CONTENT} \rangle information)$

The purpose of the type *info-carrier* is to provide a stage between specific types like *book* and more general ones like *phys-obj*, to which other concepts can be linked. For instance, a complex word like *newspaper* should have a type which implies the type *info-carrier* [6,20,24]. Note that we can easily deduce the following constraint from (6a) and (6b):

(7) $\forall (book \rightarrow phys\text{-}obj)$

This constraint will be very useful to simplify formulas where the type *book* is involved.

3.2   Coercion, Selection and Dot Objects

Let us start with the case of *read*, which has been described in [23]. The verb *read* allows for the direct selection of the dot object *book* as complement, as illustrated in (8a), but also enables coercion of its complement from type *information* in (8b) as well as from type *phys-obj* in (8c). The distinction between all these concepts can be explained as follows: although books and stories are informational in

---

[1] One of the reviewers raised the question on what grounds *phys-obj* is preferred over *information* as the primary lexical meaning facet of *book* and, more importantly, of how to decide this question for related terms like *novel* and for dot types in general. We regard this as an empirical issue which falls ultimately into the realm of psycholinguistic research. As a first approximation, we tend to rely on the information provided by monolingual dictionaries. For instance, the *Longman Dictionary of Contemporary English* tells us that a book is "a set of printed pages that are held together in a cover".

nature, a story does not need a physical realisation, whereas a book does, and although books and blackboards are physical objects, a blackboard does not necessarily contain information. The constraints for the associated types are defined in (9).

(8)   a.   John read the book.
      b.   John read the story.
      c.   John read the blackboard.

(9)   a. $\forall(story \rightarrow information)$
      b. $\forall(blackboard \rightarrow phys\text{-}obj)$
      c. $\forall(phys\text{-}obj \wedge information \rightarrow \bot)$

The semantics for *read* has to encode the direct selection of a dot object as a complement. In [23], the verb *read* is analysed with two distinct events linked by a complex relation expressing the fact that the reader first sees the object before reaching its informational content. We want to keep a similar analysis here; we build our semantic definition of *read* by taking an event node of type *reading* with two attributes, namely PERCEPTUAL-COMPONENT and MENTAL-COMPONENT, whose values are respectively of type *perception* and *comprehension*.[2] These nodes are meant to represent the decomposition of the activity of reading into two subevents, the action of looking at a physical object (the perception) and the action of processing the provided information (the comprehension). These two events linked by a non-functional temporal relation inspired from the one proposed by Pustejovsky: we call it *ordered-overlap*, and it expresses the fact that the perception starts before the comprehension and that these two subevents (typically) overlap. For the sake of simplicity, we encode the central part of this semantics into the definition of *reading* with the following constraint:

(10)
$$\forall(reading \rightarrow \exists x.\langle\text{PERC-COMP}\rangle(perception \wedge \langle ordered\text{-}overlap \rangle x) \\ \wedge \langle\text{MENT-COMP}\rangle(comprehension \wedge x))$$

Moreover, the *perception* node has an attribute STIMULUS describing the role of its object, which has to be of type *phys-obj*, and the *comprehension* node has an attribute CONTENT which refers to the information that was read. We also explicitly add in our semantics the requirement that the value of STIMULUS has a CONTENT attribute, whose value is the same for the CONTENT attribute from the *comprehension* node. Furthermore, since the argument contributed by the object can be either the stimulus of the perception (*phys-obj*) or its content, we add a disjunction of these two possibilities. We therefore obtain the formula represented in Fig. 3, with the associated elementary tree.[3] In this formula, ☐ is

---

[2] In the following, we will abbreviate these attributes by PERC-COMP and MENT-COMP, respectively.

[3] The constraint (10) should be applied here, but for reasons of space, we do not list all the conjuncts contributed by it.

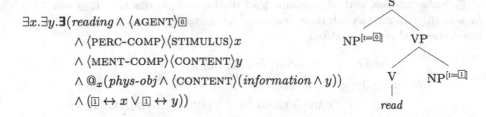

$$\exists x. \exists y. \mathbf{\exists}(reading \wedge \langle\text{AGENT}\rangle \boxed{0}$$
$$\wedge \langle\text{PERC-COMP}\rangle\langle\text{STIMULUS}\rangle x$$
$$\wedge \langle\text{MENT-COMP}\rangle\langle\text{CONTENT}\rangle y$$
$$\wedge @_x(phys\text{-}obj \wedge \langle\text{CONTENT}\rangle(information \wedge y))$$
$$\wedge (\boxed{1} \leftrightarrow x \vee \boxed{1} \leftrightarrow y))$$

Fig. 3. Semantics and elementary tree for *read*

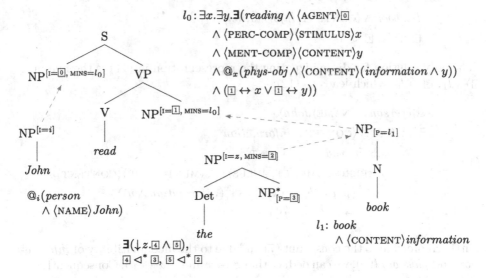

Fig. 4. Derivation for (8a)

intended to unify with a node variable when the direct object gets inserted (i.e., $\boxed{1}$ is provided as value of the feature I in the object node associated lexicalized tree for *read*), and the process of rewriting and simplifying the final formula will allow us to identify either x or y with the variable of the direct object, depending on whether this is of type *phys-obj* or *information*.

We can now use this elementary tree-frame pair to achieve a derivation for (8a), which is represented in Fig. 4. The HL formula coming with *read* is now labeled and its label is provided as potential minimal scope for quantifiers at the NP slots. Concerning the entry of *the*, we simplify here and treat is as an existential quantifier, disregarding the presuppositions it carries. The *book* formula is also labeled, and the label is made available via an interface feature P (for "proposition").[4] Due to the two scope constraints, this proposition will be part of the restriction of the quantifier (i.e., part of the subformula at $\boxed{4}$) while the *read* formula will be part of the nuclear scope, i.e., part of the subformula

---

[4] Note that in Fig. 4, we have already applied (6).

at $\boxed{5}$. Substitutions and adjunctions lead to the unifications $\boxed{0} = i$, $\boxed{1} = z$, $\boxed{2} = l_0$ and $\boxed{3} = l_1$ on the interface features. As a result, we obtain the following underspecified representation:

(11)
$$@_i(person \wedge \langle\text{NAME}\rangle\, John),\ \exists(\downarrow z.\boxed{4} \wedge \boxed{5}),$$
$$l_0 : \exists x.\exists y.\exists(reading \wedge \langle\text{AGENT}\rangle i$$
$$\wedge\ \langle\text{PERC-COMP}\rangle\langle\text{STIMULUS}\rangle x \wedge \langle\text{MENT-COMP}\rangle\langle\text{CONTENT}\rangle y$$
$$\wedge\ @_x(phys\text{-}obj \wedge \langle\text{CONTENT}\rangle(information \wedge y))$$
$$\wedge\ (z \leftrightarrow x \vee z \leftrightarrow y)),$$
$$l_1 : book \wedge \langle\text{CONTENT}\rangle information,$$
$$\boxed{4} \vartriangleleft^* l_1,\ \boxed{5} \vartriangleleft^* l_0$$

The only solution for disambiguating the representation in (11) is the mapping $\boxed{4} \mapsto l_1$, $\boxed{5} \mapsto l_0$, which leads to (12):

(12)
$$@_i(person \wedge \langle\text{NAME}\rangle\, John),$$
$$\exists(\downarrow z.book \wedge \langle\text{CONTENT}\rangle information$$
$$\wedge\ \exists x.\exists y.\exists(reading \wedge \langle\text{AGENT}\rangle i$$
$$\wedge\ \langle\text{PERC-COMP}\rangle\langle\text{STIMULUS}\rangle x \wedge \langle\text{MENT-COMP}\rangle\langle\text{CONTENT}\rangle y$$
$$\wedge\ @_x(phys\text{-}obj \wedge \langle\text{CONTENT}\rangle(information \wedge y))$$
$$\wedge\ (z \leftrightarrow x \vee z \leftrightarrow y))),$$

Furthermore, due to the constraint (7) and due to the incompatibility of *information* and *phys-obj* (9c), we can deduce that $z \leftrightarrow x$ and $\neg(z \leftrightarrow y)$. Consequently, we can simplify our formulas by omitting the $\exists x$ quantification and replacing every x with z. Putting these things together leads to the representation (13):

(13)
$$@_i(person \wedge \langle\text{NAME}\rangle\, John),$$
$$\exists(\downarrow z.book \wedge \langle\text{CONTENT}\rangle information$$
$$\wedge\ \exists y.\exists(reading \wedge \langle\text{AGENT}\rangle i \wedge \langle\text{PERC-COMP}\rangle\langle\text{STIMULUS}\rangle z$$
$$\wedge\ \langle\text{MENT-COMP}\rangle\langle\text{CONTENT}\rangle y$$
$$\wedge\ @_z(phys\text{-}obj \wedge \langle\text{CONTENT}\rangle(information \wedge y)))$$

The frame shown in Fig. 5 is a minimal model for (13) which also takes (10) into account, i.e., it is the smallest frame graph satisfying (13) and (10).

The semantic representations of (8b) and (8c) can be derived in a similar way, except that for (8b), the variable z introduced by the quantifier will be equivalent to the *information* variable y in the contribution of *read*. The interesting point in these cases is that the final semantic formula involves a node which reflects respectively that there is an implicit material on which the story is written (8b) and that implicit contents are written on the blackboard (8c). The analysis of (8a) differs from that of (8c) in that the semantics of *book* always brings a

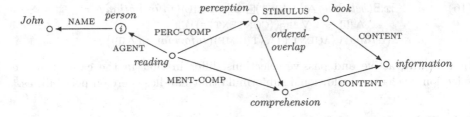

Fig. 5. Frame for (8a) *John read the book*

CONTENT attribute of type *information*, which is merged with the constraints contributed by the semantics of *read*. In (8c), by contrast, the CONTENT attribute of the blackboard is contributed solely by the verb.

It is also worth asking how to handle cases where the verb does not select a dot object as for *read*, but rather a simple type. Indeed, although the dot object *book* has the properties of physical objects and of information, there are some verbs which do not allow *book* as a complement but select a pure informational argument. These verbs actually provide no possibilities of coercion: their argument has to be of the specified type to allow a direct selection. This kind of selection is referred to as *passive* selection, in opposition to the *active* selection which enables coercion and type accommodation [23]. To understand this phenomenon, consider the following sentences (those in (15) are taken from [23]):

(14)   a.   Mary believed the story.
        b.   Mary believed the book.

(15)   a.   Mary told the story.
        b.   *Mary told the book.

The verbs *believe* and *tell* both require their argument to be of type *information*; however, the verb *believe* accepts the dot object *book* as its argument whereas *tell* does not: the sentence (15b) seems to be incorrect. Thus the examples in (14) illustrate a case of active selection, with a coercion of the complement in (14b), and those in (15) show a case of passive selection.

With our semantics for *read*, the way to build the semantics for these two verbs is quite straightforward. In comparison to *read*, we only need in each case a single node to represent the activity, respectively of *believing* and of *telling*. But the really interesting point is about the selection of the variable provided by the semantics of the argument. In the case of *read*, we had the subformula $\boxed{1} \leftrightarrow x \lor \boxed{1} \leftrightarrow y$, with $\boxed{1}$ to be unified with the variable contributed by the direct object, regardless of its type. For *believe*, we need a similar subformula that allows for the object variable to be either of type *information*, or to have a CONTENT attribute with a value of type *information*; cf. (16a). For *tell*, however, the object variable has to be unified directly with the THEME of *telling*, which is of type *information*; cf. (16b).

(16)   a. $\exists x. \exists(believing \wedge \langle \text{AGENT} \rangle \boxed{0} \wedge \langle \text{THEME} \rangle (information \wedge x)$
                $\wedge (\boxed{1} \leftrightarrow x \vee @_{\boxed{1}}(\langle \text{CONTENT} \rangle x)))$
        b. $\exists(telling \wedge \langle \text{AGENT} \rangle \boxed{0} \wedge \langle \text{THEME} \rangle (information \wedge \boxed{1}))$

In this way, active and passive selections differ in that in the case of active selection, we have an additional subformula that handles coercion possibilities.[5]

3.3   Other Cases of Coercion

Coercion is not limited to dot objects: it can occur for many other concepts with a simple type. We will discuss here a few more examples of coercion, and present ways to handle them within our framework. We will thus show that many different cases of coercion can be solved in similar ways. We start here with a sentence taken from [25]:

(17)   John left the party.

The verb *leave* requires its object to be of type *location* while in (17), the noun *party* is provided, which is of type *event* and does not carry a dot type. Here, the coercion relies on the fact that *party*, like every event, has an associated location, which is basically where the party takes place. The application of *leave* to *the party* therefore involves a transfer of meaning from the direct sense to a related one. This phenomenon is referred to as *attribute functional coercion* [25, 26] because it operates on concepts which can serve as types as well as attributes.

Our framework is capable of handling such cases without problems. Indeed, the basis of frame semantics is to work with attribute-value descriptions, and the coercion which occurs here shifts from one sense to another by following an attribute to get to the required concept type. Hence we naturally define a type *location* and an attribute LOCATION to represent the dual nature of the concept of *location*. As previously for *book*, we need to assume the general constraints in (18) to link *party* with these new elements:

(18)   a. $\forall(party \rightarrow event)$
        b. $\forall(event \rightarrow \langle \text{LOCATION} \rangle location)$
        c. $\forall(event \wedge location \rightarrow \perp)$

It remains to define the semantics for *leave* in such a way that it enables coercion to the value of the attribute LOCATION when the given argument is not of the required type. This can be done in a similar way to what we did in the case of *believe* in (16a), leading to the formula in Fig. 6, where $\boxed{1}$ is intended to be unified with the node variable from the direct object argument. By following the steps described in Sect. 3.2, we can easily produce a derivation for (17).

---

[5] As pointed out by one of the reviewers, having the attribute CONTENT in the disjunction in (16a) imposes specific constraints on the semantic structure of the argument. We leave it as a question for future research whether constraints of this type are overly restrictive when moving from selected examples to large-scale applications.

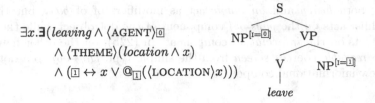

$$\exists x. \exists (leaving \land \langle\text{AGENT}\rangle \boxed{0}$$
$$\land \langle\text{THEME}\rangle (location \land x)$$
$$\land (\boxed{1} \leftrightarrow x \lor @_{\boxed{1}} (\langle\text{LOCATION}\rangle x)))$$

Fig. 6. Semantics and elementary tree for *leave*

As a starting point, we consider the yield of the syntactic unifications and the mapping of holes to formulas, which gives the following result:

$$@_i(person \land \langle\text{NAME}\rangle John),$$
(19) $$\exists (\downarrow z.party \land \exists x. \exists (leaving \land \langle\text{AGENT}\rangle i \land \langle\text{THEME}\rangle (location \land x)$$
$$\land (z \leftrightarrow x \lor @_z(\langle\text{LOCATION}\rangle x))))$$

With the constraints in (18), we can conclude that $\neg(z \leftrightarrow x)$ and consequently, we obtain the following semantics for (17):

$$@_i(person \land \langle\text{NAME}\rangle John),$$
(20) $$\exists (\downarrow z.party \land \exists x. \exists (leaving \land \langle\text{AGENT}\rangle i \land \langle\text{THEME}\rangle (location \land x)$$
$$\land @_z(\langle\text{LOCATION}\rangle x)))$$

There is only a slight difference between functional coercion of the kind just described and the treatment of dot objects shown before: frame semantics allows us to process both types of coercion phenomena in a similar way because of the underlying attribute-value structure. A further example is given by the dot object *speech*, which combines the types *event* and *information* [25]. *Speech* has the two attributes CONTENT and LOCATION, among others. More precisely, the dot type *speech* is characterized by the constraint in (21a), which, together with (18b) repeated here as (21b) gives rise to the constraint in (21c). Note that the latter constraint makes no difference between the two attributes – although they have different "levels" of origin, as CONTENT is a direct consequence of *speech* whereas LOCATION is implied by the type *event*, which is entailed by *speech*.

(21)   a. $\forall (speech \rightarrow event \land \langle\text{CONTENT}\rangle information)$
          b. $\forall (event \rightarrow \langle\text{LOCATION}\rangle location)$
          c. $\forall (speech \rightarrow \langle\text{CONTENT}\rangle information \land \langle\text{LOCATION}\rangle location)$

Two further examples, adapted from [6], are considered in (22) below. Their purpose is to show how adjectival modification which enables coercion can be handled in our framework.

(22)   a.   Mary mastered the heavy book on magic.
          b.   Mary broke every readable screen.

In (22a), both *heavy* and *on magic* act as modifiers of of *book*, but the former modifier acts on the *phys-obj* component of the dot object while the latter modifier acts on the *information* component. In (22b), on the other hand, the adjective *readable* coerces *screen* from the simple type *phys-obj* to a dot type, with a new informational component.

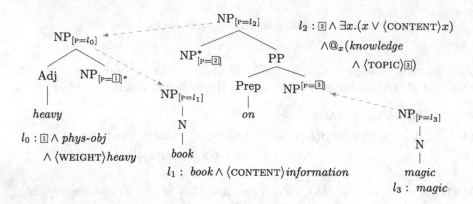

Fig. 7. Derivation for *heavy book on magic*

Let us start with the sentence in (22a). The most interesting parts of its derivation are represented in Fig. 7. We define the semantics of *heavy* by assuming that it selects directly a physical object (and so voluntarily keeping any other meaning aside). The semantics of *magic* is simply regarded as sortal for the purposes of the present example. As for *on*, its semantic representation includes a disjunction to allow for the identification with a node of the required type, using a similar technique as in the representation of *believe* above.[6] Moreover, we introduce a type *knowledge*, which is intended to be a subtype of *information*, and which has a TOPIC attribute describing what field the knowledge is about. That is, we have the constraint in (23).

(23) $\forall(knowledge \rightarrow information \land \langle \text{TOPIC} \rangle \top)$

The substitutions and adjunctions in Fig. 7 trigger unifications $\boxed{1} = l_1$, $\boxed{2} = l_0$ and $\boxed{3} = l_3$, which leads to the HL formula in (24):

(24) $book \land \langle \text{CONTENT} \rangle information \land phys\text{-}obj \land \langle \text{WEIGHT} \rangle heavy$
$\land \exists x.(x \lor \langle \text{CONTENT} \rangle x) \land @_x(knowledge \land \langle \text{TOPIC} \rangle magic)$

---

[6] The given semantic representation for *on* is considerably simplified. A more precise representation should include a selection between two effects depending on the type of the argument of *on* (unified with $\boxed{3}$ on Fig. 7), as the preposition can also occur in phrases like *the book on the table* where a physical object is involved: in this case, a more elaborated subformula with a LOCATION attribute would replace the subformula $knowledge \land \langle \text{TOPIC} \rangle \boxed{3}$.

Formula (24) can be simplified due to the fact that *book* and *knowledge* are incompatible; therefore the first element of the disjunction $x \vee \langle \text{CONTENT} \rangle x$ (which is evaluated in the *book* node) cannot be true. Consequently, we reduce the disjunction to $\langle \text{CONTENT} \rangle x$.

It is also worth noticing that the verb *master* seems to require an object of type *knowledge* and not merely *information*. Indeed, the use of this verb with another subtype of *information* as in sentences (25b) seems unacceptable, while (25a) involving a pure *knowledge* concept is fully acceptable. The sentence in (25c) shows that *master* is able to coerce at least certain types of arguments.

(25)
    a.   John mastered the theorem.
    b.   *John mastered the story.
    c.   John mastered the book.

The selectional mechanism is therefore more complex for this verb. Nevertheless, as the type *knowledge* provided by *on magic* overwrites the *information* value in the relevant example (22a), *book* has already a coerced type for its CONTENT in this case, which allows us to leave a more general analysis of *master* for future work and to assume for now the same behavior for this verb as for *believe*. A derivation for the sentence in (22a) leads thus, after unification and simplification, to the following semantic representation:

(26)
$$@_i(person \wedge \langle \text{NAME} \rangle Mary),$$
$$\exists(\downarrow z.book \wedge \langle \text{CONTENT} \rangle information \wedge phys\text{-}obj \wedge \langle \text{WEIGHT} \rangle heavy$$
$$\wedge \langle \text{CONTENT} \rangle (knowledge \wedge \langle \text{TOPIC} \rangle magic)$$
$$\wedge \exists y. \exists (mastering \wedge \langle \text{AGENT} \rangle i$$
$$\wedge \langle \text{THEME} \rangle (knowledge \wedge y) \wedge @_z(\langle \text{CONTENT} \rangle y)))$$

The case in (22b) is very similar to the coercion of *blackboard* to a dot object by *read*. The semantics of the adjective *readable* does nothing else than adding a CONTENT attribute with an *information* value to a physical object. This translates into the logical formula in (27a). Moreover, *screen* is considered as a subtype of *phys-obj*, and we assume here a simple semantics for *break*, given in (27b). Finally, (27c) recalls the semantics for *every*.[7]

(27)
    a. $\boxed{0} \wedge phys\text{-}obj \wedge \langle \text{CONTENT} \rangle information$
    b. $\exists(breaking \wedge \langle \text{AGENT} \rangle \boxed{1} \wedge \langle \text{THEME} \rangle \boxed{2})$
    c. $\forall(\downarrow x.\boxed{3} \rightarrow \boxed{4})$

The derivation for (22b) therefore leads to:

(28)
$$@_i(person \wedge \langle \text{NAME} \rangle Mary)$$
$$\wedge \forall(\downarrow x.screen \wedge \langle \text{CONTENT} \rangle information$$
$$\rightarrow \exists(breaking \wedge \langle \text{AGENT} \rangle i \wedge \langle \text{THEME} \rangle x))$$

---

[7] Lack of space prevents us from showing the associated elementary syntactic trees.

The foregoing examples have shown that our formal framework allows us to solve a large variety of coercion problems in similar ways, building on constraint-based semantic representations combining frames and HL.

4 Conclusion

In this paper, we presented a model of coercion mechanisms for the case of verbs and adjectives which select nominal arguments within a syntax-semantics interface based on frames semantics using LTAG and HL. We also provided a frame-semantic representation of Pustejovsky's dot objects which keeps the notion of referential meaning and explicitly includes the relations between the different aspects of a concept. Frame semantics is well-suited to handle such mechanisms since type shifting can simply be modeled by moving along an attribute relation from a given meaning to the coerced one. Furthermore, the approach with HL and holes semantics in the composition process allows us to implement precisely the argument selection mechanisms into the model, using a disjunction of type shifting possibilities in the logical representation of a predicate.

Another interesting point of this model is the fact that it is able to handle different cases of coercion in similar ways, thus avoiding the requirement of more complex structures when involving polysemous concepts. We also think that coercion phenomena in sentences like *Mary began the book*, in which aspectual verbs with a nominal argument are involved, could be modeled using the same kind of representation. Indeed, in Pustejovsky's analysis the underspecified information has been encoded into the lexicon by a *qualia structure*, where qualia are partial functions describing the roles that a concept can have [23]. As such, it seems possible to represent these qualia by attribute-value pairs, and modeling this kind of coercion would therefore follow the way presented in this paper. Moreover, the general constraints in HL that are used in our framework could be extended by contextual constraints as well: we would be able to change the intended qualia of a word depending on the context, and also to handle cases of metaphoric readings by adding some temporary constraints if the previous selection mechanism fails.

References

1. Abeillé, A., Rambow, O.: Tree adjoining grammar: an overview. In: Abeillé, A., Rambow, O. (eds.) Tree Adjoining Grammars: Formalisms, Linguistic Analysis and Processing, pp. 1–68. CSLI Press, Stanford (2000)
2. Areces, C., ten Cate, B.: Hybrid logics. In: Blackburn, P., Benthem, J.V., Wolter, F. (eds.) Handbook of Modal Logic, pp. 821–868. Elsevier, Amsterdam (2007)
3. Asher, N.: A Web of Words: Lexical Meaning in Context. Cambridge University Press, Cambridge (2011)
4. Asher, N.: Types, meanings and coercions in lexical semantics. Lingua **157**, 66–82 (2015)
5. Asher, N., Luo, Z.: Formalization of coercions in lexical semantics. In: Proceedings of Sinn und Bedeutung, vol. 17, pp. 63–80 (2012)

6. Asher, N., Pustejovsky, J.: Word meaning and commonsense metaphysics. Semant. Arch. (2005). http://semanticsarchive.net/
7. Asher, N., Pustejovsky, J.: A type composition logic for generative lexicon. J. Cogn. Sci. **6**, 1–38 (2006)
8. Barsalou, L.W.: Frames, concepts, and conceptual fields. In: Lehrer, A., Kittay, E.F. (eds.) Frames, Fields, and Contrasts, pp. 21–74. Lawrence Erlbaum, Mahwah (1992)
9. Bos, J.: Predicate logic unplugged. In: Dekker, P., Stokhof, M. (eds.) Proceedings of the 10th Amsterdam Colloquium, pp. 133–142 (1995)
10. Fillmore, C.J.: Frame semantics. In: Linguistics in the Morning Calm, pp. 111–137. Hanshin Publishing Co., Seoul (1982)
11. Gardent, C., Kallmeyer, L.: Semantic construction in feature-based TAG. In: Proceedings of the 10th Meeting of the European Chapter of the Association for Computational Linguistics (EACL), pp. 123–130 (2003)
12. Joshi, A.K., Schabes, Y.: Tree-adjoining grammars. In: Rozenberg, G., Salomaa, A.K. (eds.) Handbook of Formal Languages, vol. 3, pp. 69–123. Springer, Heidelberg (1997)
13. Kallmeyer, L., Joshi, A.K.: Factoring predicate argument and scope semantics: underspecified semantics with LTAG. Res. Lang. Comput. **1**(1–2), 3–58 (2003)
14. Kallmeyer, L., Lichte, T., Osswald, R., Pogodalla, S., Wurm, C.: Quantification in frame semantics with hybrid logic. In: Cooper, R., Retoré, C. (eds.) Proceedings of the ESSLLI 2015 Workshop on Type Theory and Lexical Semantics, Barcelona, Spain (2015)
15. Kallmeyer, L., Osswald, R.: Syntax-driven semantic frame composition in lexicalized tree adjoining grammars. J. Lang. Model. **1**(2), 267–330 (2013)
16. Kallmeyer, L., Osswald, R., Pogodalla, S.: Progression and iteration in event semantics - an LTAG analysis using hybrid logic and frame semantics. In: Pinón, C. (ed.) Empirical Issues in Syntax and Semantics, vol. 11 (2016)
17. Kallmeyer, L., Romero, M.: Scope and situation binding in LTAG using semantic unification. Res. Lang. Comput. **6**(1), 3–52 (2008)
18. Löbner, S.: Evidence for frames from human language. In: Gamerschlag, T., Gerland, D., Osswald, R., Petersen, W. (eds.) Frames and Concept Types. Applications in Language and Philosophy, pp. 23–67. Springer, Heidelberg (2014)
19. Nunberg, G.: The non-uniqueness of semantic solutions: polysemy. Linguist. Philos. **3**(2), 143–184 (1979)
20. Nunberg, G.: Transfers of meaning. J. Seman. **12**(2), 109–132 (1995)
21. Petersen, W.: Representation of concepts as frames. In: The Baltic International Yearbook of Cognition, Logic and Communication, vol. 2, pp. 151–170 (2006)
22. Pustejovsky, J.: The Generative Lexicon. MIT Press, Cambridge (1995)
23. Pustejovsky, J.: The semantics of lexical underspecification. Folia Linguistica **32**(3–4), 323–348 (1998)
24. Pustejovsky, J.: A survey of dot objects. Manuscript (2005)
25. Pustejovsky, J.: Coercion in a general theory of argument selection. Linguistics **49**(6), 1401–1431 (2011)
26. Pustejovsky, J., Rumshisky, A.: Mechanisms of sense extension in verbs. In: de Schryver, G.-M. (ed.) A Way with Words: Recent Advances in Lexical Theory and Analysis, pp. 67–88. Menha Publishers, Kampala (2010)
27. Vijay-Shanker, K., Joshi, A.K.: Feature structures based tree adjoining grammar. In: Proceedings of the 12th International Conference on Computational Linguistics (COLING), pp. 714–719 (1988)

Categorial Dependency Grammars with Iterated Sequences

Denis Béchet[1(✉)] and Annie Foret[2(✉)]

[1] LINA UMR CNRS 6241, Université de Nantes, Nantes, France
Denis.Bechet@univ-nantes.fr
[2] IRISA, Université de Rennes 1, Rennes, France
Annie.Foret@irisa.fr

Abstract. Some dependency treebanks use special sequences of dependencies where main arguments are mixed with separators. Classical Categorial Dependency Grammars (CDG) do not allow this construction because iterative dependency types only introduce the iterations of the same dependency. An extension of CDG is defined here that introduces a new construction for repeatable sequences of one or several dependency names. The learnability properties of the extended CDG when grammars are infered from a dependency treebank is also studied. It leads to the definition of new classes of grammars that are learnable in the limit from dependency structures.

Keywords: Categorial grammar · Dependency grammar · Iterated dependencies · Computational linguistics · Dependency treebanks · Grammatical inference · Incremental learning

1 Introduction

Dependency grammars and dependency treebanks do not always use a unique linguistic model for lists of elements. Some of them define an enumeration as a linked list of elements. Other grammars define a list as a set of dependencies that link the same word, the head of the list, to the elements of the list.

Categorial Dependency Grammars [5] (CDG) allow the second model with iterated dependency types. This construction introduces a list of dependencies with the same name and the same governor. The dependency structures (DS) in Fig. 1 shows a dependency A that is iterated on the left and on the right five times. A CDG compatible with the example could assign the type $[N \backslash A \backslash S / A^* / L / A^*]$ to the word *ran*. The dependency name A appears three times, two times as the iterative dependency type A^*. With this type, other DS are also possible: Each A^* may introduce none, one or several arguments linked to *ran* by a dependency A.

However, iterated dependency types cannot be used when a list of elements needs to be mixed with a separator like the example of Fig. 2 from corpus Sequoia [4] *"Les cyclistes et vététistes peuvent se réunir ce matin, à 9h, place*

© Springer-Verlag GmbH Germany 2016
M. Amblard et al. (Eds.): LACL 2016, LNCS 10054, pp. 34–51, 2016.
DOI: 10.1007/978-3-662-53826-5_3

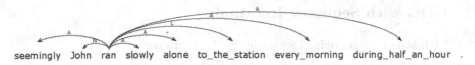

seemingly John ran slowly alone to_the_station every_morning during_half_an_hour .

Fig. 1. A dependency structure with five dependencies A

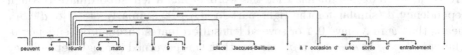

Fig. 2. A dependency structure with a list of modifiers separated by commas

Jacques-Bailleurs, à l'occasion d'une sortie d'entraînement." (fr. the cyclists and ATB bikers may meet themselves this morning, at 9, at Jacques-Bailleurs square, for a training ride)[1].

In this example, several modifiers alternate with a punctuation sign. The verb *réunir* may have type $[aff\backslash obj : obj/mod/ponct/mod/ponct/mod/ponct/mod]$. A regular expression for the part that corresponds to the modifiers and commas would be $mod(punct\,mod)^*$ or $(mod\,punct)^*mod$. It is not an iterative choice between *mod* and *ponct* like the regular expression $(mod\|ponct)^*$ but a repeatable sequence of *mod* and *ponct*. In order to formalize such structures, we propose to extend CDG types with a new construction that introduces finite sequences of dependencies. The system is an extension of classical CDG because iterated dependency types can be seen as sequence iterations where the sequence has a length of one dependency name.

We also study the learnability properties of CDG with sequence iteration when the grammar has to be infered from a dependency treebank. This concept of *identification in the limit* is due to Gold [7]. *Learning from strings* refers to hypothetical grammars generated from finite sets of strings. More generally, the hypothetical grammars may be generated from finite sets of structures defined by the target grammar. This kind of learning is called *learning from structures*. Both concepts were intensively studied (see excellent surveys in [2,8,9]). This concept lead for CDG with sequence iterations to a new class of grammar that is learnable from positive examples of dependency structures (DS).

The plan of the paper is as follows. Section 2 introduces Categorial Dependency Grammars with sequence iteration and studies their parsing properties and expressive power. The section also presents the links with linear logic, noncommutative logic and Lambek Calculus. Section 3 studies the learnability properties of such grammars from positive examples of dependency structures and defines new classes of such grammars that are learnable in this context. Section 4 presents experimental studies of sequence iterations in existent DS corpora. Section 5 concludes the paper.

---

[1] See talc2.loria.fr/deep-sequoia/sequoia-7.0/html/annodis.er_00060.html.

2   CDG with Sequence Iterations

2.1   Classical Categorial Dependency Grammars

Categorial dependency grammars can be seen as an assignment to words of first order dependency types of the form: $t = [L_m \backslash \ldots \backslash L_1 \backslash g / R_1 / \ldots / R_n]^P$. Intuitively, $w \mapsto [\alpha \backslash d \backslash \beta]^P$ means that the word w has a left subordinate through dependency d (similar for the right part $[\alpha / d / \beta]^P$). Similarly $w \mapsto [\alpha \backslash d^* \backslash \beta]^P$ means that w may have $0, 1$ or several left subordinates through dependency d. The *head type* g in $w \mapsto [\alpha \backslash g / \beta]^P$ means that w is governed through dependency g. The assignment of Example 1 determines the projective DS in Fig. 3.

Example 1.

$$
\begin{array}{ll}
in & \mapsto [c\_copul / prepos - l] \\
the & \mapsto [det] \\
beginning & \mapsto [det \backslash prepos - l] \\
was & \mapsto [c\_copul \backslash S / @fs / pred] \\
word & \mapsto [det \backslash pred] \\
. & \mapsto [@fs]
\end{array}
$$

The intuitive meaning of part P, called *potential*, is that it defines *discontinuous* dependencies of the word w. P is a string of *polarized valencies*, i.e. of symbols of four kinds: $\swarrow d$ (*left negative valency d*), $\searrow d$ (*right negative valency d*), $\nwarrow d$ (*left positive valency d*), $\nearrow d$ (*right positive valency d*). Intuitively, $v = \searrow d$ requires a subordinate through dependency d situated *somewhere* on the left, whereas the *dual* valency $\breve{v} = \swarrow d$ requires a governor through the same dependency d situated *somewhere* on the right. So together they describe the discontinuous dependency d. Similarly for the other pairs of dual valencies. For negative valencies $\swarrow d, \searrow d$ are provided a special kind of types $\#(\swarrow d)$, $\#(\searrow d)$. Intuitively, they serve to check the adjacency of a distant word subordinate through discontinuous dependency d to a *host word*. The dependencies of these types are called *anchor*. For instance, the assignment of Example 2 determines the non-projective DS in Fig. 4.

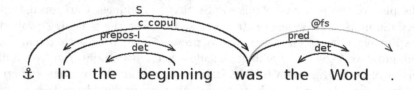

Fig. 3. Projective dependency structure.

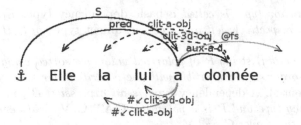

(*fr.* *she it$_{g=fem}$ to him has given*)

Fig. 4. Non-projective dependency structure.

Example 2.

$$
\begin{aligned}
elle\ &\mapsto [pred]\\
la\ &\mapsto [\#(\swarrow clit-a-obj)]^{\nearrow clit-a-obj}\\
lui\ &\mapsto [\#(\swarrow clit-3d-obj)]^{\nearrow clit-3d-obj}\\
a\ &\mapsto [\#(\swarrow clit-3d-obj)\backslash\#(\swarrow clit-a-obj)\backslash pred\backslash S/@fs/aux-a-d]\\
donnée\ &\mapsto [aux-a-d]^{\searrow clit-3d-obj\searrow clit-a-obj}\\
.\ &\mapsto [@fs]
\end{aligned}
$$

Definition 1 (CDG dependency structures). *Let* $W = a_1 \ldots a_n$ *be a list of words and* $\{d_1, \ldots, d_m\}$ *be a set of* dependency names, *with their dependency nature that can be either* local, discontinuous *or* anchor. *A graph* $D = (W, E)$ *with labeled arcs is a* dependency structure (DS) *of* W *if it has a* root, *i.e. a node* $a_i \in W$ *such that (i) for any node* $a \in W$, $a \neq a_i$, *there is a path from* a_i *to* a *and (ii) there is no arc* (a', d, a_i).[2] *An arc* $(a, d, a') \in E$ *is called* dependency d *from* a *to* a'. a *is called a* governor *of* a' *and* a' *is called a* subordinate *of* a *through* d. *The linear order on* W *is the* precedence order *on* D.

Definition 2 (CDG types). *Let* **C** *be a set of* local dependency names *and* **V** *be a set of* valency names.

The expressions of the form $\swarrow v$, $\searrow v$, $\searrow v$, $\nearrow v$, *where* $v \in$ **V**, *are called* polarized valencies. $\searrow v$ *and* $\nearrow v$ *are* positive, $\swarrow v$ *and* $\searrow v$ *are* negative; $\searrow v$ *and* $\swarrow v$ *are* left, $\nearrow v$ *and* $\searrow v$ *are* right. *Two polarized valencies with the same valency name and orientation, but with the opposite signs are* dual. *An expression of one of the forms* $\#(\swarrow v)$, $\#(\searrow v)$, $v \in$ **V**, *is called* anchor type *or just* anchor. *An expression of the form* d^* *where* $d \in$ **C**, *is called* iterated dependency type. *Local dependency names, iterated dependency types and anchor types are* primitive types.

An expression of the form $t = [L_m \backslash \ldots \backslash L_1 \backslash H / R_1 \ldots / R_n]$ *in which* $m, n \geq 0$, $L_1, \ldots, L_m, R_1, \ldots, R_n$ *are primitive types and* H *is either a* local dependency

---

[2] Evidently, every DS is connected and has a unique root.

name or an anchor type, is called a basic dependency type. $L_1, \ldots, L_m$ *and* $R_1, \ldots, R_n$ *are respectively left and right argument types of* t. *H is called the* head type *of* t.

A (possibly empty) string P of polarized valencies sorted using the standard lexicographical order $<_{lex}$ *compatible with the polarity order* $\searchangle < \searrow < \swarrow < \nearrow$, *is called a* potential. *A dependency type is an expression* B^P *in which B is a basic dependency type and P is a potential.* $\mathbf{CAT}(\mathbf{C}, \mathbf{V})$ *will denote the set of all dependency types over* $\mathbf{C}$ *and* $\mathbf{V}$.

CDG are defined using the following calculus of dependency types.[3] These rules are relativized with respect to the word positions in the sentence, which allows to interpret them as rules of construction of DS. Namely, when a type $B^{v_1 \cdots v_k}$ is assigned to the word in a position i, we encode it using the *state* $(B, i)^{(v_1, i) \cdots (v_k, i)}$. In these rules, types must be adjacent.

Definition 3 (Relativized calculus of dependency types).

$\mathbf{L}^1$. $\quad \Gamma_1\left([C], i_1\right)^{P_1}\left([C\backslash\beta], i_2\right)^{P_2}\Gamma_2 \vdash \Gamma_1\left([\beta], i_2\right)^{P_1 P_2}\Gamma_2$

$\mathbf{I}^1$. $\quad \Gamma_1\left([C], i_1\right)^{P_1}\left([C^*\backslash\beta], i_2\right)^{P_2}\Gamma_2 \vdash \Gamma_1\left([C^*\backslash\beta], i_2\right)^{P_1 P_2}\Gamma_2$

$\mathbf{\Omega}^1$. $\quad \Gamma_1\left([C^*\backslash\beta], i\right)^{P}\Gamma_2 \vdash \Gamma_1\left([\beta], i\right)^{P}\Gamma_2$

$\mathbf{D}^1$. $\quad \Gamma_1 \alpha^{P_1(\swarrow C, i_1)P(\searrow C, i_2)P_2}\Gamma_2 \vdash \Gamma_1 \alpha^{P_1 P P_2}\Gamma_2,$

$\quad$ *if the potential* $(\swarrow C, i_1)P(\searrow C, i_2)$ *satisfies the following pairing rule* **FA** *(*first available*) and where, moreover,* $i_1 < i_2$ *(non-internal constraint).*[4]

FA : $\quad$ *P has no occurrences of* $(\swarrow C, i)$ *or* $(\searrow C, i)$, *for any* i

$\mathbf{L}^1$ is the classical elimination rule. Eliminating the argument type $C \neq \#(\alpha)$ it constructs the (projective) dependency C and concatenates the potentials. $C = \#(\alpha)$ creates anchor dependencies. $\mathbf{I}^1$ derives $k > 0$ instances of C. $\mathbf{\Omega}^1$ serves in particular for the case $k = 0$. $\mathbf{D}^1$ creates *discontinuous* dependencies. It pairs and eliminates dual valencies with name C satisfying the rule **FA** to create the discontinuous dependency C.

Now, in this relativized calculus, for every proof ρ represented as a sequence of rule applications, we may define the DS $DS_x(\rho)$ *constructed* in this proof. Namely, let us consider the calculus relativized with respect to a sentence x with the set of word occurrences W. Then $DS_x(\varepsilon) = (W, \emptyset)$ is the DS constructed in the empty proof $\rho = \varepsilon$. Now, let (ρ, R) be a nonempty proof with respect to x and $(W, E) = DS_x(\rho)$. Then $DS_x((\rho, R))$ is defined as follows:
If $R = \mathbf{L}^1$ or $R = \mathbf{I}^1$, then $DS_x((\rho, R)) = (W, E \cup \{(a_{i_2}, C, a_{i_1})\})$. When C is a local dependency name, the new dependency is *local*. In the case where C is an anchor, this is an *anchor* dependency.
If $R = \mathbf{\Omega}^1$, then $DS_x((\rho, R)) = DS_x(\rho)$.
If $R = \mathbf{D}^1$, then $DS_x((\rho, R)) = (W, E \cup \{(a_{i_2}, C, a_{i_1})\})$ and the new dependency is *discontinuous*.

Definition 4 (CDG). *A categorial dependency grammar (CDG) is a system* $G = (W, \mathbf{C}, \mathbf{V}, S, \lambda)$, *where W is a finite set of words,* $\mathbf{C}$ *is a finite set of local*

---

[3] We show left-oriented rules. The right-oriented are symmetrical.

[4] This disallows internal primitive loops (the rule D^l cannot apply to a single word).

dependency names containing the selected name S (an axiom), $\mathbf{V}$ is a finite set of discontinuous dependency names and λ, called lexicon, *is a finite substitution on W such that $\lambda(a) \subset \mathbf{CAT}(\mathbf{C}, \mathbf{V})$ for each word $a \in W$. λ is extended on sequences of words W^* in the usual way.*[5]

For $G = (W, \mathbf{C}, \mathbf{V}, S, \lambda)$, a DS D and a sentence x, let $G[D, x]$ denote the relation:

$$D = DS_x(\rho) \quad \begin{array}{l} \text{where } \rho \text{ is a proof of } (t_1, 1) \cdots (t_n, n) \vdash (S, j) \\ \text{for some } n, j, \ 0 < j \leq n \text{ and } t_1 \cdots t_n \in \lambda(x). \end{array}$$

Then the language *generated by G is the set* $L(G) =_{df} \{w \mid \exists D \ G[D, w]\}$ *and the* DS-language *generated by G is the set* $\Delta(G) =_{df} \{D \mid \exists w \ G[D, w]\}$. $\mathcal{D}(CDG)$ *and $\mathcal{L}(CDG)$ will denote the families of DS-languages and languages generated by these grammars.*

Example 3. The proof in Fig. 5 shows that the DS in Fig. 4 belongs to the DS-language generated by a grammar containing the type assignments shown above for the French sentence *Elle la lui a donnée* (the word positions are not shown on types).

CDG are very expressive. Evidently, they generate all CF-languages. They can also generate non-CF languages.

Example 4. The following CDG generates the language $\{a^n b^n c^n \mid n > 0\}$ [6]:[6]

$$a \mapsto \#(\swarrow A)^{\swarrow A}, [\#(\swarrow A) \backslash \#(\swarrow A)]^{\swarrow A}$$
$$b \mapsto [B/C]^{\nwarrow A}, [\#(\swarrow A) \backslash S/C]^{\nwarrow A}$$
$$c \mapsto [C], [B \backslash C]$$

2.2   CDG with Sequences and Sequence Iterations

The extended system introduced here defines sequences and sequence iterations. An extended type $[\alpha \backslash (C_1 \bullet \cdots \bullet C_n) \backslash \beta]^P$ is viewed as a type that contains a sequence of n primitive types. It is equivalent to $[\alpha \backslash C_n \backslash \cdots \backslash C_1 \backslash \beta]^P$ (the sequence appears in the reverse order). The *starred* version of a sequence $[\alpha \backslash (C_1 \bullet \cdots \bullet C_n)^* \backslash \beta]^P$ is handled as a sequence of n primitive types that can be repeated none, once or several times. This construction with $n > 1$ is not possible with classical CDG which allows only iteration of a primitive type (the case $n = 1$). This type is equivalent to an infinite list of types:

$[\alpha \backslash \beta]^P$,
$[\alpha \backslash (C_1 \bullet \cdots \bullet C_n) \backslash \beta]^P \equiv [\alpha \backslash C_n \backslash \cdots \backslash C_1 \backslash \beta]^P$,
$[\alpha \backslash (C_1 \bullet \cdots \bullet C_n \bullet C_1 \bullet \cdots \bullet C_n) \backslash \beta]^P \equiv [\alpha \backslash C_n \backslash \cdots \backslash C_1 \backslash C_n \backslash \cdots \backslash C_1 \backslash \beta]^P$,
etc.

Definition 5. *We call* **sequence iteration** *types the expressions B^P where P is a potential, $B = [L_m \backslash \cdots \backslash L_1 \backslash H / \cdots / R_1 \cdots / R_n]$, H is either a local dependency name or an anchor type and $L_m, \ldots L_1, R_1 \ldots, R_n$ are either anchor types,*

---

[5] $\lambda(a_1 \cdots a_n) = \{t_1 \ldots t_n \mid t_1 \in \lambda(a_1), \ldots, t_n \in \lambda(a_n)\}$.
[6] One can see that a DS is not always a tree.

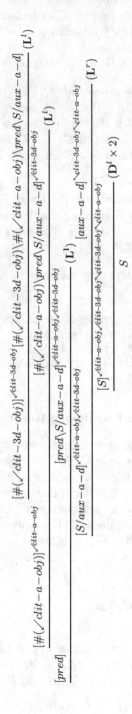

Fig. 5. Dependency structure correctness proof.

local dependency names, sequences of local dependency names or sequence iterations of local dependency names (a sequence of one local dependency name is identified to a local dependency name).

Rules for CDG with sequences and sequence iterations:

L^l. $\Gamma_1\left([C], i_1\right)^{P_1}\left([C\backslash\beta], i_2\right)^{P_2}\Gamma_2 \vdash \Gamma_1\left([\beta], i_2\right)^{P_1 P_2}\Gamma_2$

C^l. $\Gamma_1\left([(\alpha)^*\backslash\beta], i\right)^P\Gamma_2 \vdash \Gamma_1\left([\alpha\backslash(\alpha)^*\backslash\beta], i\right)^P\Gamma_2$ $(\alpha)^*$ is a sequence iteration

W^l. $\Gamma_1\left([(\alpha)^*\backslash\beta], i\right)^P\Gamma_2 \vdash \Gamma_1\left([\beta], i\right)^P\Gamma_2$ $(\alpha)^*$ is a sequence iteration

S^l. $\Gamma_1\left([(\alpha \bullet C)\backslash\beta], i\right)^P\Gamma_2 \vdash \Gamma_1\left([C\backslash\alpha\backslash\beta], i\right)^P\Gamma_2$ $(\alpha \bullet C)$ is a sequence

D^l. $\Gamma_1\, \alpha^{P_1(\swarrow C, i_1)P(\nwarrow C, i_2)P_2}\Gamma_2 \vdash \Gamma_1\, \alpha^{P_1 P P_2}\Gamma_2,$

   if the potential $(\swarrow C, i_1)P(\nwarrow C, i_2)$ satisfies **FA** and if $i_1 < i_2$

2.3 Links with Noncommutative Logic and Lambek Calculus

From a logical point of view, a CDG type B^P consists of a projective part B and a potential P. B can be seen as a logical formula in a resource sensible logic like linear logic. Because the order of formulas is also important, B can be seen either as a formula in noncommutative logic [1] or a formula in Lambek calculus [10].

In Lambek calculus, a sequence of primitive types is the product of primitive types. In the same perspective, a sequence iteration of primitive types has no equivalent in Lambek calculus.

In noncommutative logic, a type $B = [L_m\backslash\cdots\backslash L_1\backslash H/\cdots/R_1\cdots/R_n]$ can be seen as the linear type $L_m \multimap \cdots \multimap L_1 \multimap H \mathbin{\rotatebox[origin=c]{180}{$\multimap$}} R_1 \cdots \mathbin{\rotatebox[origin=c]{180}{$\multimap$}} R_n$ where $\multimap$ and $\mathbin{\rotatebox[origin=c]{180}{$\multimap$}}$ are the left and right linear implications. The sequence of primitive types $(C_1 \bullet \cdots \bullet C_n)$ is the multiplicative noncommutative product $(C_1 \odot \cdots \odot C_n)$. The following implications are valid in noncommutative logic. They justify the rules for CDG sequences:

$$(C_1 \odot \cdots \odot C_n) \multimap \beta \vdash C_n \multimap \cdots \multimap C_1 \multimap \beta$$

$$C_n \multimap \cdots \multimap C_1 \multimap \beta \vdash (C_1 \odot \cdots \odot C_n) \multimap \beta$$

The sequence iteration of primitive types $(C_1 \bullet \cdots \bullet C_n)^*$ corresponds to $?(C_1 \odot \cdots \odot C_n)$: An iteration is seen as the dual of the exponential of the multiplicative product of the primitive types. The following provable sequents justify the rules for CDG sequence iterations:

$$?(C_1 \odot \cdots \odot C_n) \multimap \beta \vdash (C_1 \odot \cdots \odot C_n \odot ?(C_1 \odot \cdots \odot C_n)) \multimap \beta$$

$$?(C_1 \odot \cdots \odot C_n) \multimap \beta \vdash \beta$$

Thus, it is possible to interpret the projective part of CDG types as a formula of noncommutative logic. The search for a valid analysis of a sentence becomes the proof search in noncommutative logic of a sequent where the formulae are one of the possible lists of types of the words through the lexicon of a grammar. This interpretation gives automatically a compositional semantic interpretation *à la Montague*.

2.4 Parsing and Expressive Power

Sequences can be seen as syntactic sugar for types. Thus, they don't change the parsing properties of languages and the expressive power of grammars. From a formal point of view, sequence iterations do not introduce new languages of string with respect to classical CDG. In fact, it is possible to emulate a sequence iteration by a simple iteration where each dependent corresponds to an element of the sequence (for instance the leftmost element of the sequence) and governs the other elements of the sequence. In contrast, sequence iterations introduce a new construction that is very common on DS corpora. For instance, the treebank Sequoia [4] models a list of elements as the alternative of an element and a punctuation mark. The introduction presents an example where the modifiers of the verb *réunir* alternate with commas: *"Les cyclistes et vététistes peuvent se réunir ce matin, à 9h, place Jacques-Bailleurs, à l'occasion d'une sortie d'entraînement."* (fr. *the cyclists and ATB bikers may meet themselves this morning, at 9, at Jacques-Bailleurs square, for a training ride).*

The parsing of CDG with sequence iterations is not very different from the parsing of classical CDG (i.e. with iterated dependency type). A sequence iteration at the leftmost position of a type $[(d_1 \bullet \cdots \bullet d_n)^* \backslash L_1 \cdots \backslash H/R_1/ \cdots]^{P_2}$ is rewritten into $[d_{n-1} \backslash \cdots \backslash d_1 \backslash (d_1 \bullet \cdots \bullet d_n)^* \backslash L_1 \cdots \backslash H/R_1/ \cdots]^{P_1 P_2}$ when the type $[d_n]^{P_1}$ is on its left (potentials $P_1 P_2$ may generate non-projective dependencies).

3 Learnability Results

The section studies the learnability properties of CDG with sequence iterations from positive examples of dependency structures (because sequences can be seen as syntactic sugar, the grammar are supposed to contain no sequence). It ends with the definition of a new family of classes of such grammars that are learnable in this context.

3.1 Inference Algorithm

A vicinity corresponds for a word to the part of a type that is used in a DS.

Definition 6 (Vicinity). *Given a DS D, the incoming and outgoing dependencies of a word w can be either local, anchor or discontinuous. For a discontinuous dependency d on a word w, we define its polarity p ($\nwarrow, \searrow, \swarrow, \nearrow$), according to its direction (left, right) and as negative if it is incoming to w, positive otherwise.*

Let D be a DS in which an occurrence of a word w has: the incoming projective dependency or anchor H (or the axiom S), the left projective dependencies or anchors $L_k, \ldots, L_1$ (in this order), the right projective dependencies or anchors $R_1, \ldots, R_m$ (in this order), and the discontinuous dependencies $d_1, \ldots, d_n \in \mathbf{V}$ with their respective polarities $p_1, \ldots, p_n$.

Then the vicinity *of w in D is the type*

$$V(w, D) = [L_1 \backslash \cdots \backslash L_k \backslash H/R_m/ \cdots /R_1]^P,$$

in which P *is a permutation of* $p_1d_1, \ldots, p_nd_n$ *in the standard lexicographical order* $<_{lex}$ *compatible with the polarity order* $\nwarrow < \searrow < \swarrow < \nearrow$.

For instance, *donnée* in Fig. 4 has the vicinity $[aux - a - d]^{\nwarrow clit-a-obj \nwarrow clit-3d-obj}$. This vicinity is nearly the same as the type of *donnée* in the lexicon because this type doesn't have a sequence iteration (or an iterated dependency type). The difference comes from the order of the polarized valencies $\nwarrow clit - a - obj$ and $\nwarrow clit - 3d - obj$ that appear in a different order. The vicinity of the verb *réunir* in Fig. 2 is $[aff \backslash obj{:}obj/mod/ponct/mod/ponct/mod/ponct/mod]$. A type that is compatible with this vicinity could be $[aff \backslash obj{:}obj/(ponct \bullet mod)^*/mod]$. In this case, the type in the lexicon and the vicinity are different.

Definition 7 (Algorithm). *Figure* 6 *presents an inference algorithm* $\mathbf{TGE}_{J-seq}^{(K)}$ *which, for every next DS in a training sequence, transforms the observed local, anchor and discontinuous dependencies of every word into a type with repeated local dependency sequences by introducing a sequence iteration for each group of at least* K *consecutive identical sequences of local dependencies. J indicates the maximum internal length of the sequences that are transformed into sequence iterations.*

Definition 8 (Generalization). *The notation* $TGen_{J-seq}^{(K)}(t_w)$, *that applies the inner loop algorithm in Fig. 6 to a type* t_w, *is extended to sets of types, lexicons and grammars, in a usual way, such that each assignment* $w \mapsto t$ *becomes* $w \mapsto TGen_{J-seq}^{(K)}(t)$

Ambiguities. Note that this process may be ambiguous. For instance, for $K = J = 2$, the generalization of $[a\backslash b\backslash a\backslash b\backslash a\backslash b\backslash a\backslash H]$ could be $[(b \bullet a)^*\backslash a\backslash H]$ or $[a\backslash(a \bullet b)^*\backslash H]$. With the same conditions on K and J, the generalization of $[b\backslash a\backslash a\backslash a\backslash a\backslash a\backslash H]$ could be $[b\backslash a^*\backslash H]$ or $[b\backslash(a \bullet a)^*\backslash a\backslash H]$. There are several ways to overcome this, such as: [**ALL mode**] adds all such types in the internal loop; or [**LML mode**] adds only the type corresponding to a leftmost longest sequence iteration with the shortest pattern. We could also consider different limiting neighbourhood conditions around the repeating pattern.

Definition 9 (LML mode). *We consider three parameters of the repeated sequence: the start position, the pattern length, the total length. In the [LML mode], the three parameters have the priorities in that order: We consider first the leftmost position as the start position, then the smallest pattern length, then the maximal number of repetitions.*

This mode is detailed by the following examples.

- $TGen_{2-seq}^{(2)}([a\backslash b\backslash a\backslash b\backslash a\backslash b\backslash a\backslash H]) = [(b \bullet a)^*\backslash a\backslash H]$ and not $[b\backslash(a \bullet b)^*\backslash H]$ because the leftmost repeated sequences for $K = J = 2$ start with the leftmost a of $[a\backslash b\backslash a\backslash b\backslash a\backslash b\backslash a\backslash H]$

Algorithm $TGE_{J-seq}^{(K)}$ (type-generalize-expand):
Input: σ, a training sequence of length N.
Output: CDG $TGE_{J-seq}^{(K)}(\sigma)$.

let $G_H = (W_H, \mathbf{C}_H, \mathbf{V}_H, S, \lambda_H)$ where $W_H := \emptyset$; $\mathbf{C}_H := \{S\}$; $\mathbf{V}_H := \emptyset$; $\lambda_H := \emptyset$;
(loop) **for** $i = 1$ to N // loop on σ
    let D such that $\sigma[i+1] = \sigma[i] \cdot D$;   // the i-th DS of σ
    let $(X, E) = D$;
    (loop) **for every** $w \in X$ // the order of the loop is not important
        $W_H := W_H \cup \{w\}$;
        let $t_w = V(w, D)$ // the vicinity of w in D
        $\mathbf{C}_H := \mathbf{C}_H \cup \{ d \mid d$ is a local dependency name of $t_w \}$
        $\mathbf{V}_H := \mathbf{V}_H \cup \{ d \mid \#(\swarrow d)$ or $\#(\searrow d)$ is an anchor type of $t_w \}$
             $\cup \{ d \mid \nwarrow d, \swarrow d, \nearrow d$ or $\searrow d$ is a polarized valency of $t_w \}$
      // – computing the generalization of t_w: $TGen_{J-seq}^{(K)}(t_w)$
        $t'_w := t_w$
        (loop) **while** $t'_w = [\alpha \backslash \delta \backslash \cdots \backslash \delta \backslash \beta]^P$
           with at least K consecutive occurrences of $\delta = d_j \cdots \backslash d_1$ $(j \leq J)$,
           $d_1, \ldots, d_j \in \mathbf{C}_H, COND_{ll}(\alpha, \delta)$ (or α not present) and $COND_{lr}(\beta, \delta)$
           $t'_w := [\alpha \backslash (d_1 \bullet \cdots \bullet d_j)^* \backslash \beta]^P$
        (loop) **while** $t'_w = [\alpha / \delta / \cdots / \delta / \beta]^P$
           with at least K consecutive occurrences of $\delta = d_j / \cdots / d_1$ $(j \leq J)$,
           $d_1, \ldots, d_j \in \mathbf{C}_H, COND_{rl}(\alpha, \delta)$ and $COND_{rr}(\beta, \delta)$ (or β not present)
           $t'_w := [\alpha / (d_1 \bullet \cdots \bullet d_j)^* / \beta]^P$
      // – the final t'_w defines $TGen_{J-seq}^{(K)}(t_w)$
      $\lambda_H(w) := \lambda_H(w) \cup \{t'_w\}$;   // expansion
end end

where $COND_{ll}(\alpha, \delta) = \alpha$ does not end in δ
     $COND_{lr}(\beta, \delta) = \beta$ does not start with $\delta \backslash$
     $COND_{rl}(\alpha, \delta) = \alpha$ does not end in $/\delta$
     $COND_{rr}(\beta, \delta) = \beta$ does not start with δ

Fig. 6. Inference algorithm $\mathbf{TGE}_{J-seq}^{(K)}$; the inner loop defines $TGen_{J-seq}^{(K)}(t_w)$ on types.

– $TGen_{2-seq}^{(2)}([H/a/a/a/a/a]) = [H/a^*]$ and not $[H/(a \bullet a)^*]$ because the sequences for a^* and $(a \bullet a)^*$ both start with the leftmost a in $[H/a/a/a/a/a]$ but the pattern length of a^* is one (the smallest) and the pattern length of $(a \bullet a)^*$ is two.

– $TGen_{2-seq}^{(2)}([H/a/b/a/b/a/b/a]) = [H/(b \bullet a)^*/a]$ and not $[H/(b \bullet a)^*/a/b/a]$ because for $K = J = 2$ even if there are two repeated sequences starting at the leftmost a with a pattern length of two $(b \bullet a)$ that are $a/b/a/b$ and $a/b/a/b/a/b$, the maximal number of repetitions is three and corresponds to $a/b/a/b/a/b$.

3.2 Algorithm Properties

Some Terminology. The following definitions are introduced for ease of writing.

Definition 10 (argument-form). *By an* argument-form *we mean a part of a type with the form* $L_m\backslash \ldots \backslash L_1\backslash$ *or the form* $/R_1 \ldots /R_n$ *where each* L_i, R_i *is a possible argument in a CDG type (in short an* argument-form *is a writing fragment on one side in a CDG type).*

Definition 11 (Component). *By a* star-component *in a type or an argument-form* t, *we mean any* $x^*\backslash$ *or* $/x^*$ *occurring in the writing of* t. *By a* primitive component *in a type or an argument-form* t, *we mean any* $x^*\backslash$, $/x^*$, $d\backslash$, *or* $/d$ *where* d *is a local dependency name or an anchor type, occurring in the writing of* t. *These notions are extended to the form without* $\backslash$ *or* $/$.

Definition 12 (Parallel Decomposition). *If* t' *is the result of the algorithm* $TGen_{J-seq}^{(K)}$ *on* $t = [L_1\backslash \cdots \backslash H/\cdots /R_1]^P$ *in the LML mode, we can decompose in parallel:* $t = [\alpha_1 \cdots H \cdots \alpha_N]^P$ *and* $t' = [\beta_1 \cdots H \cdots \beta_N]^{P'}$ *where* $P' = sort(P)$, *each* α_i *is an argument-form,* β_i *is a primitive component and:*

$$\beta_1 = TGen_{J-seq}^{(K)}(\alpha_1) \ldots \beta_j = TGen_{J-seq}^{(K)}(\alpha_j) \ldots and \ \beta_N = TGen_{J-seq}^{(K)}(\alpha_N)$$

The pair $(\alpha_1 \ldots \alpha_N, \beta_1 \ldots \beta_N)$ *defines the parallel decomposition of* (t, t') *in the LML mode; we call* (α_i, β_i) *a* block *and we say that each index* i selects *block* (α_i, β_i) *in the decomposition.*

Construction and Key Lemmas

Definition 13 (Expansion). *For any type* t, *we define its* full expansion $FE(t)$ *as the set of types obtained from* t *by erasing or by replacing its star-components* x^* $(d^*$ *or* $(d_1 \bullet d_2)^*$ *when* $J = 2)$ *by any successive repetitions of* x.

Note. This set is infinite when there is at least one star-component, but is used as an intermediate for proofs. It corresponds to the possible vicinities that can be associated to a word in a DS.

Definition 14 (Expansion of Rank K'). *For any* t, *type or argument-form, we define its* full expansion of rank K', $FE^{K'}(t)$, *as the set of types obtained from* t *by erasing or by replacing all its star-components* x^* *by any successive repetitions of* x *not more than* K' *times.*

Lemma 1. *Let* $K > 1$, $J = 1$ *or* 2 *and* $K' \geq K + 1$. *For any type* t:

$$TGen_{J-seq}^{(K)}(FE^{K'}(t)) = TGen_{J-seq}^{(K)}(FE^{K+1}(t)) \tag{1}$$

Proof. We show (1). Obviously $TGen_{J-seq}^{(K)}(FE^{K+1}(t)) \subseteq TGen_{J-seq}^{(K)}$ $(FE^{K'}(t))$. We show the converse for $J = 2$ ($J = 1$ is a subcase of $J = 2$). Suppose $t_1 \in FE^{K'}(t_0)$, let $t_2 = TGen_{J-seq}^{(K)}(t_1)$ and let α_j, β_j, for $1 \leq j \leq N$ denote the parallel decomposition of (t_1, t_2) in the LML mode. We discuss by induction on the construction of t_0, considering the parallel decomposition.

We consider the leftmost star-component x^* in t_0 repeated more than $K + 1$ times in t_1. We show that we can replace it by t_1' with only $K + 1$ repetitions of this pattern instead (unchanged elsewhere).

- If $|x| = 1$, then x^* of t_0 corresponds to $d\backslash d\backslash \cdots d\backslash$ or $/d/d\backslash \cdots d$ in t_1.

(1.1) If this argument-form of t_1 (and x^* of t_0) corresponds to a unique block i in the parallel decomposition of (t_1, t_2), then α_i contains more than $K + 1$ x and $\beta_i = x^*$; in that case, we define t_1' by replacing in α_i all the repetition of x with only $K + 1$ repetitions of x. In this case, x^* of t_0 corresponds to $K + 1$ x in t_1' and the algorithm yields the same type.

(1.2) If the argument-form corresponds to several adjacent blocks in the parallel decomposition of (t_1, t_2), the leftmost x is the end of a block i with $\beta_i = (x \bullet d_1)^*$ and the others are in the block $i + 1$ with $\beta_{i+1} = x^*$. α_{i+1} contains at least K x. We define t_1' by replacing in α_{i+1} all the repetition of x by only K repetitions of x. In this case, x^* of t_0 corresponds to $K + 1$ x in t_1' which yields the same type (algorithm output).

- If $|x| = 2$, then x is the succession of d_1 and d_2 ($x = d_2 \bullet d_1$ and it corresponds to $d_1\backslash d_2\backslash d_1 \cdots \backslash d_1\backslash d_2\backslash$ or $/d_1/d_2/d_1 \cdots /d_1/d_2)$:

(2.1) If x^* of t_0 corresponds to a unique block i, in that case, as in (1.1), we define t_1' by replacing in α_i the repetition of d_1 and d_2 with $K + 1$ repetitions of d_1 and d_2. In this case, x^* of t_0 corresponds to $K + 1$ x in t_1' which yields the same type (algorithm output).

(2.2) if $d_1 \neq d_2$ and x^* corresponds to several adjacent blocks in the parallel decomposition of (t_1, t_2) starting at block i, this means that in the LML mode the leftmost d_1 corresponds to the end of block i, the rightmost d_2 correspond to the beginning of block $i + 2$ and the other local dependency names $d_2, d_1, d_2 \ldots, d_2, d_1$ correspond to block $i + 1$ with $\beta_{i+1} = (d_2 \bullet d_1)^*$. We define t_1' by replacing in α_{i+1} the repetition of d_2 and d_1 with K repetitions of d_2 and d_1. In this case, x^* of t_0 corresponds to $K + 1$ x in t_1' which yields the same type (algorithm output).

(2.3) if $d_1 = d_2$, we have the same cases as in (1.1) and (1.2) but with more than $2K + 2$ local dependency names.

We can repeat this process until no expansion is made more than $K + 1$ times, hence the converse inclusion.

For example, if $t_0 = a\backslash a\backslash (b \bullet a)^*\backslash b\backslash b\backslash H$, with $J = 2, K = 2, K' = 4$: the decomposition for $t_1 = a\backslash a\backslash a\backslash b\backslash a\backslash b\backslash a\backslash b\backslash b\backslash b\backslash H$ (with $K' = K + 2$ repetitions) can be compared to that of $t_1' = a\backslash a\backslash a\backslash b\backslash a\backslash b\backslash a\backslash b\backslash b\backslash b\backslash H$ with $K + 1$ repetitions (we recall that the display order is reverted for internal sequence as arguments):

$$TGen_{2-seq}^{(2)} \begin{vmatrix} \alpha_1 = a\backslash a\backslash a\backslash \\ \beta_1 = a^* \end{vmatrix} \begin{vmatrix} \alpha_2 = b\backslash a\backslash b\backslash a\backslash b\backslash a\backslash \\ \beta_2 = (a \bullet b)^* \end{vmatrix} \begin{vmatrix} \alpha_3 = b\backslash b\backslash b\backslash \\ \beta_3 = b^* \end{vmatrix} \begin{Vmatrix} t_1 \\ t_2 \end{Vmatrix}$$

$$TGen_{2-seq}^{(2)} \begin{vmatrix} \alpha_1 = a\backslash a\backslash a\backslash \\ \beta_1 = a^* \end{vmatrix} \begin{vmatrix} \alpha_2 = b\backslash a\backslash b\backslash a\backslash \\ \beta_2 = (a \bullet b)^* \end{vmatrix} \begin{vmatrix} \alpha_3 = b\backslash b\backslash b\backslash \\ \beta_3 = b^* \end{vmatrix} \begin{Vmatrix} t_1' \\ t_2 \end{Vmatrix}$$

Note that $a\backslash a\backslash a\backslash b\backslash a\backslash b\backslash b\backslash b\backslash$, with K repetitions only, yields a different decomposition.

Corollary 1. *Let $K > 1$ and $J = 1$ or 2. For any type t the result of the algorithm $TGen_{J-seq}^{(K)}$ on the full extension of t is a finite set and is the same set as the result of this algorithm on $FE^{K+1}(t)$.*

The definitions of FE^K and FE^{K+1} are extended to sets, lexicons and grammars in the usual way.

Lemma 2. *Let $K > 1$ and $J = 1$ or 2. Let G be a CDG with sequence iterations. We have:*

(1) all vicinities of words in DS of $\Delta(G)$ belong to some $FE(t)$, where t is assigned by G.

(2) if σ is a finite sequence in $\Delta(G)$, then $\Delta(TGE^{(K)}_{J-seq}(\sigma)) \subseteq \Delta(G')$ where G' is $TGE^{(K)}_{J-seq}$ on $FE^{K+1}(G)$

Proof. If G generates $D \in \sigma$ where a word w occurs with a vicinity t_w, for which G uses the assignment $w \mapsto t$ in the derivation, then t_w must be in $FE(t)$. Finally, we use Corollary 1 relating $FE(t)$ to $FE^{K+1}(t)$.

Theorem 1 (Convergence). *Let $K > 1$ and $J = 1$ or 2. Let G be any CDG. The algorithm $TGE^{(K)}_{J-seq}$ stabilizes on every training sequence in $\Delta(G)$ to a grammar with assignments in $TGE^{(K)}_{J-seq}$ on $(FE^{K+1}(G))$.*

Proof. We have (1) $TGE^{(K)}_{J-seq}(\sigma[i]) \subseteq TGE^{(K)}_{J-seq}(\sigma[i+1]) \subseteq \ldots$ As observed in Lemma 2, the vicinities for the words of the DS in σ belong to $FE(G)$. If we had an infinite chain of types $t'_i = TGE^{(K)}_{J-seq}(t_i)$, with assignments $w_i \mapsto t'_i$ in $TGE^{(K)}_{J-seq}(\sigma[i])$, but not in $TGE^{(K)}_{J-seq}(\sigma[i-1])$ (we could consider one such chain concerning a same word w as the lexicon of G is finite) ; now all t_i also belong to some $FE^{K_i}(G)$, then if $K' > K + 1$, there exists t''_i in $FE^{K+1}(G)$, such that $t'_i = TGE^{(K)}_{J-seq}(t''_i)$, we can thus view the set of t'_i as the result of $TGE^{(K)}_{J-seq}$ on a subset of $FE^{K+1}(G)$; obviously $FE^{K+1}(G)$ is finite, we would then have a contradiction.

Therefore for any G and any $K > 1$:
$$\exists N, \forall N' \geq N \; TGE^{(K)}_{J-seq}(\sigma[N']) = TGE^{(K)}_{J-seq}(\sigma[N])$$
Furthermore, if $w \mapsto t' \in TGE^{(K)}_{J-seq}(\sigma[N])$ there exists $w \mapsto t'' \in FE^{K+1}(G)$, such that $t' = TGE^{(K)}_{J-seq}(t'')$: in that sense the assignments in $TGE^{(K)}_{J-seq}(\sigma[N])$ are in $TGE^{(K)}_{J-seq}$ on $(FE^{K+1}(G))$.

Proposition 1. *Let $K > 1$ and $J = 1$ or 2.*
If G is a CDG and σ is a sequence in $\Delta(G)$ then

(1) $TGE^{(K)}_{J-seq}(\sigma[i]) \subseteq TGE^{(K)}_{J-seq}(\sigma[i+1])$ *monotonicity/incrementality*
(2) $\sigma[i] \subseteq \Delta(TGE^{(K)}_{J-seq}(\sigma[i]))$ *expansivity*
(3) $\Delta(TGE^{(K)}_{J-seq}(\sigma[i])) \subseteq \Delta(G')$ where G' is $TGE^{(K)}_{J-seq}$ on $FE^{K+1}(G)$

Proof. (1) holds by definition of the algorithm (that expands the lexicon); (2) can be shown by adapting the derivation ; (3) follows from a preceeding lemma.

3.3   A Family of Learnable Classes

Definition 15. *Two grammars are said* strongly equivalent *if they generate the same dependency structure language. The* strong equivalence criterion*:*

(i) *G is strongly equivalent to* $TGE^{(K)}_{J-seq}$ *on* $FE^{K+1}(G)$ *defines the subclass written* $CCDG^{K}_{J-seq}$ *of grammars satisfying (i).*

Theorem 2. *Let* $K > 1$ *and* $J = 1$ *or* 2*. The algorithm* $TGE^{(K)}_{J-seq}$ *learns the class of CDG satisfying the strong equivalence criterion (i), from labelled dependency structures.*

Proof. From Proposition 1(1): $TGE^{(K)}_{J-seq}(\sigma[i]) \subseteq TGE^{(K)}_{J-seq}(\sigma[i+1]) \subseteq \ldots$
The stabilization property holds (Theorem 1):
$$\exists N, \forall N' \geq N \; TGE^{(K)}_{J-seq}(\sigma[N']) = TGE^{(K)}_{J-seq}(\sigma[N])$$
Then by Proposition 1(2): $\Delta(G) \subseteq \Delta(TGE^{(K)}_{J-seq}(\sigma[N]))$,
and using (i) and Proposition 1(3): $\Delta(G) \subseteq \Delta(TGE^{(K)}_{J-seq}(\sigma[N])) \subseteq \Delta(G)$.
Therefore for any grammar, such that (i) we get the convergence to a grammar generating the same structure language.

Observe that this class does not impose a bound on the number of types associated to a word (in contrast to k-valued grammars). The learnability for $J = 1$ was studied in [3], with a special case of our algorithm.

4   Extended CDG and Dependency Treebanks

From Dependency Treebanks to Vicinities. Our workflow applies to data in the Conll format[7]. The CDG potentials in this section are considered as empty[8].

For each governor unit in each corpus we have computed (using MySQL and Camelis[9]): (1) its *vicinity* in the root simplified form $[l_1 \backslash \ldots \backslash l_n \backslash root / r_m / \ldots / r_1]$ (where l_1 to l_n on the left and r_1 to r_m on the right are the successive dependency names from that governor), then (2) its generalization as *star-vicinity*, replacing consecutive repetitions of d_k on a same side with d_k^*; and (3) its generalization as *vicinity_2seq* following the LML mode of the algorithm in Fig. 6 for J = K = 2.

Our development allows to mine repetitions and to call several kinds of viewers: we use the item/word description interactive viewer camelis and the sentence parse conll viewer [11] or grew[10].

Figure 7 on its left, shows the *root simplified vicinities* computed on corpus Sequoia; the resulting file has been loaded as an interactive information context, in Camelis; this tool manages three synchronised windows: the current query is on the top, selecting the objects on the right, their properties can be browsed in the multi-facets index on the left.

---

[7] http://universaldependencies.org/format.html.
[8] this complies with Sequoia data, but may be a simplification for some other corpora.
[9] www.irisa.fr/LIS/softwares.
[10] http://talc2.loria.fr/grew/.

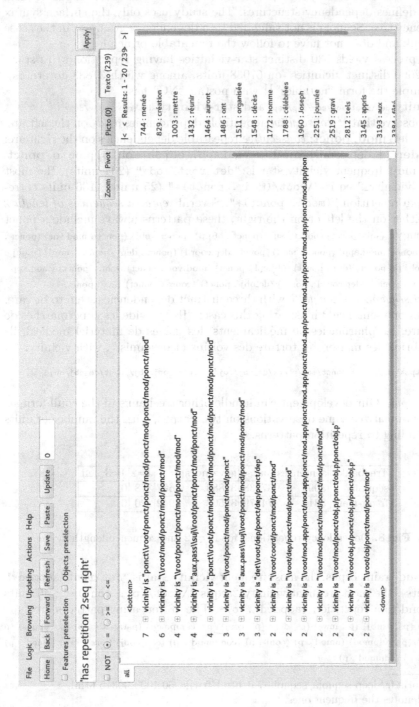

Fig. 7. Simplified vicinities computed on corpus Sequoia

Results on the French corpus Sequoia. We consider a version of corpus Sequoia [4] that defines dependency structures. The study uses only the surface syntax dependency tree. Sequoia is not validated by a dependency grammar in the sense of Mel'čuk and does not have to follow the repeatable principle.

The process yields 530 distinct star-vicinities having repetition(s) (a star), among 2660 distinct vicinities (on 67038 units, among which 37883 governors). For example the form "notables"[11] with postag "NC" has:

vicinity `det\root/mod/mod/dep/dep` and star-vicinity `det\root/mod*/dep*` .

We observe that:[12] *consecutive* repeatable dependencies $d_1.d_1$ on the left are: aff, **dep**, det, **mod**, **ponct**; *consecutive* repeatable dependencies on the right are: **coord**, **dep** (+ dep.coord), **mod** (+ mod.app), obj:obj+obj.p, p_obj.o, **ponct**

The most frequent vicinity_star is `"det\root/mod*"` (204 units), the most frequent vicinity_2seq is `"\root/(mod . ponct)*"` (25 units), 166 units correspond to a repetition `"(mod . ponct)*"`. Several *repeated sequences of length 2* occur, either on the left or on the right, these patterns always include a ponct dependency: (suj . ponct) (ponct . suj) (ponct . obj.p) (ponct . obj) (ponct . mod.voc) (ponct . mod.rel) (ponct . mod.app) (ponct . mod) (ponct . dep.coord) (ponct . dep) (ponct . coord) (p_obj.o . ponct) (obj.p . ponct) (obj . ponct) (obj.cpl . ponct) (mod.voc . ponct) (mod . ponct) (mod.app . ponct) (dep . ponct) (dep.coord . ponct) (de_obj . ponct) (coord . ponct) (ats . ponct)

Repeated sequences of length 3, with three distinct dependencies seem to be rare. We found one sentence[13] illustrating this case: "Ils ont vidé les supermarchés de nourriture, les pharmacies de médicaments, les usines de matériel médical, ils ont cambriolé les maisons et torturé des voisins et des amis.", with vicinity:

`"aux.tps\\suj\\root/ponct/mod/ponct/de_obj/obj/ponct/de_obj/obj/ponct/de_obj/obj"`

Other corpus. Our development can handle other treebanks in the conll format. Table 4 summarizes some observations on two corpus, with the number of units corresponding to repetition patterns.

Treebank	sentences	units	governors	J=1	J=2	J=3, left
sequoia	3099	67038	37883	1667	378	0
fr-ud-train	3312	74979	33568	1942	220	0

Fig. 8. Dependency repetitions, for K = 2 and sequence length J

In the fr-ud-train corpus, the most frequent vicinity_star is `"det\root/adpmod*"` (194 units), the most frequent vicinity_2seq is `"\root/(p . conj)*"` ; 45 units correspond to a repetition (adpmod . p)*. The 18 repeating patterns are: (p . parataxis) (p . nsubj) (p . mwe) (p . dep) (p . conj) (p . compmod) (parataxis . p) (p . appos) (p . advmod) (p . adpmod) (nmod . p) (conj . p) (compmod . p) (cc . conj) (aux . neg) (amod . p) (advmod . p) (adpmod . p)

---

[11] talc2.loria.fr/deep-sequoia/sequoia-7.0/html/frwiki_50.1000_00315.html.
[12] bold denotes the frequent ones.
[13] talc2.loria.fr/deep-sequoia/sequoia-7.0/html/frwiki_50.1000_00091.html.

5   Conclusion

In this paper, we have extended classical Categorial Dependency Grammars with a new construction to handle repeatable sequences of several dependencies. The work was motivated by the observation of such patterns. We have proposed a learning algorithm. A version of this algorithm has been implemented and applied to some treebanks (in Conll). Some design and computational variants are possible depending on the *repetition principle reading*. On the formal side, further analysis could consider richer patterns. On the experimental side, other treebanks could be explored as well. It would also be interesting to reconsider these notions in other formalisms or application domains.

References

1. Abrusci, V., Ruet, P.: Non-commutative logic i: the multiplicative fragment. Ann. Pure Appl. Logic **101**(1), 29–64 (1999)
2. Angluin, D.: Inductive inference of formal languages from positive data. Inf. Control **45**, 117–135 (1980)
3. Béchet, D., Dikovsky, A., Foret, A.: Two models of learning iterated dependencies. In: Groote, P., Nederhof, M.-J. (eds.) FG 2010-2011. LNCS, vol. 7395, pp. 17–32. Springer, Heidelberg (2012). doi:10.1007/978-3-642-32024-8_2
4. Candito, M., Perrier, G., Guillaume, B., Ribeyre, C., Fort, K., Seddah, D., de la Clergerie, E.: Deep syntax annotation of the sequoia french treebank. In: Proceedings of LREC, pp. 2298–2305. European Language Resources Association (ELRA), May 2014
5. Dekhtyar, M., Dikovsky, A., Karlov, B.: Categorial dependency grammars. Theoret. Comput. Sci. **579**, 33–63 (2015)
6. Dikovsky, A.: Dependencies as categories. In: Kruijff, G.J.M., Duchier, D. (eds.) COLING 2004 Recent Advances in Dependency Grammar, pp. 82–89. COLING, Geneva, Switzerland, 28 August 2004. http://aclweb.org/anthology/W04-1512
7. Gold, E.M.: Language identification in the limit. Inf. Control **10**, 447–474 (1967)
8. de la Higuera, C.: Grammatical Inference: Learning Automata and Grammars. Cambridge University Press, New York (2010)
9. Kanazawa, M.: Learnable Classes of Categorial Grammars. Studies in Logic, Language and Information. FoLLI & CSLI, Stanford (1998)
10. Lambek, J.: On the calculus of syntactic types. In: Jakobson, R. (ed.) Structure of Languages and its Mathematical Aspects, pp. 166–178. American Mathematical Society, Providence (1961)
11. Rosa, R.: Terminal-based CoNLL-file viewer (2014). http://hdl.handle.net/11234/1-1456, LINDAT/CLARIN digital library at Institute of Formal and Applied Linguistics, Charles University in Prague

Implementing Variable Vectors in a CCG Parser

Daisuke Bekki[1,2,3,4($\boxtimes$)] and Ai Kawazoe[4]

[1] Faculty of Core Research, Ochanomizu University,
2-1-1 Ohtsuka, Bunkyo-ku, Tokyo 112-8610, Japan
bekki@is.ocha.ac.jp
[2] CREST, Japan Science and Technology Agency (JST),
4-1-8 Honcho, Kawaguchi, Saitama 332-0012, Japan
[3] National Institute of Advanced Industrial Science and Technology (AIST),
1-1-1 Higashi, Tsukuba, Ibaraki 305-8561, Japan
[4] National Institute of Informatics (NII),
2-1-2 Hitotsubashi, Chiyoda-ku, Tokyo 101-8430, Japan

Abstract. This article addresses problems that arise from the use of category variables T in combinatory categorial grammars (CCGs), in particular, that they require extension of semantic languages with *variable vectors* in a form such as $\lambda \overline{x}.M$ or $M \overline{x}$. As a solution to such problems, we introduce a technique for implementing variable vectors within the context of *lightblue*, a Japanese CCG parser implemented within the Haskell programming language with a dependent type semantics (DTS) representation.

Keywords: Combinatory categorial grammar · Variable vectors

1 Introduction

1.1 Category Variables T and Variable Vectors $\lambda \overline{x}$ or $M \overline{x}$

Type-raising rules in combinatory categorial grammars (CCGs) typically introduce *category variables*, often represented by a bold letter T together with an index $\boxed{i}$ ($i \in \mathbb{N}$) to distinguish between category variables (Steedman 2000).

$$
\begin{aligned}
(1) \qquad X : a &\Longrightarrow_{>T} \quad T^{\boxed{i}}/(T^{\boxed{i}} \backslash X) : \lambda p.pa \\
X : a &\Longrightarrow_{<T} \quad T^{\boxed{i}} \backslash (T^{\boxed{i}}/X) : \lambda p.pa
\end{aligned}
$$

The theoretical status of the type-raising rules can be understood in at least two different ways. The first perspective is that the two rules in (1) are not, themselves, rules; rather, they are rule schema, with T being meta-level variables. Instantiating T with an actual syntactic category then defines a countably

---

My sincere thanks to Koji Mineshima and Ribeka Tanaka for many helpful discussions. I also thank the developers of JUMAN++ and KNP, in particular, Sadao Kurohashi, Daisuke Kawahara, Hajime Morita and Yuta Hayashibe, for sharing information. I also thank the anonymous reviewers of LACL2016 for many insightful comments. This research is partially supported by JST, CREST.

M. Amblard et al. (Eds.): LACL 2016, LNCS 10054, pp. 52–67, 2016.
DOI: 10.1007/978-3-662-53826-5_4

infinite set of rules. Let us call this perspective, in which T is not an object-level expression, *category variable as meta-variable*. The second, *category variable as type variable*, understands the whole syntactic calculus of CCG as a weak polymorphic type theory (i.e., polymorphism without quantification over T), where category variables T are object-level expressions.

Under either view, T may be instantiated by, for example, a functional category with an arbitrary number of arguments, but their semantic representation in (1), $\lambda p.pa$, is invariant. This is because only p's first argument, which is to be filled with a, matters; the number of other argument slots is irrelevant.

However, there are two situations where the invariance of semantic representation for T cannot be maintained. Those cases require the use of lambda calculus with terms whose argument slots are both "expandable" and "shrinkable." Variable vectors $\overline{x}$ that appear in terms of the form $\lambda \overline{x}.M$ or $M\overline{x}$ are typical tools for this purpose (Steedman 2000). The first situation is coordination between type-raised NPs. A typical lexical entry for quantifier NPs in English, such as *every N* and *some N* (with the accusative case), has the syntactic category $T\backslash(T/NP)$, and these entries participate in the coordinated structure in the following way.

(2)
$$\frac{T\backslash(T/NP):f \quad CONJ:\wedge \quad T\backslash(T/NP):g}{T\backslash(T/NP):\lambda \overline{x}.(f\overline{x} \wedge g\overline{x})}_{\langle\Phi\rangle}$$

The resulting semantic representation, $\lambda \overline{x}.(f\overline{x} \wedge g\overline{x})$, shows the two usages of, and the necessity of, a variable vector $\overline{x}$; when used with λ in the form of $\lambda \overline{x}.M$, it is a function that takes an arbitrary number of arguments, enough that the corresponding syntactic category becomes S. When used with a function f (or g), in the form $f\overline{x}$ (or $g\overline{x}$), it is a result of applying a sequence of arguments taken by the corresponding binder $\lambda \overline{x}$, and the sequence preserves the order of arguments.

The reason why variable vectors are needed in (2), unlike in (1), is that $\wedge$ is a truth function that conjoins only propositions. In (2), $f\overline{x}$ and $g\overline{x}$ are safely conjoined by the truth function $\wedge$ because the corresponding category for $f\overline{x}$ and $g\overline{x}$ is S, and this ensures that their semantic type is proposition. However, the number of arguments needed for f and g to become propositions depends on the syntactic category that instantiates T. Therefore, we need a variable vector $\overline{x}$ for semantic representations in cases where we want to not specify the number of arguments. For example, the following two CCG derivations instantiate (1), which uniformly describes the coordination calculus therein.

(3)
$$\frac{\dfrac{\dfrac{\text{John}}{NP:j}}{\dfrac{S\backslash NP\backslash(S\backslash NP/NP)}{:\lambda p.pj}}^{<T} \quad \dfrac{\dfrac{\text{and}}{CONJ}}{:\wedge} \quad \dfrac{\dfrac{\text{Mary}}{NP:m}}{\dfrac{S\backslash NP\backslash(S\backslash NP/NP)}{:\lambda p.pm}}^{<T}}{\begin{array}{c} S\backslash NP\backslash(S\backslash NP/NP) \\ :\lambda p.\lambda x.(((\lambda p.pj)p)x \wedge ((\lambda p.pm)p)x) \\ \twoheadrightarrow_\beta \lambda p.\lambda x.(pjx \wedge pmx) \end{array}}_{\langle\Phi\rangle}$$

The example (3) shows a coordinated structure between two direct objects in English, for which both $\boldsymbol{T} \equiv S \backslash NP$ and $\overline{x} \equiv (p, x)$; namely, $\lambda \overline{x}. \equiv \lambda p. \lambda x.$ and $f\overline{x} \equiv ((fp)x)$.

$$
(4) \quad
\cfrac{
\cfrac{
\cfrac{\text{John}}{NP : j}
}{
\begin{array}{c} S \backslash NP/NP \backslash (S \backslash NP/NP/NP) \\ : \lambda p.pj \end{array}
} {}^{<T}
\quad
\cfrac{\text{and}}{\begin{array}{c} CONJ \\ : \wedge \end{array}}
\quad
\cfrac{
\cfrac{\text{Mary}}{NP : m}
}{
\begin{array}{c} S \backslash NP/NP \backslash (S \backslash NP/NP/NP) \\ : \lambda p.pm \end{array}
} {}^{<T}
}{
\begin{array}{c} S \backslash NP/NP \backslash (S \backslash NP/NP/NP) \\ \lambda p.\lambda y.\lambda x.((((\lambda p.pj)p)y)x \wedge (((\lambda p.pm)p)y)x) \\ \twoheadrightarrow_\beta \lambda p.\lambda y.\lambda x.(pjyx \wedge pmyx) \end{array}
} {}^{\langle \Phi \rangle}
$$

The example (4) shows a coordinated structure between two indirect objects in English, for which both $\boldsymbol{T} \equiv S \backslash NP/NP$ and $\overline{x} \equiv (p, y, x)$; namely, $\lambda \overline{x}. \equiv \lambda p.\lambda y.\lambda x.$, $f\overline{x} \equiv (((fp)y)x)$.

The second situation in which variable vectors are needed is semantic representation of quantifiers.

$$
(5) \quad
\begin{array}{l}
every \vdash \boldsymbol{T} \backslash (\boldsymbol{T}/NP)/N : \lambda n.\lambda p.\forall x.nx \to (px)\overline{x} \\
some \vdash \boldsymbol{T} \backslash (\boldsymbol{T}/NP)/N : \lambda n.\lambda p.\exists x.nx \wedge (px)\overline{x}
\end{array}
$$

Since truth functions $\to$ and $\wedge$ conjoin only propositions, $(px)\overline{x}$ must be a propositional term. However, the number of arguments needed to make p into a proposition depends on the instantiation of the syntactic category $\boldsymbol{T}/NP$, and thus must be underspecified, as seen in (5).

1.2   Problem of Variable Vectors

A more precise definition of $\overline{x}$ ($\lambda \overline{x}$ and $M\overline{x}$) could be given as follows:[1]

Definition 1 (Variable vectors).

$$
\lambda \overline{x}.M[\ldots f\overline{x} \ldots] \stackrel{def}{\equiv}
\begin{cases}
M[\ldots f \ldots] \\
\lambda x.\lambda \overline{x}.M[\ldots (fx)\overline{x} \ldots] & where\ x \notin fv\,(M[\ldots f\overline{x} \ldots])
\end{cases}
$$

The problem with this pseudo-definition is that it is "defined" in nondeterministic style but the choice is determined by the form of the corresponding syntactic category. For example, in (2), the value of $|\overline{x}|$ (namely, how many arguments are needed for f and g) depends on the instantiation of $\boldsymbol{T}$. In (5), $|\overline{x}|$ (how many arguments are needed for px) again depends on the instantiation of $\boldsymbol{T}$. The strategy adopted here is that one should take the second choice in Definition 1 n times when $|\overline{x}| = n$, but this determination requires a reference to the corresponding syntactic category.

The fact that the definition of variable vectors depends on the corresponding syntactic category is problematic when trying to situate the definition within

---

[1] In the literature, Bekki (2010) defines variable vectors as in Definition 1 (which is slightly adapted for this article), among others. $M[\ldots N \ldots]$ indicates a term M that contains N as a subterm.

the formal theory of CCG. Moreover, implementation is complicated by the definition because variable vectors are not a notion that can be defined entirely within the pure lambda calculus.

Another problem that arises with variable vectors is that they are not sub-terms from the perspective of a lambda calculus without them. In (4), for example, $\bar{x} \equiv (p, y, x)$, namely, $\lambda \bar{x}. \equiv \lambda p.\lambda y.\lambda x.$ and $f\bar{x} \equiv (((fp)y)x)$; however, neither $\lambda p.\lambda y.\lambda x.$ nor pyx is a subterm from the viewpoint of the syntax of lambda calculus.

Thus, translation from lambda calculus extended to include variable vectors to lambda calculus without them is no longer possible by simple substitution of variable vectors with a lambda term. Instead, translation requires transformation of a syntactic tree of lambda calculus. In other words, variable vectors underspecify the syntactic structure that surrounds them.

1.3   Previous Work on Category Variables

Presently available implementations of CCG parsers, such as the C&C parser (Clark and Curran 2004), EasyCCG (Lewis and Steedman 2014), and Jigg (Noji and Miyao 2016), have avoided implementation of category variables.

As is well-known, a näive top-down CCG parser is not guaranteed to terminate, because of the existence of the following (infinite) path:

$$\cfrac{\cfrac{\cfrac{\vdots}{S/X_1/X_2} \quad X_2}{S/X_1} \, {}^{>} \quad X_1}{S} \, {}^{>}$$

Likewise, a näive bottom-up CCG parser with type-raising rules would not terminate for the following infinite path.[2]

$$\cfrac{\cfrac{\cfrac{X}{T^{\boxed{1}}/(T^{\boxed{1}}\backslash X)} \, {}^{>T}}{T^{\boxed{2}}/(T^{\boxed{2}}\backslash(T^{\boxed{1}}/(T^{\boxed{1}}\backslash X)))} \, {}^{>T}}{\vdots} \, {}^{>T}$$

Linguistically, adopting type-raising rules allows the extraction of wh-phrases from complex NP islands (Ozaki and Bekki 2012). When regarding CCG as a

---

[2] Another issue concerning the introduction of category variables is the need for unification between syntactic categories, which tends to be slow. We will not discuss how to avoid this in the implementation of *lightblue*, as such discussion is beyond the scope of this article.

substructural combinatory logic, the type-raising rules correspond to the $\mathcal{C}^*$-combinator and thus strengthen the deduction theorem (i.e., extractability).[3]

Thus, there are both computational and linguistic motivations for not adopting the type-raising rules, and category variables in CCG parsers.

However, this is not sufficient reason to annihilate category variables altogether, since a categorial grammar may employ category variables without adopting the type-raising rules. This is a reasonable choice, particularly because category variables are useful for packing ambiguous but syntactically similar lexical items. Moreover, the computational effect of lexical packing is evident in some languages, such as Japanese, in which all arguments appear on the left-hand side of predicates (and thus, any quantificational NP can be given a single lexical item by using variable vectors).

1.4   *Lightblue*: A Robust CCG Parser with DTS

Lightblue is a wide-coverage CCG parser for Japanese, implemented in the Haskell programming language, which outputs semantic representations in terms of dependent type semantics (DTS; Bekki (2014), Bekki and Mineshima (2016)).

DTS is a proof-theoretic discourse semantics of natural language based on dependent type theory ((Martin-Löf (1984), Nordström et al. (1990)), which extends TTG (Ranta 1994) with *underspecified terms* (notation $@_i^A$), through which anaphora resolution and presupposition binding are calculated via type checking and proof search in dependent type theory.

There are few Japanese parsers, other than *lightblue*, that yield (logical) semantic representations. Exceptions are Haruniwa (Tsaiwei et al. 2014) and ccg2lambda (Martínez Gómez et al. 2016). Both separate syntactic and semantic parsing; the former yields syntactic trees as output and the latter transforms them into semantic representations. However, since their lexicons are automatically obtained from large-scale corpora, they do not allow a semantics developer to add, delete, or modify a single lexical item, which is a standard way to improve lexicalized grammar, during the process of grammar development.

Parser development in *lightblue* is purely lexicalized; the data type for a lexical entry is a triple of a phonetic form, a CCG syntactic category, and a DTS preterm, so a semantics developer has direct access to each lexical entry. The *lightblue* lexicon has about 994,416 lexical entries for open words obtained from the dictionary of JUMAN, a part-of-speech tagger and morphological analyzer (Morita et al. 2015), which is automatically obtained (and distilled) from the world wide web, plus 758 lexical entries for closed words excerpted from Bekki (2010).

The CCG part of *lightblue* can make use of category variables, and the DTS part can make use of variable vectors, which significantly reduces the number of items in lexicons. *Lightblue* can also use empty categories, which gives it more flexibility, but that will be discussed elsewhere, and not in this paper.

---

[3] Evaluation of strengthened extractability in CCG depends on whether the complex NP constraint is inherently syntactic or arises from choices made about parsing or other extra-syntactic processes. Our discussion here assumes the former. However, if one assumes the latter, then adopting type-raising rules or a categorial grammar with stronger extractability would be essential.

2   DTS with Variable Vectors in de Bruijn Notation

As a semantic theory, *lightblue* employs DTS in de Bruijn notation, in order to avoid variable name clash or, alternatively, α conversion everywhere. The standard implementation of lambda calculus in de Bruijn notation is widely known from Pierce (2005), the techniques of which can be naturally extended to DTS.

2.1   Syntax and Reduction

The syntax of DTS with variable names extended with two constructors for variable vectors is given as follows[4], where the constructors $\lambda \overline{x}.M$ and $M\overline{x}$ give binders and bindees of variable vectors, respectively.

Definition 2 (Syntax of DTS with variable names).

$$\Lambda :: = x \mid c \mid \text{type} \mid \text{kind} \mid (x{:}\Lambda) \to \Lambda \mid \lambda x.\Lambda \mid \Lambda\Lambda$$
$$\mid \begin{bmatrix} x{:}\Lambda \\ \Lambda \end{bmatrix} \mid (\Lambda, \Lambda) \mid \pi_i(\Lambda) \mid @_i^\Lambda \mid \lambda \overline{x}.\Lambda \mid \Lambda\overline{x}$$

In de Bruijn notation, a variable name is replaced with a non-negative integer i, which is bound by the ith binder that takes scope over it. The syntax is defined as described below, where $j \in \{1, 2\}$.

Definition 3 (Syntax of DTS in de Bruijn notation).

$$\Lambda :: = i \mid c \mid \text{type} \mid \text{kind} \mid \Lambda \to \Lambda \mid \lambda\Lambda \mid \Lambda\Lambda$$
$$\mid \begin{bmatrix} \Lambda \\ \Lambda \end{bmatrix} \mid (\Lambda, \Lambda) \mid \pi_j(\Lambda) \mid @_i^\Lambda \mid \overline{\lambda}\Lambda \mid \Lambda\overline{i}$$

The syntax of Definition 3 is naturally implemented by the following Haskell data type.

```
data Preterm =
    Var Int |              -- ^ Variables
    Con Text |             -- ^ Constant symbols
    Type |                 -- ^ The sort \"type\"
    Kind |                 -- ^ The sort \"kind\"
    Pi Preterm Preterm |   -- ^ Dependent function types (or Pi types)
    Lam Preterm |          -- ^ Lambda abstractions
    App Preterm Preterm |  -- ^ Function Applications
    Sigma Preterm Preterm | -- ^ Dependent product types (or Sigma types)
    Pair Preterm Preterm | -- ^ Pairs
    Proj Selector Preterm | -- ^ (First and second) Projections
    Asp Int Preterm |      -- ^ The asperand terms (or underspecified terms)
    Lamvec Preterm |       -- ^ Lambda abstractions of a variable vector
    Appvec Int Preterm |   -- ^ Function applications with a variable vector
```

---

[4] The full version of *lightblue* employs, besides those constructors given in Definition 3, the intensional equality type, the natural number type, and the enumeration types in Nordström et al. (1990). For brevity, these are omitted.

Semantic representations of (a nominative version of) quantifiers *every* and *some* in DTS are given as follows (Bekki and Mineshima 2016):

$$(6) \quad \lambda n.\lambda p.\lambda \overline{x} \left(u{:}\begin{bmatrix} x{:}\textbf{entity} \\ nx(\lambda x.\top) \end{bmatrix} \right) \to p(\pi_1(u))\overline{x}$$

$$(7) \quad \lambda n.\lambda p.\lambda \overline{x} \begin{bmatrix} u{:}\begin{bmatrix} x{:}\textbf{entity} \\ nx(\lambda x.\top) \end{bmatrix} \\ p(\pi_1(u))\overline{x} \end{bmatrix}$$

With the data type `Preterm`, these representations are described in Haskell code as follows.

(8)   (Lam (Lam (Lamvec (Pi (Sigma (Con "entity'") (App (App (Var 3) (Var 0)) (Lam Top))) (Appvec 1 (App (Var 2) (Proj Fst (Var 0))))))))

(9)   (Lam (Lam (Lamvec (Sigma (Sigma (Con "entity") (App (App (Var 3) (Var 0)) (Lam Top))) (Appvec 1 (App (Var 2) (Proj Fst (Var 0)))))))))

The definitions of free variables and substitution for `Preterm` are obtained by extending their standard lambda calculus definitions, from Pierce (2005), with `Pi`, `Sigma`, `Asp`, `Lamvec`, and `Appvec`. The form `subst m l i` is understood as `m[l/i]`, that is, the preterm m in which an index i is substituted by a preterm l.

```
subst :: Preterm -> Preterm -> Int -> Preterm
subst preterm l i = case preterm of
  Var j       -> if i == j then l else Var j
  Pi a b      -> Pi (subst a l i) (subst b (shiftIndices l 1 0) (i+1))
  Lam m       -> Lam (subst m (shiftIndices l 1 0) (i+1))
  App m n     -> App (subst m l i) (subst n l i)
  Sigma a b   -> Sigma (subst a l i) (subst b (shiftIndices l 1 0) (i+1))
  Pair m n    -> Pair (subst m l i) (subst n l i)
  Proj s m    -> Proj s (subst m l i)
  Asp j m     -> Asp j (subst m l i)
  Lamvec m    -> Lamvec (subst m (shiftIndices l 1 0) (i+1))
  Appvec j m  -> Appvec j (subst m l i)
  m           -> m
```

The essence of the definition of `subst` lies in the use of the `shiftIndices` function. The form `shiftIndices m d i` appears in the cases of `Pi`, `Lam`, `Sigma`, and `Lamvec` above and executes *d-place shift* (Pierce 2005). Namely, it adds d to every index within m that is greater than i, whose role is to accommodate all indices to the new environment in which m is placed. The `shiftIndices` function is recursively defined as follows.

```
shiftIndices :: Preterm -> Int -> Int -> Preterm
shiftIndices preterm d i = case preterm of
  Var j       -> if j >= i
                 then Var (j+d)
                 else Var j
  Pi a b      -> Pi (shiftIndices a d i) (shiftIndices b d (i+1))
  Lam m       -> Lam (shiftIndices m d (i+1))
```

```
App m n      -> App (shiftIndices m d i) (shiftIndices n d i)
Sigma a b    -> Sigma (shiftIndices a d i) (shiftIndices b d (i+1))
Pair m n     -> Pair (shiftIndices m d i) (shiftIndices n d i)
Proj s m     -> Proj s (shiftIndices m d i)
Asp j m      -> Asp j (shiftIndices m d i)
Lamvec m     -> Lamvec (shiftIndices m d (i+1))
Appvec j m   -> if j >= i
                   then Appvec (j+d) (shiftIndices m d i)
                   else Appvec j (shiftIndices m d i)
m            -> m
```

For example, $(\lambda M)[L/i]$ reduces to $(\lambda(M[\text{shiftIndices } L\ 1\ 0/i+1])$, adding 1 to all indices greater than or equal to 0 (i.e., all indices) in L, since L, a preterm to replace, is going though one λ.

The reason for the restriction "greater than or equal to 0" is that indices in L that are bound within L must stay intact. For example, L, a preterm to replace, goes through two λs, then indices less than or equal to i+1 should remain unchanged.

$$(\lambda\lambda M)[L/i] = \lambda((\lambda M)[\text{shiftIndices } L\ 1\ 0/i+1]$$
$$= \lambda\lambda(M[\text{shiftIndices } (\text{shiftIndices } L\ i\ 0)\ 1\ 0/i+2]$$

Note that the definitions of Pi and Sigma add 1 only in the nuclear scope part (not in the restriction part). This reflects that in constructions such as $(x{:}A) \to B$ and $\begin{bmatrix} x{:}A \\ B \end{bmatrix}$, A is outside the scope of x.

The above definition of subst is then used to define beta reduction of preterms as follows, which also uses the shiftIndice function.

```
betaReduce :: Preterm -> Preterm
betaReduce preterm = case preterm of
  Pi a b     -> Pi (betaReduce a) (betaReduce b)
  Lam m      -> Lam (betaReduce m)
  App m n    -> case betaReduce m of
    Lam v -> betaReduce (shiftIndices (subst v (shiftIndices n 1 0) 0) (-1) 0)
    e        -> App e (betaReduce n)
  Sigma a b  -> Sigma (betaReduce a) (betaReduce b)
  Pair m n   -> Pair (betaReduce m) (betaReduce n)
  Proj s m   -> case betaReduce m of
    Pair x y -> case s of
                  Fst -> x
                  Snd -> y
    e        -> Proj s e
  Asp i m -> Asp i (betaReduce m)
  Lamvec m   -> Lamvec (betaReduce m)
  Appvec i m -> Appvec i (betaReduce m)
  m          -> m
```

2.2   Expanding and Shrinking of Variable Vectors

The expanding and shrinking operations of variable vectors consist of the three primitive functions addLambda, deleteLambda, and replaceLambda.

addLambda i m works on the form Appvec j f (within m), which is bound by the ith binder from the position of the addLambda i m (= the jth binder from the position of Appvec j f), and replaces it with Appvec j (App (addLambda i m) (Var (j+1))). In the second choice in Definition 1, this operation is used for replacing $M[\ldots f\overline{x}\ldots]$ with $M[\ldots (fx)\overline{x}\ldots]$ when we replace $\lambda\overline{x}$ with $\lambda x.\lambda\overline{x}$[5].

```
addLambda :: Int -> Preterm -> Preterm
addLambda i preterm = case preterm of
  Var j | j > i      -> Var (j+1)
        | j < i      -> Var j
        | otherwise -> Error
  Pi a b      -> Pi (addLambda i a) (addLambda (i+1) b)
  Lam m       -> Lam (addLambda (i+1) m)
  App m n     -> App (addLambda i m) (addLambda i n)
  Sigma a b   -> Sigma (addLambda i a) (addLambda (i+1) b)
  Pair m n    -> Pair (addLambda i m) (addLambda i n)
  Proj s m    -> Proj s (addLambda i m)
  Asp j m     -> Asp j (addLambda i m)
  Lamvec m    -> Lamvec (addLambda (i+1) m)
  Appvec j m | j > i     -> Appvec (j+1) (addLambda i m)
             | j < i     -> Appvec j (addLambda i m)
             | otherwise -> Appvec j (App (addLambda i m) (Var (j+1)))
  m          -> m
```

deleteLambda i m works on the form Appvec j f (within m, under the same conditions as addLambda), and replaces it with deleteLambda i m. In other words, deleteLambda deletes the occurrence(s) of a corresponding variable vector from m.

```
deleteLambda :: Int -> Preterm -> Preterm
deleteLambda i preterm = case preterm of
  Var j | j > i      -> Var (j-1)
        | j < i      -> Var j
        | otherwise -> Error
  Pi a b      -> Pi (deleteLambda i a) (deleteLambda (i+1) b)
  Lam m       -> Lam (deleteLambda (i+1) m)
  App m n     -> App (deleteLambda i m) (deleteLambda i n)
  Sigma a b   -> Sigma (deleteLambda i a) (deleteLambda (i+1) b)
  Pair m n    -> Pair (deleteLambda i m) (deleteLambda i n)
  Proj s m    -> Proj s (deleteLambda i m)
  Asp j m     -> Asp j (deleteLambda i m)
  Lamvec m    -> Lamvec (deleteLambda (i+1) m)
  Appvec j m | j > i     -> Appvec (j-1) (deleteLambda i m)
             | j < i     -> Appvec j (deleteLambda i m)
             | otherwise -> deleteLambda i m
  m          -> m
```

---

[5] The Error is used here for simplifying the exposition. It is a constant symbol that represent an error in the actual code of *lightblue*.

replaceLambda i m works on the form Appvec j f (within m, under the same conditions as addLambda and deleteLambda), and replaces it with App (replaceLambda i m) (Var j). Namely, it replaces the occurrence(s) of a corresponding variable vector with a variable.

```
replaceLambda :: Int -> Preterm -> Preterm
replaceLambda i preterm = case preterm of
  Pi a b     -> Pi (replaceLambda i a) (replaceLambda (i+1) b)
  Lam m      -> Lam (replaceLambda (i+1) m)
  App m n    -> App (replaceLambda i m) (replaceLambda i n)
  Sigma a b -> Sigma (replaceLambda i a) (replaceLambda (i+1) b)
  Pair m n   -> Pair (replaceLambda i m) (replaceLambda i n)
  Proj s m   -> Proj s (replaceLambda i m)
  Asp j m    -> Asp j (replaceLambda i m)
  Lamvec m   -> Lamvec (replaceLambda (i+1) m)
  Appvec j m | i == j      -> App (replaceLambda i m) (Var j)
             | otherwise -> Appvec j (replaceLambda i m)
  m          -> m
```

Using these three functions gives the following definition of variable vectors.

$$\lambda\overline{x}.M[\ldots f\overline{x}\ldots] \overset{def}{\equiv} \begin{cases} M[\ldots f,\ldots] \\ \lambda x.\lambda\overline{x}.M[\ldots (fx)\overline{x}\ldots] \end{cases}$$

can be represented by the following (pseudo-)Haskell code:

$$\text{Lamvec } M[\ldots \text{Appvec } j \ f \ldots] \overset{def}{\equiv} \begin{cases} \text{deleteLambda 0 M} \\ \text{Lam (Lamvec (addLambda 0 M))} \end{cases}$$

Here, $f\overline{x}$ in $M[\ldots f\overline{x}\ldots]$ is replaced with f by deleteLambda 0 M, and is replaced with $(fx)\overline{x}$ by addLambda 0 M. Note that a condition that $x \notin fv(M[\ldots f\overline{x}\ldots])$ in Definition 1 is no longer necessary under de Bruijn notation.

2.3   Interaction Between Category and Lambda Terms

The remaining task is to provide a function that takes a pair comprising a syntactic category and a preterm in DTS and returns a preterm within which variable vectors are expanded or shrunk as needed. The transvec function, defined as follows, does this job, transforming variable vectors within a given preterm by adjusting the preterm's number of arguments using the tree functions of the last section.

```
transvec :: Cat -> Preterm -> Preterm
transvec c preterm = case c of
  SL x _ -> case preterm of
            Lam m     -> Lam (transvec x m)
            Lamvec m -> Lam (transvec x (Lamvec (addLambda 0 m)))
```

```
              m          -> m -- Var, Con, App, Proj, Asp, Appvec
BS x _ -> case preterm of
              Lam m    -> Lam (transvec x m)
              Lamvec m -> Lam (transvec x (Lamvec (addLambda 0 m)))
              m          -> m -- Var, Con, App, Proj, Asp, Appvec
NP _   -> case preterm of
              Lamvec m -> deleteLambda 0 m
              m          -> m
S _ -> case preterm of
              Lam (Lamvec m) -> Lam (deleteLambda 0 m)
              Lamvec (Lam m) -> deleteLambda 0 (Lam m)
              Lamvec m -> Lam (replaceLambda 0 m)
              m          -> m
N -> case preterm of
              Lam (Lam (Lamvec m)) -> Lam (Lam (deleteLambda 0 m))
              Lam (Lamvec (Lam m)) -> Lam (deleteLambda 0 (Lam m))
              Lamvec (Lam (Lam m)) -> deleteLambda 0 (Lam (Lam m))
              Lamvec (Lam (Lam m)) -> Lam (replaceLambda 0 (Lam m))
              Lam (Lamvec m) -> Lam (Lam (replaceLambda 0 m))
              Lamvec m -> Lam (Lam (replaceLambda 0 (addLambda 0 m)))
              m          -> m
_ -> preterm
```

By using `transvec` functions, we can implement a CCG parser with category variables in the syntax and variable vectors in the semantics.

3   Some Examples

Let us demonstrate some parse results of *lightblue*. The first example is a Japanese verb phrase consisting of a quantifier (in the object position) and a transitive verb.

(10)   subete-no ningen -ni   aw   -u
       every   man   cm-DAT meet PRES

       (lit.) 'meet every man'

The following derivation is an output of *lightblue* given the phrase *subete-no ningen-ni* (except for a minor modification that replaces names of constant symbols in Japanese with equivalents in English).

$$\cfrac{\cfrac{\text{subete-no}}{\begin{array}{l} T^{\boxed{1}}_{S_{v:5:k|v:5:s|+:\boxed{2}}} \quad /(T^{\boxed{1}}_{S_{v:5:k|v:5:s|+:\boxed{2}}} \quad \backslash NP_{nc})/N \\ {\scriptstyle neg|cont|+:\boxed{3}} \qquad\qquad {\scriptstyle neg|cont|+:\boxed{3}} \\ {\scriptstyle \pm t:\boxed{4},\pm p:\boxed{5},\pm n:\boxed{6}} \quad {\scriptstyle \pm t:\boxed{4},\pm p:\boxed{5},\pm n:\boxed{6}} \\[4pt] :\lambda x_0.\lambda x_1.\lambda \overline{x_2}.\left(u_3:\begin{bmatrix} x_4:\textbf{entity} \\ x_0(x_4)\,(\lambda x_5.\top) \end{bmatrix}\right) \to x_1\,(\pi_1\,(u_3))\,\overline{x_2} \end{array}}(534) \qquad \cfrac{\text{ningen}}{\begin{array}{l} N \\ :\lambda x_0.\lambda x_1.\begin{bmatrix} s_2:\textbf{state} \\ u_3:\textbf{man}\,(s_2,x_0) \\ x_1(s_2) \end{bmatrix} \end{array}}(CN)}{\begin{array}{l} T^{\boxed{1}}_{S_{v:5:k|v:5:s|+:\boxed{2}}} \quad /(T^{\boxed{1}}_{S_{v:5:k|v:5:s|+:\boxed{2}}} \quad \backslash NP_{nc}) \\ {\scriptstyle neg|cont|+:\boxed{3}} \qquad\qquad {\scriptstyle neg|cont|+:\boxed{3}} \\ {\scriptstyle \pm t:\boxed{4},\pm p:\boxed{5},\pm n:\boxed{6}} \quad {\scriptstyle \pm t:\boxed{4},\pm p:\boxed{5},\pm n:\boxed{6}} \\[4pt] :\lambda x_0.\lambda \overline{x_1}.\left(u_2:\begin{bmatrix} x_3:\textbf{entity} \\ \begin{bmatrix} s_4:\textbf{state} \\ \textbf{man}\,(s_4,x_3) \end{bmatrix} \end{bmatrix}\right) \to x_0\,(\pi_1\,(u_2))\,\overline{x_1} \end{array}}>$$

$$\text{subete-no ningen}$$

$$
\underset{\substack{\underline{T}_{S_{v:5:k|v:5:s|+\boxed{2}}}^{\boxed{1}}\\ neg|cont|+\boxed{3}\\ \pm t\cdot\boxed{4},\pm p\cdot\boxed{5},\pm n\cdot\boxed{6}}}{}\quad \underset{\substack{/(\underline{T}_{S_{v:5:k|v:5:s|+\boxed{2}}}^{\boxed{1}}\\ neg|cont|+\boxed{3}\\ \pm t\cdot\boxed{4},\pm p\cdot\boxed{5},\pm n\cdot\boxed{6}}}{}\quad \backslash NP_{nc}) \overset{(above)}{}
$$

$$
:\lambda x_0.\lambda\overline{x_1}.\left(u_2:\begin{bmatrix}x_3:\textbf{entity}\\ s_4:\textbf{state}\\ \textbf{man}\,(s_4,x_3)\end{bmatrix}\right)\to x_0\,(\pi_1\,(u_2))\,\overline{x_1}
$$

$$
\underset{\substack{\underline{T}_S^{\boxed{1}}\\ neg|cont|+\boxed{3}\\ \pm t\cdot\boxed{4},\pm p\cdot\boxed{5},\pm n\cdot\boxed{6}}}{}\quad\underset{\substack{/(\underline{T}_S^{\boxed{1}}\\ neg|cont|+\boxed{3}\\ \pm t\cdot\boxed{4},\pm p\cdot\boxed{5},\pm n\cdot\boxed{6}}}{}\;\backslash NP_{ni})\backslash NP_{nc})\quad\overset{(524)}{}
$$

$$
:\lambda x_0.\lambda x_1.x_1(x_0)
$$

$$
\underset{\substack{\underline{T}_{S_{v:5:k|v:5:s|+\boxed{2}}}^{\boxed{1}}\\ neg|cont|+\boxed{3}\\ \pm t\cdot\boxed{4},\pm p\cdot\boxed{5},\pm n\cdot\boxed{6}}}{}\quad\underset{\substack{/(\underline{T}_{S_{v:5:k|v:5:s|+\boxed{2}}}^{\boxed{1}}\\ neg|cont|+\boxed{3}\\ \pm t\cdot\boxed{4},\pm p\cdot\boxed{5},\pm n\cdot\boxed{6}}}{}\;\backslash NP_{ni})\quad\overset{>}{}
$$

$$
:\lambda x_0.\lambda\overline{x_1}.\left(u_2:\begin{bmatrix}x_3:\textbf{entity}\\ s_4:\textbf{state}\\ \textbf{man}\,(s_4,x_3)\end{bmatrix}\right)\to x_0\,(\pi_1\,(u_2))\,\overline{x_1}
$$

In the above derivation, T is a category variable. The upper scripts of T, such as $\boxed{1}$, indicate that structure sharing is taking place, and each of their scopes is local. Namely, $\boxed{1}$ within the phrase *subete-no* and $\boxed{1}$ within the phrase *ni* do not share their structures with each other. The lower script of T shows the "final-output category" of T. For example, T_S must be unified with a function category that ends as S, or S itself (T_S is equal to $S|\$$ in the standard CCG notation).

The lower script attached to syntactic categories other than T shows their syntactic features. For example, NP_{ni} is an NP with *ni*-feature (i.e., an NP marked with the dative case), and NP_{nc} is an NP with no case. A feature such as $v:5:k$ shows that it is a verb that belongs to the conjugation series $5k$. The symbol $|$ is a disjunction (or a union) between syntactic features so that unification between syntactic categories returns their intersection if it is not empty. Features such as $neg|cont$ show conjugation forms (i.e., the negation form or the continuous form). Binary features $\pm t, \pm p, \pm n$ represent past/nonpast, polite/non-polite, and negated/non-negated, respectively. Details of these syntactic features in Japanese are described in Bekki (2010). The number on the right-hand side of a lexical item corresponds to an entry number in Bekki (2010).

The notation of DTS that *lightblue* adopts follows that of Bekki and Mineshima (2016). Although the internal representations of DTS preterms are in de Bruijn notation, they are transformed into DTS preterms with variable names when visualized. The point of the above derivation lies in the last step, where function application takes place: the left node is *every man*, a quantifier NP. How many arguments it would take after merging with a predicate is underspecified by T, and thus its semantic representation contains a variable vector. Meanwhile, *ni* is a dative case marker (which is semantically various) whose whole syntactic category $T/(T\backslash NP_{ni})\backslash NP_{nc}$ unifies with the $T\backslash NP_{nc}$ part of the quantifier NP. As a result, beta reduction between their semantic representations ends up with the following preterm.

$$
\left(\lambda x_0.\lambda\overline{x_1}.\left(u_2:\begin{bmatrix}x_3:\textbf{entity}\\ s_4:\textbf{state}\\ \textbf{man}\,(s_4,x_3)\end{bmatrix}\right)\to x_0\,(\pi_1(u_2))\,\overline{x_1}\right)(\lambda x_0.\lambda x_1.x_1(x_0))
$$

$$\twoheadrightarrow_\beta \lambda \overline{x_1}. \left(u_2 : \begin{bmatrix} x_3 : \mathbf{entity} \\ \begin{bmatrix} s_4 : \mathbf{state} \\ \mathbf{man}\,(s_4, x_3) \end{bmatrix} \end{bmatrix} \right) \to (\lambda x_0.\lambda x_1.x_1(x_0))\,(\pi_1(u_2))\,\overline{x_1}$$

$$\twoheadrightarrow_\beta \lambda \overline{x_1}. \left(u_2 : \begin{bmatrix} x_3 : \mathbf{entity} \\ \begin{bmatrix} s_4 : \mathbf{state} \\ \mathbf{man}\,(s_4, x_3) \end{bmatrix} \end{bmatrix} \right) \to (\lambda x_1.x_1(\pi_1(u_2)))\overline{x_1}$$

This is a `Lamvec` construction, and the number of arguments that it would take is underdetermined. However, since its corresponding syntactic category (which could be read off from the derivation) is $T/(T\backslash NP_{ni})$, we can tell that it takes at least one more argument (of syntactic category $T\backslash NP_{ni}$) and then would take an arbitrary number of arguments. Thus, the variable vector should be expanded one step and become $\lambda x.\lambda \overline{x_1} \ldots$, and the inner $\overline{x_1}$ should be replaced with the sequence $(x_0, \overline{x_1})$. The `transvec` function ensures this behavior. First, the form of syntactic category $T/(T\backslash NP_{ni})$ and the preterm match the following line of `transvec`.

```
SL x _ -> case preterm of
        Lamvec m -> Lam (transvec x (Lamvec (addLambda 0 m)))
```

Here, `m` corresponds to the following subterm.

$$\left(u_2 : \begin{bmatrix} x_3 : \mathbf{entity} \\ \begin{bmatrix} s_4 : \mathbf{state} \\ \mathbf{man}\,(s_4, x_3) \end{bmatrix} \end{bmatrix} \right) \to (\lambda x_1.x_1(\pi_1(u_2)))\overline{x_1}$$

for which `addLambda 0 m` is evaluated, adding a variable x_0 at the position of $\overline{x}$.

$$\left(u_2 : \begin{bmatrix} x_3 : \mathbf{entity} \\ \begin{bmatrix} s_4 : \mathbf{state} \\ \mathbf{man}\,(s_4, x_3) \end{bmatrix} \end{bmatrix} \right) \to ((\lambda x_1.x_1(\pi_1(u_2)))x_0)\overline{x_1}$$

This beta-reduces to the following preterm.

$$\left(u_2 : \begin{bmatrix} x_3 : \mathbf{entity} \\ \begin{bmatrix} s_4 : \mathbf{state} \\ \mathbf{man}\,(s_4, x_3) \end{bmatrix} \end{bmatrix} \right) \to x_0(\pi_1(u_2))\overline{x_1}$$

In the case where this form is again a `Lamvec` construction, the *transvec* function recursively applies the *transvec* function to it, making reference to the syntactic category T, which does not do anything for this case. After that, one more lambda operator is added to the top (which binds x_0), and the whole result is wound up with the following form.

$$\lambda x_0.\lambda \overline{x_1}. \left(u_2 : \begin{bmatrix} x_3 : \mathbf{entity} \\ \begin{bmatrix} s_4 : \mathbf{state} \\ \mathbf{man}\,(s_4, x_3) \end{bmatrix} \end{bmatrix} \right) \to x_0\,(\pi_1\,(u_2))\,\overline{x_1}$$

This subterm will then merge with the transitive verb *aw-u* ("meet"). The number of arguments of *aw-u* is two (a nominative NP and a dative NP). Thus, the final syntactic category becomes $S\backslash NP_{ga}$, as shown below.

$$\frac{\begin{array}{c}\text{aw}\\ S_{v:5:w}\backslash NP_{ga}\backslash NP_{ni}\\ \text{stem}\end{array}}{}\,(JCon)\quad \frac{\begin{array}{c}\text{u}\\ S_{v:5:w|v:5:TOW\cdot\boxed{1}}\,|\,S_{v:5:w|v:5:TOW\cdot\boxed{1}}\\ \text{term}|\text{attr}\quad\quad\quad\text{stem}\end{array}}{}\,(130)$$

$$\frac{\begin{array}{c}[\text{subete-no ningen}]\text{-ni}\\ T^{\boxed{1}}_{\substack{S_{v:5:k|v:5:s|+\boxed{2}}\\ neg|cont|+\boxed{3}\\ \pm t\boxed{4},\pm p\boxed{5},\pm n\boxed{6}}}/(T^{\boxed{1}}_{\substack{S_{v:5:k|v:5:s|+\boxed{2}}\\ neg|cont|+\boxed{3}\\ \pm t\boxed{4},\pm p\boxed{5},\pm n\boxed{6}}}\backslash NP_{ni})\end{array}\quad\begin{array}{c}:\lambda x_0.\lambda x_1.\lambda x_2.\begin{bmatrix}e_3\text{:event}\\ u_4\text{:meet}(e_3,x_1,x_0)\\ x_2(e_3)\end{bmatrix}\end{array}\quad\begin{array}{c}:\lambda x_0.x_0\end{array}}{}$$

$$:\lambda x_0.\lambda\overline{x_1}.\left(u_2\colon\begin{bmatrix}x_3\text{:entity}\\ \begin{bmatrix}s_4\text{:state}\\ \mathbf{man}\,(s_4,x_3)\end{bmatrix}\end{bmatrix}\right)\rightarrow x_0\,(\pi_1\,(u_2))\,\overline{x_1}\qquad :\lambda x_0.\lambda x_1.\lambda x_2.\begin{bmatrix}e_3\text{:event}\\ u_4\text{:meet}(e_3,x_1,x_0)\\ x_2(e_3)\end{bmatrix}$$

$$\frac{\quad\quad S_{v:5:w\cdot\boxed{1}}\backslash NP_{ga}}{\text{term}|\text{attr}\cdot\boxed{4}}$$

$$:\lambda x_0.\lambda x_1.\left(u_2\colon\begin{bmatrix}x_3\text{:entity}\\ \begin{bmatrix}s_4\text{:state}\\ \mathbf{man}\,(s_4,x_3)\end{bmatrix}\end{bmatrix}\right)\rightarrow\begin{bmatrix}e_3\text{:event}\\ u_4\text{:meet}\,(e_3,x_0,\pi_1\,(u_2))\\ x_1(e_3)\end{bmatrix}$$

The point here is that $T\backslash NP_{ni}$ on the left-hand side unifies with $S\backslash NP_{ga}\backslash NP_{ni}$ on the right-hand side. Therefore, T, the result of this merge operation, must be $S\backslash NPga$, with two more arguments (one for NP_{ni} and another for a continuation). This tells the `transvec` function to transform $\lambda\overline{x_1}$ in the semantic representation into two λs, by the `replacelambda` function. First, β-reduction proceeds as follows:

$$\left(\lambda x_0.\lambda\overline{x_1}.\left(u_2\colon\begin{bmatrix}x_3\text{:entity}\\ \begin{bmatrix}s_4\text{:state}\\ \mathbf{man}\,(s_4,x_3)\end{bmatrix}\end{bmatrix}\right)\rightarrow x_0\,(\pi_1\,(u_2))\,\overline{x_1}\right)\left(\lambda x_0.\lambda x_1.\lambda x_2.\begin{bmatrix}e_3\text{:event}\\ u_4\text{:meet}\,(e_3,x_1,x_0)\\ x_2(e_3)\end{bmatrix}\right)$$

$$\rightarrow_\beta\lambda\overline{x_1}.\left(u_2\colon\begin{bmatrix}x_3\text{:entity}\\ \begin{bmatrix}s_4\text{:state}\\ \mathbf{man}\,(s_4,x_3)\end{bmatrix}\end{bmatrix}\right)\rightarrow\left(\lambda x_0.\lambda x_1.\lambda x_2.\begin{bmatrix}e_3\text{:event}\\ u_4\text{:meet}\,(e_3,x_1,x_0)\\ x_2(e_3)\end{bmatrix}\right)(\pi_1\,(u_2))\,\overline{x_1}$$

$$\rightarrow_\beta\lambda\overline{x_1}.\left(u_2\colon\begin{bmatrix}x_3\text{:entity}\\ \begin{bmatrix}s_4\text{:state}\\ \mathbf{man}\,(s_4,x_3)\end{bmatrix}\end{bmatrix}\right)\rightarrow\left(\lambda x_1.\lambda x_2.\begin{bmatrix}e_3\text{:event}\\ u_4\text{:meet}\,(e_3,x_1,(\pi_1\,(u_2)))\\ x_2(e_3)\end{bmatrix}\right)\overline{x_1}$$

Then, a pair $S\backslash NP_{ga}$ and the above preterm match the following line of the *transvec* function, in the same way as the previous example.

```
SL x _ -> case preterm of
            Lamvec m -> Lam (transvec x (Lamvec (addLambda 0 m)))
```

Applying `addLambda 0` and putting `Lamvec` on top yields the following preterm.

$$\lambda\overline{x_1}.\left(u_2\colon\begin{bmatrix}x_3\text{:entity}\\ \begin{bmatrix}s_4\text{:state}\\ \mathbf{man}\,(s_4,x_3)\end{bmatrix}\end{bmatrix}\right)\rightarrow\left(\lambda x_1.\lambda x_2.\begin{bmatrix}e_3\text{:event}\\ u_4\text{:meet}\,(e_3,x_1,(\pi_1\,(u_2)))\\ x_2(e_3)\end{bmatrix}\right)x_0\overline{x_1}$$

$$\rightarrow_\beta\lambda\overline{x_1}.\left(u_2\colon\begin{bmatrix}x_3\text{:entity}\\ \begin{bmatrix}s_4\text{:state}\\ \mathbf{man}\,(s_4,x_3)\end{bmatrix}\end{bmatrix}\right)\rightarrow\left(\lambda x_2.\begin{bmatrix}e_3\text{:event}\\ u_4\text{:meet}\,(e_3,x_0,(\pi_1\,(u_2)))\\ x_2(e_3)\end{bmatrix}\right)\overline{x_1}$$

This is sent to the recursive call of *transvec*, making reference to the corresponding syntactic category S. This time, it matches with the following line of `transvec`.

```
S _ -> case preterm of
            Lamvec m -> Lam (replaceLambda 0 m)
```

Application of the `replaceLambda` function replaces the variable vector $\overline{x_1}$ with a variable x_1, and putting another λ on top yields the following preterm, as expected.

$$\lambda x_0.\lambda x_1.\left\{u_2:\begin{bmatrix}x_3:\mathbf{entity}\\\begin{bmatrix}s_4:\mathbf{state}\\\mathbf{man}\,(s_4,x_3)\end{bmatrix}\end{bmatrix}\right\}\rightarrow\left(\lambda x_2.\begin{bmatrix}e_3:\mathbf{event}\\\begin{bmatrix}u_4:\mathbf{meet}\,(e_3,x_0,(\pi_1\,(u_2)))\\x_2(e_3)\end{bmatrix}\end{bmatrix}\right)x_1$$

$$\twoheadrightarrow_\beta\ \lambda x_0.\lambda x_1.\left(u_2:\begin{bmatrix}x_3:\mathbf{entity}\\\begin{bmatrix}s_4:\mathbf{state}\\\mathbf{man}\,(s_4,x_3)\end{bmatrix}\end{bmatrix}\right)\rightarrow\begin{bmatrix}e_3:\mathbf{event}\\\begin{bmatrix}u_4:\mathbf{meet}\,(e_3,x_0,(\pi_1\,(u_2)))\\x_1(e_3)\end{bmatrix}\end{bmatrix}$$

4   Conclusion and Future Work

While the use of category variables in CCG offers advantages such as packing ambiguous syntactic candidates during parsing, we have also seen that it requires semantic language to be extended with variable vectors, whose formalization and implementation have not been straightforward so far.

In this article, we introduced *lightblue*, a Japanese CCG parser implementation equipped with variable vectors and a mechanism for expanding and shrinking them, according to the corresponding syntactic categories. Since the semantic language of *lightblue* is DTS, which is based on dependent type theory, a natural extension of simply typed lambda calculus, its implementation by a functional programming language is straightforward and natural. The implementation consists of three primitive functions, `addLambda`, `deleteLambda`, and `replaceLambda`, that expand and shrink variable vectors, together with the `transvec` function that transforms a preterm, making reference to a corresponding syntactic category, by choosing and applying an appropriate definition in Definition 1 according to the number of arguments that the category anticipates.

Although the formalism and implementation of variable vectors in *lightblue* provides a solution to the problems that we addressed, some issues remain to be pursued. For example, future research should prove a version of completeness of the `transvec` function regarding how the normal form is defined for typed lambda calculus or dependent type theory with variable vectors. In each step of applying combinatory rules, does the `transvec` function always transform a given preterm to a normal form? Does the `transvec` function terminate, and under what conditions? These remain as open issues, which we believe provides an attractive research topic regarding syntactic–semantic transparency in combinatory (or other) categorial grammar(s).

References

Bekki, D.: Nihongo-Bunpoo-no Keisiki-Riron - Katuyootaikei, Toogohantyuu, Imigoo-sei - (trans. 'Formal Japanese Grammar: the conjugation system, categorial syntax, and compositional semantics'). Kuroshio Publisher, Tokyo (2010)

Bekki, D.: Representing anaphora with dependent types. In: Asher, N., Soloviev, S.V. (eds.) LACL 2014. LNCS, vol. 8535, pp. 14–29. Springer, Heiderburg (2014)

Bekki, D., Mineshima, K.: Context-passing and underspecification in dependent type semantics. In: Chatzikyriakidis, S., Luo, Z. (eds.) Type-Theoretical Semantics: Current Perspectives. Springer, Heidelberg (2016)

Clark, S., Curran, J.R.: Parsing the WSJ using CCG and log-linear models. In: The Proceedings of the 42nd Annual Meeting on Association for Computational Linguistics, pp. 103–110. Association for Computational Linguistics (2004)

Lewis, M., Steedman, M.: A* CCG parsing with a supertag-factored model. In: The Proceedings of the 2014 Conference on Empirical Methods in Natural Language Processing (EMNLP), Doha, Qatar, pp. 990–1000. Association of Computational Linguistics (2014)

Martin-Löf, P.: Intuitionistic Type Theory, vol. 17. Bibliopolis, Naples (1984). Sambin, G. (ed.)

Martínez Gómez, P., Mineshima, K., Miyao, Y., Bekki, D.: ccg2lambda: a computational semantics system. In: The Proceedings of the Association of Computational Linguistics (ACL 2016), Berlin, pp. 85–90 (2016)

Morita, H., Kawahara, D., Kurohashi, S.: Morphological analysis for unsegmented languages using recurrent neural network language model. In: The Proceedings of Conference on Empirical Methods in Natural Language Processing (EMNLP 2015), pp. 2292–2297 (2015)

Noji, H., Miyao, Y.: Jigg: a framework for an easy natural language processing pipeline. In: The Proceedings of the 54th Association of Computational Linguistics, pp. 103–108 (2016)

Nordström, B., Petersson, K., Smith, J.: Programming in Martin-Löf's Type Theory. Oxford University Press, Oxford (1990)

Ozaki, H., Bekki, D.: Extractability as the deduction theorem in subdirectional combinatory logic. In: Béchet, D., Dikovsky, A. (eds.) LACL 2012. LNCS, vol. 7351, pp. 186–200. Springer, Heidelberg (2012). doi:10.1007/978-3-642-31262-5_13

Pierce, B.C.: Advanced Topics in Types and Programming Languages. The MIT Press, Cambridge (2005)

Ranta, A.: Type-Theoretical Grammar. Oxford University Press, Oxford (1994)

Steedman, M.J.: The Syntactic Process (Language, Speech, and Communication). The MIT Press, Cambridge (2000)

Tsaiwei, F., Butler, A., Yoshimoto, K.: Parsing Japanese with a PCFG treebank grammar. In: The Proceedings of the Twentieth Meeting of the Association for Natural Language Processing, Sapporo, pp. 432–435 (2014)

On Classical Nonassociative Lambek Calculus

Wojciech Buszkowski[✉]

Faculty of Mathematics and Computer Science,
Adam Mickiewicz University, Poznań, Poland
buszko@amu.edu.pl

Abstract. CNL, intoduced by de Groote and Lamarche [11], is a conservative extension of Nonassociative Lambek Calculus (NL) by a De Morgan negation $^\sim$, satisfying $A^\sim/B \Leftrightarrow A\backslash B^\sim$. [11] provides a fine theory of proof nets for CNL and shows cut elimination and polynomial decidability. Here the purely proof-theoretic approach of [11] is enriched with algebras and phase spaces for CNL. We prove that CNL is a strongly conservative extension of NL, CNL has the strong finite model property, the grammars based on CNL (also with assumptions) generate the context-free languages, and the finitary consequence relation for CNL is decidable in polynomial time.

Keywords: Lambek calculus · Phase space · Sequent system · Type grammar

1 Introduction

NL, due to Lambek [13], admits formulas built from variables and the connectives $\otimes, \backslash, /$. The axioms and the rules are as follows.

$$\text{(NL-id) } A \Rightarrow A$$

$$(\otimes \Rightarrow) \ \frac{\Gamma[(A,B)]\Rightarrow C}{\Gamma[A\otimes B]\Rightarrow C} \quad (\Rightarrow \otimes) \ \frac{\Gamma \Rightarrow A \quad \Delta \Rightarrow B}{(\Gamma,\Delta)\Rightarrow A\otimes B}$$

$$(\backslash \Rightarrow) \ \frac{\Gamma[B]\Rightarrow C \quad \Delta \Rightarrow A}{\Gamma[(\Delta,A\backslash B)]\Rightarrow C} \quad (\Rightarrow \backslash) \ \frac{(A,\Gamma)\Rightarrow B}{\Gamma \Rightarrow A\backslash B}$$

$$(/ \Rightarrow) \ \frac{\Gamma[A]\Rightarrow C \quad \Delta \Rightarrow B}{\Gamma[(A/B,\Delta)]\Rightarrow C} \quad (\Rightarrow /) \ \frac{(\Gamma,B)\Rightarrow A}{\Gamma \Rightarrow A/B}$$

$$\text{(NL-cut) } \frac{\Gamma[A]\Rightarrow B \quad \Delta \Rightarrow A}{\Gamma[\Delta]\Rightarrow B}$$

This is a sequent system for NL. Sequents are of the form $\Gamma \Rightarrow A$, where A is a formula and Γ is a formula structure. Formula structures are defined recursively: (i) all formulas are formula structures, (ii) if Γ and Δ are formula structures, then (Γ, Δ) is a formula structure. Formula structures represent the elements of the free groupoid generated by formulas. A context $\Gamma[]$ is a formula structure

© Springer-Verlag GmbH Germany 2016
M. Amblard et al. (Eds.): LACL 2016, LNCS 10054, pp. 68–84, 2016.
DOI: 10.1007/978-3-662-53826-5_5

containing one special formula x. $\Gamma[\Delta]$ denotes the substitution of Δ for x in $\Gamma[\,]$. We reserve A, B, C, D for formulas and Γ, Δ, Θ for formula structures.

NL is strongly complete with respect to residuated groupoids (see Sect. 2 for the definition). Recall that a logic (in the form of a sequent system) is strongly complete with respect to a class of (ordered) algebras $\mathcal{C}$, if the following equivalence holds: $\Gamma \Rightarrow A$ is provable in this logic from the set of sequents Φ if and only if, for any algebra from $\mathcal{C}$ and any valuation μ, $\Gamma \Rightarrow A$ is true for μ whenever all sequents from Φ are true for μ. The right-hand side of this equivalence expresses the semantic entailment: $\Gamma \Rightarrow A$ follows from Φ in $\mathcal{C}$. For systems considered here, $\Gamma \Rightarrow A$ is *true* for μ, if $\mu(\Gamma) \leq \mu(A)$.

NL1 is NL admitting empty antecedents of sequents and containing the constant 1, the axiom (a-1) $\Rightarrow$ 1 and the rules:

$$(1 \Rightarrow) \quad \frac{\Gamma[\Delta] \Rightarrow A}{\Gamma[(1, \Delta)] \Rightarrow A}, \quad \frac{\Gamma[\Delta] \Rightarrow A}{\Gamma[(\Delta, 1)] \Rightarrow A}.$$

NL1 is strongly complete with respect to residuated unital groupoids.

Classical Nonassociative Lambek Calculus (CNL) can be presented as an extension of NL with negation $\sim$, admitting the axioms $A^{\sim\sim} \Leftrightarrow A$, $A^\sim / B \Leftrightarrow A \backslash B^\sim$ and the transposition rule:

$$\frac{A \Rightarrow B}{B^\sim \Rightarrow A^\sim}.$$

Here $A \Leftrightarrow B$ replaces two sequents: $A \Rightarrow B$ and $B \Rightarrow A$. In [11], CNL is presented as a Schütte style (i.e. one-sided) sequent system in language $\otimes, \oplus, \sim$, where $A \oplus B$ is equivalent to $(B^\sim \otimes A^\sim)^\sim$. So $\oplus$ corresponds to the operation 'par' in linear logics. We do not follow the popular notation of Girard [10], but replace it with a notation used in substructural logics [9]. CNL is a nonassociative variant of Cyclic Noncommutative MALL [15], but it lacks the multiplicative units.

In Sect. 2 we define CNL-algebras, i.e. the ordered algebras corresponding to CNL. We also define phase spaces, appropriate for nonassociative logics without units. We show that CNL-algebras arise from symmetric phase spaces, satisfying a compatibility condition.

In Sect. 3 we present CNL as a dual Schütte style system, which seems closer to the syntax of NL and the framework of type grammars. We discuss the strong completeness of CNL with respect to CNL-algebras and phase spaces. In particular, we outline a model-theoretic proof of cut elimination, similar to those for different substructural logics (see [9] for a discussion). Theorem 2 states that CNL is a strongly conservative extension of NL; we give a model-theoretic proof. At the end we briefly discuss analogous results for related logics: CNL1, i.e. CNL with constants 1 and 0, CNL and CNL1 with $\vee, \wedge$, and others.

In Sect. 4 we prove an interpolation lemma for CNL (with assumptions), analogous to the interpolation lemma for NL [4,8]. Using this lemma, we prove the strong finite model property (SFMP) for CNL (see [9] for the definition), the context-freeness of the languages generated by CNL-grammars and the polynomial time decidability of the consequence relation for CNL. These results remain true for CNL1. At the end, we discuss their status for other logics.

The size limits do not allow us to study CNL^-, i.e. the variant of CNL with two negations $\sim, ^-$, satisfying $A^{\sim -} \Leftrightarrow A$, $A^{-\sim} \Leftrightarrow A$, $A^\sim/B \Leftrightarrow A\backslash B^-$ and the transposition rules. CNL^- is a nonassociative variant of Noncommutative MALL [1], also called Classical Bilinear Logic in [14]; again it lacks units. The corresponding algebras are briefly discussed in Sect. 2. We only note here that CNL^- does not have SFMP. $A^\sim \Rightarrow A^-$ entails $A^- \Rightarrow A^\sim$ in finite CNL^--algebras, since $a^\sim < a^-$ enforces the infinite chain $a < a^{\sim\sim} < a^{\sim\sim\sim\sim} < \ldots$; there exist infinite CNL^--algebras such that $a^\sim < a^-$, for some element a.

2   Algebras and Phase Spaces

The algebraic models of NL are residuated groupoids $\mathbf{M} = (M, \otimes, \backslash, /, \leq)$ such that $(M, \leq)$ is a nonempty poset and $\otimes, \backslash, /$ are binary operations on M, satisfying:

$$\text{(RES)} \quad a \otimes b \leq c \text{ iff } b \leq a\backslash c \text{ iff } a \leq c/b,$$

for all $a, b, c \in M$. The models of NL1 are residuated unital groupoids, i.e. residuated groupoids containing the unit element for $\otimes$ (denoted by 1). It follows that $1\backslash a = a$, $a/1 = a$.

A pair $\sim, ^-$ of unary operations on a poset $(P, \leq)$ is called *an involutive pair of negations*, if for all $a, b \in P$ the following conditions are satisfied:

(TR) if $a \leq b$ then $b^\sim \leq a^\sim$ and $b^- \leq a^-$,
(DN) $a^{-\sim} = a$, $a^{\sim -} = a$;

if $\sim$ equals $^-$, then $\sim$ is called *a De Morgan negation* (then $a^{\sim\sim} = a$).

The models of CNL are residuated groupoids $\mathbf{M}$ with a De Morgan negation $\sim$ satisfying the compatibility condition:

(COM) for all $a, b, c \in M$, if $a \otimes b \leq c$ then $c^\sim \otimes a \leq b^\sim$.

We refer to these algebras as *CNL-algebras*. Unital CNL-algebras (i.e. with the unit for $\otimes$) are called *CNL1-algebras*.

In any CNL-algebra the following conditions are equivalent: $a \otimes b \leq c$, $c^\sim \otimes a \leq b^\sim$, $b \otimes c^\sim \leq a^\sim$. On the basis of other axioms, (COM) is equivalent to:

(TR') $a\backslash b^\sim = a^\sim/b$ for all $a, b \in M$,

and either of the following transposition laws: $a\backslash b = a^\sim/b^\sim$, $a/b = a^\sim\backslash b^\sim$.

In any CNL-algebra one defines *the dual product*: $a \oplus b = (b^\sim \otimes a^\sim)^\sim$. The following equations hold:

$$a\backslash b = a^\sim \oplus b, \quad a/b = a \oplus b^\sim.$$

Consequently, $\oplus, \backslash, /$ are definable in terms of $\otimes, \sim$.

In any CNL1-algebra one defines: $0 = 1^\sim$. Then, $1 = 0^\sim$, 0 is the unit for $\oplus$, and $a^\sim = a\backslash 0 = 0/a$.

CNL$^-$-algebras are residuated groupoids $\mathbf{M}$ with an involutive pair of negations, satisfying:

(COM⁻) for all $a, b, c \in M$, if $a \otimes b \leq c$ then $c^- \otimes a \leq b^-$ and $b \otimes c^\sim \leq a^\sim$.

Unital CNL⁻-algebras are referred to as *CNL1⁻-algebras*. In any CNL⁻-algebra the three conditions in (COM⁻) are equivalent. (COM⁻) is equivalent to:

(TR") $a^\sim/b = a\backslash b^-$, for all $a, b \in M$.

Hence in any CNL1⁻-algebra, $1^\sim = 1^-$. One defines $0 = 1^\sim$ and obtains: $a^\sim = a\backslash 0$, $a^- = 0/a$.

CNL⁻-algebras (resp. CNL-algebras) are term equivalent to (resp. cyclic) involutive p.o. groupoids [9].

The equation $(a^- \otimes b^-)^\sim = (a^\sim \otimes b^\sim)^-$ is valid in CNL⁻-algebras. One defines $a \oplus b = (b^- \otimes a^-)^\sim$ and obtains:

$$a\backslash b = a^\sim \oplus b, \ a/b = a \oplus b^-.$$

Consequently, $\oplus, \backslash, /$ are definable in terms of $\otimes, ^\sim, ^-$.

These algebras can be constructed from *phase spaces*, i.e. structures $(M, \cdot, R)$ such that $(M, \cdot)$ is a groupoid and $R \subseteq M^2$. We focus on *symmetric* phase spaces (R is symmetric).

A *closure operation* on a poset $(P, \leq)$ is a map $C : P \mapsto P$, satisfying: (C1) $x \leq C(x)$, (C2) if $x \leq y$ then $C(x) \leq C(y)$, (C3) $C(C(x)) \leq C(x)$, for all $x, y \in P$. A *nucleus* on a p.o. groupoid $(M, \cdot, \leq)$ is a closure operation C on $(M, \leq)$, satisfying: (C4) $C(x) \cdot C(y) \leq C(x \cdot y)$. If **M** is a residuated groupoid, then C is a nucleus on $(M, \cdot, \leq)$ iff C is a closure operation on $(M, \leq)$ and satisfies: (C4') $x\backslash y$ and y/x are C-closed for any $x \in M$ and any C-closed $y \in M$. Recall that x is *C-closed*, if $C(x) = x$.

Let $R \subseteq M^2$. For $X \subseteq M$, one defines:

$$X^\sim = \{a \in M : \forall_{b \in X} R(b, a)\}, \ X^- = \{a \in M : \forall_{b \in X} R(a, b)\}.$$

The maps $^\sim, ^-$ are a Galois connection on $\mathcal{P}(M)$: $X \subseteq Y^\sim$ iff $Y \subseteq X^-$. Consequently, $X \subseteq Y$ entails $Y^\sim \subseteq X^\sim$ and $Y^- \subseteq X^-$. The maps $\phi_R(X) = X^{-\sim}$ and $\psi_R(X) = X^{\sim-}$ are closure operations on $(\mathcal{P}(M), \subseteq)$. It follows that X is ϕ_R-closed (resp. ψ_R-closed) iff $X = Y^\sim$ (resp. $X = Y^-$) for some Y.

Proposition 1. *The following conditions are equivalent. (i) $X^\sim = X^-$ for all $X \subseteq M$, (ii) R is symmetric: $R(a, b)$ entails $R(b, a)$, for all $a, b \in M$.*

Let $(M, \cdot, R)$ be a symmetric phase space. Then, $\phi_R = \psi_R$. By M_R we denote the family of ϕ_R-closed subsets of M. Clearly $^\sim$ is a De Morgan negation on $(M_R, \subseteq)$.

Let $(M, \cdot, R)$ be a phase space. For $X, Y \subseteq M$, one defines: $X \cdot Y = \{a \cdot b : a \in X, b \in Y\}$, $X\backslash Y = \{y \in M : X \cdot \{y\} \subseteq Y\}$, $X/Y = \{x \in M : \{x\} \cdot Y \subseteq X\}$. $\mathcal{P}(\mathcal{M})$ with $\cdot, \backslash, /, \subseteq$ is a residuated groupoid. Let C be a nucleus on $(\mathcal{P}(M), \cdot, \subseteq)$. Then, $(M_C, \otimes^C, \backslash^C, /^C, \subseteq)$ is a residuated groupoid, where M_C is the family of C-closed subsets of M, $X \otimes^C Y = C(X \cdot Y)$, and $\backslash, /$ are the operations defined on $\mathcal{P}(M)$, restricted to M_C. If $\cdot$ is associative (resp. commutative), then $\otimes^C$ is associative (resp. commutative). If 1 is the unit for $\cdot$ in M, then $C(\{1\})$ is the unit for $\otimes^C$ (see e.g. [9]).

By *a phase space for CNL* we mean a symmetric phase space $(M, \cdot, R)$, satisfying the compatibility condition:

(COM-R) for all $a, b, c \in M$, $R(a \cdot b, c)$ iff $R(a, b \cdot c)$.

Phase spaces for CNL^- are defined in a similar way except that the symmetry of R is replaced with $\phi_R = \psi_R$.

Proposition 2. *For any phase space, (COM-R) holds if and only if, for all* $X, Y, Z \subseteq M$, $X \cdot Y \subseteq Z^\sim$ *iff* $Z \cdot X \subseteq Y^-$.

Proof. We show ($\Rightarrow$). $X \cdot Y \subseteq Z^\sim$ is equivalent to $\forall_{x \in X} \forall_{y \in Y} \forall_{z \in Z} R(z, x \cdot y)$, and $Z \cdot X \subseteq Y^-$ to iff $\forall_{z \in Z} \forall_{x \in X} \forall_{y \in Y} R(z \cdot x, y)$. Both statements are equivalent, by (COM-R). For ($\Leftarrow$), take $X = \{b\}$, $Y = \{c\}$, $Z = \{a\}$. Now $\{b\} \cdot \{c\} \subseteq \{a\}^\sim$ iff $R(a, b \cdot c)$, and $\{a\} \cdot \{b\} \subseteq \{c\}^-$ iff $R(a \cdot b, c)$. $\qquad\square$

Corollary 1. *For any phase space, (COM-R) holds if and only if, for all* $Y, Z \subseteq M$, $Z^\sim / Y = Z \backslash Y^-$.

Theorem 1. *Let* $(M, \cdot, R)$ *be a phase space for CNL. Then* M_R, *ordered by* $\subseteq$, *with operations* $\otimes^{\phi_R}$ *and* $\backslash, /, \sim$, *restricted to* M_R, *is a CNL-algebra.*

Proof. First, we show that ϕ_R satisfies (C4'). Using (COM-R), we show that $\{a\} \backslash \{b\}^\sim = \{b \cdot a\}^\sim$ and $\{a\}^\sim / \{b\} = \{b \cdot a\}^\sim$ for all $a, b \in M$. We have: $c \in \{a\} \backslash \{b\}^\sim$ iff $a \cdot c \in \{b\}^\sim$ iff $R(b, a \cdot c)$ iff $R(b \cdot a, c)$ iff $c \in \{b \cdot a\}^\sim$. The second equation is proved similarly (use the symmetry of R). This yields $X \backslash Y^\sim = (Y \cdot X)^\sim$ and $X^\sim / Y = (Y \cdot X)^\sim$, for all $X, Y \subseteq M$, by the well-known distribution laws: $\cdot$ distributes over infinite joins in both arguments, $\backslash$ (resp. $/$) distributes over infinite meets in the second (resp. first) argument and converts joins into meets in the first (resp. second) argument, and $\sim$ converts joins into meets. So for $X = \{a_i\}_{i \in I}$, $Y = \{b_j\}_{j \in J}$ we have:

$$X \backslash Y^\sim = \bigcap_{i \in I} \bigcap_{j \in J} \{a_i\} \backslash \{b_j\}^\sim = \bigcap_{i \in I} \bigcap_{j \in J} \{b_j \cdot a_i\}^\sim = (Y \cdot X)^\sim.$$

Let $X \subseteq M$, $Z \in M_R$. Then $Z = Y^\sim$ for some Y. Hence $X \backslash Z = (Y \cdot X)^\sim$ belongs to M_R, and similarly for Z/X.

Since ϕ_R is a nucleus on $(\mathcal{P}(M), \cdot, \subseteq)$, then M_R with $\otimes^{\phi_R}, \backslash, /, \subseteq$ is a residuated groupoid. Since R is symmetric, $\sim$ is a De Morgan negation on M_R. (TR') $X \backslash Y^\sim = X^\sim / Y$, for $X, Y \in M_R$, has been shown in the preceding paragraph; (TR') also follows from Corollary 1. $\qquad\square$

If $(M, \cdot, R)$ is a phase space for CNL, then the CNL-algebra constructed above is referred to as *the complex algebra* of the phase space. Worthy of noting, every CNL-algebra **M** is isomorphic to a subalgebra of the complex algebra of the phase space $(M, \otimes, R)$, where R is defined by: $R(a, b)$ iff $a \leq b^\sim$. Let $[a]^\downarrow$ denote the principal downset in $(M, \leq)$ generated by a, i.e. $[a]^{\downarrow} = \{x \in M : x \leq a\}$. Then, $[a]^{\downarrow} = \{a^{\sim}\}^{\sim}$. (Here $\sim$ is used in two meanings: the inner one as an operation in $\mathbf{M}$, the outer one as an operation on $\mathcal{P}(M)$.) The so-defined R is symmetric and $[a]^{\downarrow} \in M_R$. The map $h(a) = [a]^{\downarrow}$ is the required isomorphism. We omit the proof.

REMARK 1. In fact, for any symmetric phase space $(M, \cdot, R)$, (COM-R) holds if and only if ϕ_R is a nucleus and (TR') (equivalently (COM)) holds in the complex algebra.

A *unital phase space* is a structure $(M, \cdot, 1, R)$ such that $(M, \cdot, 1)$ is a unital groupoid and $R \subseteq M^2$. A *phase space for* CNL1 is a unital phase space $(M, \cdot, 1, R)$ such that $(M, \cdot, R)$ is a phase space for CNL. The analogue of Theorem 1 remains true. Now $\phi_R(\{1\})$ is the unit for $\otimes^{\phi_R}$ in the complex algebra.

For unital phase spaces, (COM-R) implies:

(Eq-R) $R(a, b)$ iff $R(1, a \cdot b)$ iff $R(a \cdot b, 1)$.

R can be represented by a set $O \subseteq M$, satisfying:

(COM-O) for all $a, b, c \in M$, $a \cdot (b \cdot c) \in O$ iff $(a \cdot b) \cdot c \in O$.

For $R \subseteq M^2$, we define $O_R = \{a \in M : R(1, a)\}$, and for $O \subseteq M$, we define $R_O = \{(a, b) \in M^2 : a \cdot b \in O\}$. By (Eq-R), $R_{O_R} = R$ and $O_{R_O} = O$. Furthermore, R satisfies (COM-R) iff O_R satisfies (COM-O). So there is a one-one correspondence between relations $R \subseteq M^2$ satisfying (COM-R) and sets $O \subseteq M$ satisfying (COM-O). Therefore, unital phase spaces, satisfying (COM-R), can also be defined as structures $(M, \cdot, 1, O)$ such that $(M, \cdot, 1)$ is a unital groupoid and $O \subseteq M$ satisfies (COM-O). This resembles the standard definitions of phase spaces for linear logics [1,10,15].

REMARK 2. If 1 is not present, then we can define $O_R = \{a \cdot b : R(a, b)\}$ and R_O as above, but this only yields the inclusions: $R \subseteq R_{O_R}$ and $O_{R_O} \subseteq O$. O satisfies (COM-O) iff R_O satisfies (COM-R). On the other hand, if O_R satisfies (COM-O), then R satisfies (COM-R), but the converse implication fails. If, however, $(M, \cdot)$ is a free groupoid, then there is a one-one correspondence between relations $R \subseteq M^2$ and sets $O \subseteq M$ such that each element of O is of the form $x \cdot y$, for some $x, y \in M$. Also R satisfies (COM-R) iff O_R satisfies (COM-O).

Let $(M, \cdot, 1, O)$ be a unital phase space. The symmetry of R_O is equivalent to *the cyclic law* for O:

(Cy) for all $a, b \in M$, if $a \cdot b \in O$ then $b \cdot a \in O$.

Accordingly, a phase space for CNL1 can be defined as a unital phase space $(M, \cdot, 1, O)$, satisfying (COM-O) and (Cy). Observe that $X^{\sim} = X \backslash O = O/X$, for any $X \subseteq M$. We denote $\phi_O = \phi_{R_O}$, and similarly for ψ_O, M_O. O is ϕ_O-closed, since $O = \{1\}^{\sim}$. So $O \in M_O$; also $O^{\sim}$ is the unit for $\otimes^{\phi_O}$ and O is the unit for the dual product. If M does not contain 1, then O, even satisfying (COM-O) and (Cy), need not belong to M_O.

EXAMPLE 1. Consider the phase space $(M, \cdot, O)$ such that $M = \Sigma^+$, $\cdot$ is the concatenation of strings, and O is the set of all strings of length 1. Clearly O satisfies (COM-O) and (Cy). So the complex algebra of $(M, \cdot, R_O)$ is a CNL-algebra. We have $\emptyset^\sim = \Sigma^+$ and $X^\sim = \emptyset$ for $X \neq \emptyset$. Therefore $M_O = \{\emptyset, \Sigma^+\}$ and $O \notin M_O$.

EXAMPLE 2. We construct a phase space $(M, +, R)$ such that $(M, +)$ is a commutative semigroup, $R \subseteq M^2$ is symmetric and satisfies (COM-R), but $R \neq R_O$, for any $O \subseteq M$. Let M consist of all pairs of positive integers. For $a, b \in M$, $a = (a_1, a_2)$, $b = (b_1, b_2)$, we set $a + b = (a_1 + b_1, a_2 + b_2)$. Let R consist of all $(a, b) \in M^2$ such that neither a, nor b is of the form $x + y$, for any $x, y \in M$. Clearly R is symmetric and satisfies (COM-R). Assume $R = R_O$ for some $O \subseteq M$. Since $R((1, 2), (2, 1))$, then $(3, 3) \in O$. We have $(3, 3) = (1, 1) + (2, 2)$, which yields $R((1, 1), (2, 2))$. This contradicts the definition of R, since $(2, 2) = (1, 1) + (1, 1)$. $\square$

This example shows that the notion of a phase space with a relation R is essentially wider than that with a set O for the non-unital spaces, even based on (commutative) semigroups. Therefore the former may also be useful in the theory of associative linear logics with no multiplicative units (not only in language, but in the corresponding algebras). Clearly (COM-O) (resp. (Cy)) holds for any $O \subseteq M$, if $\cdot$ is associative (resp. commutative).

3   Logics

We present a dual Schütte style system for CNL. Formulas are built from variables $p, q, \ldots$, negated variables $p^\sim, q^\sim, \ldots$, and connectives $\otimes, \oplus$. A, B, C, D range over formulas. By $\mathcal{S}$ we denote the free groupoid generated by all formulas. Γ, Δ, Θ range over elements of $\mathcal{S}$. These elements are represented as formula structures. The groupoid operation is: $\Gamma \cdot \Delta = (\Gamma, \Delta)$.

In CNL, sequents are formula-structures, containing at least two formulas; the set of all sequents is denoted by $\mathcal{S}^{(2)}$. So the distinction between quasi-sequents and sequents in [11] corresponds to our distinction between formula-structures and sequents. In axioms and rules of our systems (and after the provability symbol $\vdash$) we omit outer parentheses, e.g. we write $\vdash \Gamma, \Delta$ for $\vdash (\Gamma, \Delta)$. The axioms and the rules of CNL are as follows.

$$\text{(id)} \ p, p^\sim$$

$$\text{(r-}\otimes\text{)} \ \frac{(A, B), \Gamma}{A \otimes B, \Gamma} \quad \text{(r-}\oplus\text{)} \ \frac{A, \Gamma \quad B, \Delta}{A \oplus B, (\Delta, \Gamma)}$$

$$\text{(r-sym)} \ \frac{\Gamma, \Delta}{\Delta, \Gamma} \quad \text{(r-com)} \ \frac{(\Gamma, \Delta), \Theta}{\Gamma, (\Delta, \Theta)}$$

(r-$\otimes$), (r-$\oplus$) are the introduction rules for connectives, and (r-sym), (r-com) are the structural rules (expressing the symmetry of R and the condition (COM-R) in phase spaces for CNL).

We write $\Gamma \sim \Delta$, if Δ can be derived from Γ by finitely many applications of (r-sym), (r-com). Clearly $\sim$ is an equivalence relation (but not a congruence in $\mathcal{S}$).

Proposition 3. *For any sequent $\Gamma' \in \mathcal{S}^{(2)}$, containing one marked formula $\underline{A}$, there exists a unique $\Delta' \in \mathcal{S}$ such that $\Gamma' \sim (\underline{A}, \Delta')$.*

Proof. We describe an algorithm which reduces Γ' to some sequent $(\underline{A}, \Delta')$. We underline the substructure containing $\underline{A}$. The reduction rules are as follows.

$$(R1)\ (\Gamma, \underline{\Delta}) \to (\underline{\Delta}, \Gamma)$$
$$(R2)\ ((\underline{\Gamma}, \Delta), \Theta) \to (\underline{\Gamma}, (\Delta, \Theta))$$
$$(R3)\ ((\Gamma, \underline{\Delta}), \Theta) \to (\underline{\Delta}, (\Theta, \Gamma))$$

Each reduction step can be executed by applying at most three instances of (r-sym), (r-com). This procedure is deterministic. If we run it on a sequent $\Gamma' \in \mathcal{S}^{(2)}$, then the algorithm terminates in finitely many steps and yields $(\underline{A}, \Delta')$. The uniqueness of Δ', satisfying $\Gamma' \sim (\underline{A}, \Delta')$, follows from the fact:

(F1) if Γ' reduces to $(\underline{A}, \Delta')$ and $\Theta' \sim \Gamma'$, then Θ' reduces to $(\underline{A}, \Delta')$.

The proof of (F1) has two parts: (I) one proves it for Θ' resulting from Γ' by one application of (r-sym) or (r-com), (II) one proves (F1) by induction on the number of applications of (r-sym), (r-com) leading from Γ' to Θ'. We skip details.

Now assume that $\Gamma' \sim (\underline{A}, \Delta)$ and $\Gamma' \sim (\underline{A}, \Delta')$. Then $(\underline{A}, \Delta) \sim (\underline{A}, \Delta')$. By (F1), $(\underline{A}, \Delta)$ reduces to $(\underline{A}, \Delta')$. Since the algorithm stops on sequents of this form, then $\Delta = \Delta'$. □

EXAMPLE 3. Take $\Gamma' = ((B, (C', \underline{A})), (C, D))$. The reduction looks as follows:

$$\Gamma' \to_{R3} ((C', \underline{A}), ((C, D), B)) \to_{R3} (\underline{A}, (((C, D), B), C')).$$

Due to Proposition 3, the introduction rules can be restricted to the left-most occurrences of formulas in sequents, as above.

We say that a reduction of Γ' to $(\underline{A}, \Delta')$ *preserves* a substructure Θ of Γ', if Θ can be replaced by a variable in the whole reduction. The reduction in Example 3 preserves (C, D).

Lemma 1. *Assume that Γ' reduces to $(\underline{A}, \Delta')$ and Θ is a substructure of Γ', which does not contain $\underline{A}$. Then, the reduction preserves Θ.*

Proof. Let Γ_1 result from Γ' after one has replaced Θ by a new variable p. By Proposition 3, Γ_1 reduces to a sequent $(\underline{A}, \Delta_1)$. Now we substitute Θ for p in the whole reduction, which yields the reduction of Γ' to a sequent $(\underline{A}, \Delta)$. We have $\Delta = \Delta'$, since the algorithm is deterministic. Consequently, the reduction of Γ' to $(\underline{A}, \Delta')$ preserves Θ. □

Let $\mathbf{M}$ be a CNL-algebra. *A valuation in* $\mathbf{M}$ is a homomorphism of the free algebra of CNL-formulas into $\mathbf{M}$ such that $\mu(p^\sim) = \mu(p)^\sim$, for any (non-negated) variable p. The valuation μ is extended for sequents, by setting: $\mu((\Gamma, \Delta)) = \mu(\Gamma) \otimes \mu(\Delta)$. The sequent (Γ, Δ) is *true* for μ in $\mathbf{M}$, if $\mu(\Gamma) \leq \mu(\Delta)^\sim$. A sequent is *valid* in $\mathbf{M}$, if it is true for all valuations in $\mathbf{M}$.

The above system of CNL is *weakly complete*: the provable sequents are precisely the sequents valid in all CNL-algebras. Since the system is cut-free, its weak completeness entails the cut-elimination theorem (see below). Soundness is easy. The proof of completeness is a routine modification of similar proofs for different substructural logics, tracing back to Lafont [12]; see [9] for a wider discussion. Since for CNL and its variants no proof can be found in the literature, we give some details. We write $\vdash \Gamma$ if Γ is provable in CNL.

In metalanguage, one defines $A^\sim$ for any formula A:

$$(p^\sim)^\sim = p$$

$$(A \otimes B)^\sim = B^\sim \oplus A^\sim \quad (A \oplus B)^\sim = B^\sim \otimes A^\sim$$

By formula induction, one proves $A^{\sim\sim} = A$ and $\mu(A^\sim) = \mu(A)^\sim$, for any formula A and any valuation μ in $\mathbf{M}$. Also $\vdash A, A^\sim$, for any A.

It is convenient to write $\Gamma \Rightarrow A$ for the sequent $(\Gamma, A^\sim)$; due to (r-sym), it is deductively equivalent to $(A^\sim, \Gamma)$. Clearly $\Gamma \Rightarrow A$ is true for μ in $\mathbf{M}$, if $\mu(\Gamma) \leq \mu(A)$. We define $[A] = \{\Gamma \in \mathcal{S} : \vdash \Gamma \Rightarrow A\}$.

We consider the phase space $(M, \cdot, R)$ such that $(M, \cdot) = (\mathcal{S}, \cdot)$ and $R = \{(\Gamma, \Delta) \in \mathcal{S}^2 : \vdash \Gamma, \Delta\}$. Since $(M, \cdot)$ is a free groupoid, R can be replaced by the set $O_R = \{(\Gamma, \Delta) \in \mathcal{S} : R(\Gamma, \Delta)\}$ (see Remark 2 in Sect. 2). Due to (r-com), (r-sym), R is symmetric and satisfies (COM-R). By Theorem 1, M_R with inclusion and $\otimes^{\phi_R}, \backslash, /, ^\sim$ is a CNL-algebra. For any formula A, we have: $[A] = \{A^\sim\}^\sim$. So $[A]$ is ϕ_R-closed for any formula A.

We define a valuation μ in M_R:

$$\mu(p) = [p] = \{p^\sim\}^\sim, \ \mu(p^\sim) = \mu(p)^\sim. \tag{1}$$

By formula induction, one proves:

$$A \in \mu(A) \subseteq [A], \text{ for any formula } A. \tag{2}$$

We only consider the case: $A \otimes B$. Since $A \in \mu(A)$, $B \in \mu(B)$, then $(A, B) \in \mu(A) \cdot \mu(B) \subseteq \mu(A \otimes B)$. We use the fact:

(F2) if $(A, B) \in X$ and X is ϕ_R-closed then $A \otimes B \in X$.

Let $X = Y^\sim$, $(A, B) \in X$. Then, for all $\Gamma \in Y$, $\vdash (A, B), \Gamma$, hence $\vdash A \otimes B, \Gamma$, by (r-$\otimes$). So $A \otimes B \in X$. Consequently $A \otimes B \in \mu(A \otimes B)$.

We show $\mu(A \otimes B) \subseteq [A \otimes B]$. Since $[A \otimes B]$ is ϕ_R-closed, it suffices to show $\mu(A) \cdot \mu(B) \subseteq [A \otimes B]$. Let $\Gamma \in \mu(A)$, $\Delta \in \mu(B)$. Then, $\Gamma \in [A]$, $\Delta \in [B]$, hence $\vdash A^\sim, \Gamma, \vdash B^\sim, \Delta$. By (r-$\oplus$), $\vdash (A \otimes B)^\sim, (\Gamma, \Delta)$, which yields $(\Gamma, \Delta) \in [A \otimes B]$.

Now assume $\nvdash \Gamma, \Delta$. By Proposition 3, there exists a sequent $(A, \Theta) \sim (\Gamma, \Delta)$. Then $\nvdash A, \Theta$, hence $\Theta \notin [A^\sim]$. By (2), $\Theta \notin \mu(A^\sim)$ and $\Theta \in \mu(\Theta)$. Consequently (A, Θ) is not true for μ in the complex algebra of $(M, \cdot, R)$. It follows that (Γ, Δ) is not true, since the set of true sequents is invariant under $\sim$. This finishes the proof of weak completeness.

The sequents valid in CNL-algebras are closed under the cut rule:

$$(\text{cut}) \frac{\Gamma[A] \quad A^\sim, \Delta}{\Gamma[\Delta]} \ .$$

Therefore (cut) is admissible in the cut-free system of CNL. By Proposition 3, this rule can also be formulated in the form:

$$(\text{cut'}) \frac{A, \Gamma \quad A^\sim, \Delta}{\Delta, \Gamma} \ .$$

The system of CNL with (cut') is *strongly complete* with respect to CNL-algebras: the sequents provable from a set of assumptions Φ are precisely those which follow from Φ in CNL-algebras.

Let $f(\Gamma)$ be the formula arising from Γ after one has replaced each comma by $\otimes$. Every sequent (Γ, Δ) is deductively equivalent to $(f(\Gamma), f(\Delta))$. This is easy to prove with applying (cut); for the cut-free system one can use the reversibility of (r-$\otimes$). Therefore, without lost of generality, we assume that all sequents in Φ are of the form (A, B).

In the proof of strong completeness, one constructs the complex algebra of $(\mathcal{S}, \cdot, R)$, where $R = \{(\Gamma, \Delta) \in \mathcal{S} : \Phi \vdash \Gamma, \Delta\}$. Now $[A] = \{\Gamma \in \mathcal{S} : \Phi \vdash \Gamma \Rightarrow A\}$, and μ is defined by (1).

In the presence of (cut'), the inclusion in (2) can be replaced by $\mu(A) = [A]$; so $A \in \mu(A)$ may be omitted. We use the fact:

(F3) if X is ϕ_R-closed, $A \in X$ and $\Phi \vdash \Gamma \Rightarrow A$, then $\Gamma \in X$.

This is needed to prove that all sequents from Φ are true for μ in the complex algebra. Let $(A, B) \in \Phi$. Then, $A \in [B^\sim]$, hence $[A] \subseteq [B^\sim]$, by (F3). Consequently $\mu(A) \subseteq \mu(B^\sim) = \mu(B)^\sim$.

REMARK 3. We have shown in Sect. 2 that not every phase space $(M, \cdot, R)$ can be replaced by $(M, \cdot, O)$. The above proof shows that CNL is strongly complete with respect to phase spaces of the latter form, satisfying (COM-O) and (Cy) (even based on free groupoids). It follows that every CNL-algebra is isomorphic to a subalgebra of the complex algebra of some space $(M, \cdot, O)$ such that $(M, \cdot)$ is a free groupoid.

The connectives $\backslash, /$ can be defined by: $A \backslash B = A^\sim \oplus B$, $A / B = A \oplus B^\sim$. Each NL-sequent $\Gamma \Rightarrow A$ can be treated as a CNL-sequent $\Gamma \Rightarrow A$, i.e. $(A^\sim, \Gamma)$. We prove that CNL with (cut') is a *strongly conservative extension* of NL with (NL-cut). The weak conservativeness was proved in [11] by proof-theoretic methods.

Theorem 2. *Let Φ be a set of NL-sequents (of the form $C \Rightarrow D$), and let $\Gamma \Rightarrow A$ be an NL-sequent. Then, $\Phi \vdash_{NL} \Gamma \Rightarrow A$ iff $\Phi \vdash_{CNL} \Gamma \Rightarrow A$.*

Proof. The only-if part is easy. The easiest proof uses the strong completeness, hence soundness, of NL with respect to residuated groupoids and the strong completeness of CNL with respect to CNL-algebras. We prove the if-part.

We consider the free groupoid $(M, \cdot)$ generated by all NL-formulas and formally negated NL-formulas $A^\sim$, i.e. A with superscript $\sim$. The elements of M are represented as formula-structures, as above. We define $O \subseteq M$ as the smallest set which contains all $(A^\sim, \Gamma)$ such that $\Phi \vdash_{NL} \Gamma \Rightarrow A$ and is closed under (r-sym), (r-com). Clearly each element of O contains at least two formulas and exactly one negated formula.

We consider the complex algebra M_O, i.e. M_R for $R = R_O$. Since O satisfies (COM-O) and (Cy), M_O is a CNL-algebra, by Theorem 1.

For any NL-formula A, we define $[A] = \{\Gamma : \Phi \vdash_{NL} \Gamma \Rightarrow A\}$. We show $[A] = \{A^\sim\}^\sim$. Clearly $[A] \subseteq \{A^\sim\}^\sim$, by the definition of O. We prove $\{A^\sim\}^\sim \subseteq [A]$. Let $\Gamma \in \{A^\sim\}^\sim$. Then $(A^\sim, \Gamma) \in O$. By the definition of O, there exists a NL-sequent $\Delta \Rightarrow A$ such that $\Phi \vdash_{NL} \Delta \Rightarrow A$ and $(A^\sim, \Gamma) \sim (A^\sim, \Delta)$. By Proposition 3, $\Gamma = \Delta$ (take $A^\sim$ as the marked formula). Consequently $\Gamma \in [A]$.

So all sets $[A]$ are ϕ_O-closed. We define μ by (1). By formula induction we show $\mu(A) = [A]$ for any NL-formula A. This is obvious for p.

The cases $A \backslash B$, A/B are treated in the same way as in analogous proofs for NL. Let us consider $A \backslash B$. Assume $\Gamma \in \mu(A \backslash B)$. Since $A \in \mu(A)$, then $(A, \Gamma) \in \mu(B)$. So $(A, \Gamma) \in [B]$, which yields $\Gamma \in [A \backslash B]$, by $(\Rightarrow \backslash)$. Assume $\Gamma \in [A \backslash B]$. By the reversibility of $(\Rightarrow \backslash)$ in NL, $(A, \Gamma) \in [B]$. Let $\Delta \in \mu(A)$. Then $\Delta \in [A]$, which yields $(\Delta, \Gamma) \in [B]$, by (NL-cut). So $(\Delta, \Gamma) \in \mu(B)$ for any $\Delta \in \mu(A)$, and consequently $\Gamma \in \mu(A \backslash B)$.

The case $A \otimes B$ needs (F2), (F3), which remain true for NL-formulas. We prove (F2). Let $X = Y^\sim$, $(A, B) \in X$. Then, $(\Gamma, (A, B)) \in O$ for any $\Gamma \in Y$. We fix $\Gamma \in Y$. Let $C^\sim$ be the only negated formula in Γ; we treat $C^\sim$ as the marked formula. By Proposition 3, there is a unique Δ such that $(\Gamma, (A, B)) \sim (C^\sim, \Delta)$. By the construction of O, $\Phi \vdash_{NL} \Delta \Rightarrow C$. By Lemma 1, the reduction of $(\Gamma, (A, B))$ to $(C^\sim, \Delta)$ preserves (A, B), hence $\Delta = \Theta[(A, B)]$. Accordingly $\Phi \vdash_{NL} \Theta[A \otimes B] \Rightarrow C$, by $(\otimes \Rightarrow)$, hence $(C^\sim, \Theta[A \otimes B]) \in O$. Clearly $(\Gamma, A \otimes B) \sim (C^\sim, \Theta[A \otimes B])$. Consequently $(\Gamma, A \otimes B) \in O$. This yields $A \otimes B \in X$. (F3) can be proved in a similar way ($\vdash$ in (F3) means $\vdash_{NL}$).

We prove $[A \otimes B] \subseteq \mu(A \otimes B)$. Since $A \in \mu(A)$, $B \in \mu(B)$, then $(A, B) \in \mu(A) \cdot \mu(B) \subseteq \mu(A \otimes B)$. By (F2), $A \otimes B \in \mu(A \otimes B)$. Hence $[A \otimes B] \subseteq \mu(A \otimes B)$, by (F3). We prove $\mu(A \otimes B) \subseteq [A \otimes B]$. Since $[A \otimes B]$ is ϕ_O-closed, it suffices to show $\mu(A) \cdot \mu(B) \subseteq [A \otimes B]$, which amounts to $[A] \cdot [B] \subseteq [A \otimes B]$. This holds, by $(\Rightarrow \otimes)$.

Now assume $\Phi \nvdash_{NL} \Gamma \Rightarrow A$. Then $\Gamma \in \mu(\Gamma)$, $\Gamma \notin [A] = \mu(A)$, and consequently $\Gamma \Rightarrow A$ is not true for μ. Let $C \Rightarrow D \in \Phi$. $\mu(C) \subseteq \mu(D)$ follows from $[C] \subseteq [D]$. Therefore $\Gamma \Rightarrow A$ does not follow from Φ in CNL-algebras. Consequently $\Phi \nvdash_{CNL} \Gamma \Rightarrow A$.   □

The results of this section can be extended for several richer logics. Proofs are similar, and we omit them.

First, we consider CNL with $\sim$ in the language. So formulas are built from variables and $\otimes, \oplus, \sim$. One adds the rules:

$$(\text{r-}\sim\sim) \, \frac{A, \Gamma}{A^{\sim\sim}, \Gamma}$$

$$(\text{r-}\otimes^\sim) \, \frac{A^\sim, \Gamma \quad B^\sim, \Delta}{(A \otimes B)^\sim, (\Gamma, \Delta)} \quad (\text{r-}\oplus^\sim) \, \frac{(B^\sim, A^\sim), \Gamma}{(A \oplus B)^\sim, \Gamma} \, .$$

This system is equivalent to the former one in a strong sense. Every formula with $\sim$ can be translated into a formula without $\sim$ (except its occurrences at variables), using the metalanguage definition of $\sim$, given above. The translation can be extended for sequents and sets of sequents. Γ is provable from Φ in CNL with $\sim$ if and only if the translation of Γ is provable from the translation of Φ in CNL without $\sim$.

CNL1 is obtained by adding the constants $1, 0$, treated as atomic formulas, and:

$$(\text{a-}0) \, 0, \quad (\text{r-}1) \, \frac{\Gamma}{1, \Gamma} \, .$$

The new axiom (a-0) introduces a sequent containing only one formula. We define sequents as all elements of $\mathcal{S}$. The set of formula-structures is defined as the free unital groupoid $\mathcal{S}_1 = \mathcal{S} \cup \{\lambda\}$, where λ satisfies $\Gamma \cdot \lambda = \Gamma = \lambda \cdot \Gamma$. One may imagine λ as the 'empty structure'. Γ and Δ may be empty in (r-$\otimes$), (r-$\oplus$).

For CNL1 without $\sim$, the metalanguage negation is defined as above, with: $1^\sim = 0$, $0^\sim = 1$. Given a CNL1-algebra and a valuation μ, one sets $\mu(\lambda) = 1$. A sequent $\Gamma \in \mathcal{S}$ is said to be *true* for μ, if $\mu(\Gamma) \leq 0$. For sequents (Γ, Δ) this amounts to the former definition of a true sequent.

CNL1 (in both versions) admits cut elimination, since the cut-free system is weakly complete with respect to CNL1-algebras. With (cut') it is strongly complete. CNL1 is a strongly conservative extension of NL1.

CNL* is obtained from CNL1 by dropping 1 and 0. Since CNL* is strongly complete with respect to CNL1-algebras, then CNL1 is a strongly conservative extension of CNL*. Notice that CNL* is stronger than CNL; $p \otimes p^\sim$ is provable in CNL*, by (id) and (r-$\otimes$), but not in CNL. In CNL1-algebras this law expresses $a \otimes a^\sim \leq 0$, which lacks sense in CNL-algebras without 0. The axiom (id) expresses $a \leq a$, which holds in all ordered algebras.

In the completeness proofs, the underlying unital groupoid is $(\mathcal{S}_1, \cdot, \lambda)$ and O consists of all provable sequents. Then O satisfies (COM-O) and (Cy), hence the complex algebra is a CNL1-algebra. (1) is extended by: $\mu(0) = O$, $\mu(1) = \phi_O(\{\lambda\})$.

If C is a closure operation on a complete lattice, then the C-closed sets are closed under infinite meets. So they form a complete lattice. The results of this section can be extended to CNL and CNL1 with lattice connectives $\vee, \wedge$, satisfying the lattice laws. These logics may be called Full CNL and Full CNL1 (FCNL and FCNL1) by analogy with FNL, i.e. NL with $\vee, \wedge$. FCNL (resp. FCNL1) is a strongly conservative extension of FNL (resp. FNL1).

For FCNL, the connectives are $\otimes, \oplus, \vee, \wedge$. One adds three new rules.

$$(\text{r-}\wedge) \ \frac{A, \Gamma}{B \wedge A, \Gamma} \quad \frac{A, \Gamma}{A \wedge B, \Gamma} \quad (\text{r-}\vee) \ \frac{A, \Gamma \ \ B, \Gamma}{A \vee B, \Gamma}$$

In the complex algebra of $(M, \cdot, R)$ (an arbitrary phase space) one defines: $X \wedge Y = X \cap Y$, $X \vee Y = \phi_R(X \cup Y)$. With these operations the complex algebra of a phase space for CNL is a lattice-ordered CNL-algebra. We refer to these algebras as FCNL-algebras. FCNL1-algebras are defined in a similar way.

These results remain true for associative and/or commutative CNL-algebras and CNL1-algebras. The associative FCNL1-algebras are the algebras of Cyclic Noncommutative MALL [15]; the commutative and associative FCNL1-algebras are the algebras of MALL [10]. The completeness results were proved in these papers. The fact that Cyclic Noncommutative MALL is a (weakly) conservative extension of FL1 was proved in Abrusci [2] by a tedious proof-theoretic argument. This can be proved like Theorem 2, which yields the strong conservativeness.

4   Main Results

We need an extended subformula property for $\Phi \vdash_{CNL} \Gamma$. Let T be a set of formulas. $\mathcal{S}_T$ consists of all $\Gamma \in \mathcal{S}$ such that every formula in Γ belongs to T. A T-sequent is a sequent $\Gamma \in \mathcal{S}^2 \cap \mathcal{S}_T$. A T-proof is a formal proof from Φ in CNL which consists of T-sequents only. We write $\Phi \vdash^T_{CNL} \Gamma$, if there exists a T-proof of Γ from Φ in CNL. We write $\vdash$ for $\vdash_{CNL}$ and $\vdash^T$ for $\vdash^T_{CNL}$. We define $[A]^T = \{\Gamma \in \mathcal{S}_T : \Phi \vdash^T \Gamma \Rightarrow A\}$.

Lemma 2. *Let T be a set of formulas, closed under subformulas and $\sim$. Let Φ be a set of T-sequents of the form (A, B). For any T-sequent Γ_0, $\Phi \vdash \Gamma_0$ if and only if $\Phi \vdash^T \Gamma_0$.*

Proof. The if-part is obvious. For the only-if part, we consider the phase space $(M, \cdot, R)$ such that $M = \mathcal{S}_T$, $\cdot$ is defined as in Sect. 3, and $R = \{(\Gamma, \Delta) \in (\mathcal{S}_T)^2 : \Phi \vdash^T \Gamma, \Delta\}$. Clearly R is symmetric and satisfies (COM-R). So the complex algebra of $(M, \cdot, R)$ is a CNL-algebra. We define: $\mu(p) = [p]^T = \{p^\sim\}^\sim$ for $p \in T$; the values of μ for $p \notin T$ may be arbitrary. One proves: $\mu(A) = [A]^T$ for any $A \in T$, by the same argument as in Sect. 3. Consequently, if $\Phi \vdash^T \Gamma_0$ does not hold, then Γ_0 is not true for μ, but all sequents in Φ are true for μ. Therefore $\Phi \vdash \Gamma_0$ does not hold. $\qquad\square$

Corollary 2. *Let T be the smallest set of formulas, containing all formulas occurring in Φ or Γ and being closed under subformulas and $\sim$. If $\Phi \vdash \Gamma$, then $\Phi \vdash^T \Gamma$.*

We prove an interpolation lemma for CNL: every proper substructure Δ of a provable sequent Γ can be replaced by a formula (*an interpolant*) from a finite set.

Lemma 3. *Let T, Φ, Γ_0 be as in Lemma 2, and let Δ_0 be a substructure of Γ_0, $\Delta_0 \neq \Gamma_0$. We write $\Gamma_0 = \Theta_0[\Delta_0]$. If $\Phi \vdash^T \Gamma_0$, then there exists $D \in T$ such that $\Phi \vdash^T D^\sim, \Delta_0$ and $\Phi \vdash^T \Theta_0[D]$.*

Proof. Assume $\Phi \vdash^T \Gamma_0$. We proceed by induction on T-proofs from Φ. If Δ_0 is a formula, then $D = \Delta_0$. So the thesis holds, if Γ_0 is an axiom (id) or belongs to Φ. We assume that Δ_0 is not a formula.

Case: (r-$\otimes$). Then D is the same as in the premise.

Case: (r-$\oplus$). $1°$. $\Delta_0 = (\Delta, \Gamma)$. Then $D = (A \oplus B)^\sim$. $2°$. Δ_0 is a substructure of Γ or Δ. Then D is as in the appropriate premise.

Cases: (r-sym). D is as in the premise.

Case: (r-com) downwards. If $\Delta_0 = (\Delta, \Theta)$, then $D = D_1^\sim$, where D_1 is the interpolant of Γ in the premise. Otherwise D is the interpolant of Δ_0 in the premise. (r-com) upwards is treated in a similar way.

Case: (cut'). D is as in the appropriate premise. $\quad\square$

There are two important consequences of Lemma 3.

Theorem 3. *CNL has the strong finite model property (SFMP).*

Proof. Let Φ be a finite set of sequents of the form (A, B). We show that for any sequent Γ, if $\Phi \vdash_{CNL} \Gamma$ does not hold, then there exist a finite CNL-algebra $\mathbf{M}$ and a valuation μ in $\mathbf{M}$ such that all sequents from Φ are true for μ, but Γ is not true for μ.

Assume $\Phi \nvdash \Gamma$. Let T be defined as in Corollary 2. Clearly T is finite and $\Phi \nvdash^T \Gamma$. Let $\mathbf{M}$ be the complex algebra constructed in the proof of Lemma 2, and let μ be defined as there. It suffices to show that $\mathbf{M}$ is finite, this means: there are only finitely many ϕ_R-closed sets, i.e. sets of the form $X^\sim$, for $X \subseteq \mathcal{S}_T$. We have $X^\sim = \bigcap_{\Gamma \in X} \{\Gamma\}^\sim$. So it suffices to show that there are only finitely many sets of the form $\{\Gamma\}^\sim$.

Let $\Delta \in \{\Gamma\}^\sim$. Then, $\Phi \vdash^T \Gamma, \Delta$. By Lemma 3, $\Phi \vdash^T \Gamma, D$, for some $D \in T$ such that $\Phi \vdash^T D^\sim, \Delta$. We have: $D \in \{\Gamma\}^\sim$ and $\Delta \in [D]^T$. By (F3) (precisely: its version for T-sequents and T-proofs), $[D]^T \subseteq \{\Gamma\}^\sim$. Consequently, $\{\Gamma\}^\sim$ is the union of some family of sets $[D]^T$, for $D \in T$. There are only finitely many sets $[D]^T$ such that $D \in T$, which yields our claim. $\quad\square$

By a *CNL-grammar* we mean a triple $G = (\Sigma, I, A_0)$ such that Σ is a nonempty, finite alphabet, I is a map from Σ to the family of finite sets of CNL-formulas, and A_0 is a CNL-formula. For any $\Gamma \in \mathcal{S}$, we define a sequence of formulas $s(\Gamma)$: $s(A) = A$, $s((\Gamma, \Delta)) = s(\Gamma)s(\Delta)$, i.e. the concatenation of $s(\Gamma)$ and $s(\Delta)$. We say that G assigns A to the string $a_1 \ldots a_n$ ($a_i \in \Sigma$), if there exists $\Gamma \in \mathcal{S}$ such that $(A^\sim, \Gamma)$ is provable, $s(\Gamma) = A_1 \ldots A_n$ and $A_i \in I(a_i)$ for $i = 1, \ldots, n$. Here 'provable' means 'provable in CNL'. We also consider grammars based on CNL augmented with finitely many assumptions; then 'provable' means 'provable from Φ in CNL', where Φ is the set of assumptions. *The language of G is the set of all $x \in \Sigma^+$ such that G assigns A_0 to x.*

Theorem 4. *Let Φ be a finite set of sequents. Let G be a CNL-grammar based on CNL augmented with the assumptions from Φ. Then, the language of G is a context-free language.*

Proof. Fix a grammar $G = (\Sigma, I, A_0)$. Let T be the smallest set of formulas which contains A_0 and all formulas appearing in Φ, I and is closed under sub-formulas and $\sim$. Clearly T is finite. Let $(A_0^\sim, \Gamma)$ be provable, $\Gamma \in \mathcal{S}_T$. Let (A, B) be a substructure of Γ; so $\Gamma = \Theta[(A, B)]$. By Lemma 3, there exists $D \in T$ such that $(D^\sim, (A, B))$ and $(A_0^\sim, \Theta[D])$ are provable. Accordingly, every $\Gamma \in \mathcal{S}_T$ such that $(A_0^\sim, \Gamma)$ is provable can be derived (as a derivation tree) from A_0 by means of context-free rules: $A \mapsto B$ (resp. $A \mapsto B, C$) such that $A, B, C \in T$ and $B \Rightarrow A$ (resp. $(B, C) \Rightarrow A$) is provable. The language of G is generated by the context-free grammar with the terminal alphabet Σ, the nonterminal alphabet T, the start symbol A_0, and the production rules as above plus $A \mapsto a$ for $A \in I(a)$. $\qquad\square$

Conversely, every ϵ-free context-free language is generated by some CNL-grammar (without assumptions). This follows from Theorem 2 and the fact that every ϵ-free context-free language is generated by an NL-grammar [3].

Theorem 3 implies the decidability of the finitary consequence relation for CNL. We prove that it is decidable in polynomial time. [11] shows the polynomial time decidability of CNL.

Theorem 5. *The relation $\Phi \vdash \Gamma$, for finite sets Φ and $\Gamma \in \mathcal{S}^{(2)}$, is decidable in polynomial time.*

Proof. A sequent $\Gamma \in \mathcal{S}^{(2)}$ is said to be *restricted*, if it is of the form (A, B), $(A, (B, C))$ or $((A, B), C)$. So (id) and all sequents from Φ are restricted. Fix a finite set Φ and $\Gamma_0 \in \mathcal{S}^{(2)}$. Let T be defined as in Corollary 2 (for $\Gamma = \Gamma_0$).

By CNL_r^T we denote the system whose axioms and rules are those of CNL with (cut'), limited to restricted T-sequents. Clearly there are finitely many restricted T-sequents. All sequents provable in CNL_r^T from Φ can be determined in polynomial time (in the size of $\Phi \cup \{\Gamma_0\}$).

By CNL_Φ^T we denote the system whose axioms are all sequents provable in CNL_r^T from Φ and the only inference rule is (cut) (now admitting unrestricted T-sequents). Notice that (cut) is not the same as (cut'). Observe that every restricted T-sequent provable in CNL_Φ^T must be provable in CNL_r^T from Φ (if the conclusion of (cut) is restricted, then the premises are restricted; also (cut) limited to restricted T-sequents is derivable in CNL_r^T). We prove:

$$\Phi \vdash_{CNL}^T \Gamma \text{ iff } \Gamma \text{ is provable in } \mathrm{CNL}_\Phi^T.$$

($\Leftarrow$) is obvious. For ($\Rightarrow$), we observe that CNL_Φ^T has the interpolation property: if $\Theta_0[\Delta_0]$ is provable and $\Delta_0 \neq \Theta_0[\Delta_0]$, then there exists $D \in T$ such that $(D^\sim, \Delta_0)$ and $\Theta_0[D]$ are provable.

First, one proves this property for CNL_r^T with the assumptions from Φ in the same way as Lemma 3. For rules (r-$\oplus$), (r-com) one uses the fact that $(A, A^\sim)$, for $A \in T$, is provable in CNL_r^T.

Second, one shows this property for CNL_Φ^T by induction on derivations based on (cut), which is easy. The only interesting case is the following: $\Theta_0[\Theta_1[\Delta]]$ arises by (cut) from $A^\sim, \Delta$ and $\Theta_0[\Theta_1[A]]$, and $\Delta_0 = \Theta_1[\Delta]$. Then, the interpolant of Δ_0 equals the interpolant of $\Theta_1[A]$ in $\Theta_0[\Theta_1[A]]$.

Third, one shows that all rules of CNL, restricted to T-sequents, are admissible in CNL_Φ^T. We only consider (r-sym). Let (Γ, Δ) be provable in CNL_Φ^T. By interpolation, there exist $C \in T$, $D \in T$ such that (C, D), $(C^\sim, \Gamma)$, $(D^\sim, \Delta)$ are provable in CNL_Φ^T. Since (C, D) is provable in CNL_r^T from Φ, then (D, C) is provable in CNL_r^T from Φ, and consequently, (Δ, Γ) is provable in CNL_Φ^T, by two applications of (cut). This yields $(\Rightarrow)$.

By Lemma 2, $\Phi \vdash_{CNL} \Gamma_0$ if and only if Γ_0 is provable in CNL_Φ^T. In particular, for a restricted Γ_0, Γ_0 is provable in CNL from Φ if and only if Γ_0 is provable in CNL_r^T. $\square$

We have noted in Sect. 1 that CNL^- does not have SFMP. The status of Theorems 4 and 5 for CNL^- remains an open problem. They are true for the pure CNL^- (i.e. $\Phi = \emptyset$); the proof will be given in another paper.

Chvalovsky [7] proves that the consequence relation for FNL is undecidable. Since FCNL is a strongly conservative extension of FNL, then the consequence relation for FCNL is undecidable (hence SFMP fails). On the other hand, the analogues of Theorems 3 and 4 hold for DFCNL, i.e. FCNL admitting the distributive laws for $\vee, \wedge$, like for DFNL and its variants [5, 6].

References

1. Abrusci, V.M.: Phase semantics and sequent calculus for pure noncommutative classical linear propositional logic. J. Symb. Log. **56**, 1403–1451 (1991)
2. Abrusci, V.M.: Classical conservative extensions of Lambek calculus. Stud. Logica. **71**, 277–314 (2002)
3. Buszkowski, W.: Generative capacity of nonassociative Lambek calculus. Bull. Pol. Acad. Sci. Math. **34**, 507–516 (1986)
4. Buszkowski, W.: Lambek calculus with nonlogical axioms. In: Casadio, C., Scott, P.J., Seely, R. (eds.) Language and Grammar. Studies in Mathematical Linguistics and Natural Language, pp. 77–93. CSLI Publications, Stanford (2005)
5. Buszkowski, W.: Interpolation and FEP for logics of residuated algebras. Log. J. IGPL **19**, 437–454 (2011)
6. Buszkowski, W., Farulewski, M.: Nonassociative Lambek calculus with additives and context-free languages. In: Grumberg, O., Kaminski, M., Katz, S., Wintner, S. (eds.) Languages: From Formal to Natural. LNCS, vol. 5533, pp. 45–53. Springer, Heidelberg (2009). doi:10.1007/978-3-642-01748-3_4
7. Chvalovsky, K.: Undecidability of consequence relation in full nonassociative Lambek calculus. J. Symb. Log. **80**, 567–576 (2015)
8. Farulewski, M.: Finite embeddability property for residuated groupoids. Rep. Math. Log. **43**, 25–42 (2008)
9. Galatos, N., Jipsen, P., Kowalski, T., Ono, H.: Residuated Lattices: An Algebraic Glimpse at Substructural Logics. Elsevier, Amsterdam (2007)
10. Girard, J.-Y.: Linear logic. Theoret. Comput. Sci. **50**, 1–102 (1987)

11. de Groote, P., Lamarche, F.: Classical non-associative Lambek calculus. Stud. Logica. **71**, 355–388 (2002)
12. Lafont, Y.: The finite model property of various fragments of linear logic. J. Symb. Log. **62**, 1202–1208 (1997)
13. Lambek, J.: On the calculus of syntactic types. In: Jakobson, R. (ed.) Structure of Language and Its Mathematical Aspects, pp. 166–178. AMS, Providence (1961)
14. Lambek, J.: Cut elimination for classical bilinear logic. Fundamenta Informaticae **22**, 53–67 (1995)
15. Yetter, D.N.: Quantales and (non-commutative) linear logic. J. Symb. Log. **55**, 41–64 (1990)

Proof Assistants for Natural Language Semantics

Stergios Chatzikyriakidis[1,2(✉)] and Zhaohui Luo[3]

[1] Department of Philosophy, Linguistics and Theory of Science,
University of Gothenburg, Gothenburg, Sweden
stergios.chatzikyriakidis@gu.se
[2] Open University of Cyprus, Nicosia, Cyprus
[3] Department of Computer Science, Royal Holloway,
University of London, London, UK
zhaohui@hotmail.ac.uk

Abstract. In this paper we discuss the use of interactive theorem provers (also called proof assistants) in the study of natural language semantics. It is shown that these provide useful platforms for NL semantics and reasoning on the one hand, and allow experiments to be performed on various frameworks and new theories, on the other. In particular, we show how to use Coq, a prominent type theory based proof assistant, to encode type theoretical semantics of various NL phenomena. In this respect, we can encode the NL semantics based on type theory for quantifiers, adjectives, common nouns, and tense, among others, and it is shown that Coq is a powerful engine for checking the formal validity of these accounts as well as a powerful reasoner about the implemented semantics. We further show some toy semantic grammars for formal semantic systems, like the Montagovian Generative Lexicon, Type Theory with Records and neo-Davidsonian semantics. It is also explained that experiments on new theories can be done as well, testing their validity and usefulness. Our aim is to show the importance of using proof assistants as useful tools in natural language reasoning and verification and argue for their wider application in the field.

Keywords: Type theory · Proof assistants · Reasoning · Formal semantics · Coq

1 Introduction

Interactive theorem provers (also called proof assistants) have come a long way since they were first introduced in the late 60's as tools to formalise mathematics (cf., the AUTOMATH project [3]). Today, a number of state-of-the-art proof assistants exist and their uses have been proven fruitful both in formalisation

S. Chatzikyriakidis—Partially supported by Centre for Linguistic Theory and Studies in Probability, University of Gothenburg.

Z. Luo—Partially supported by EU COST Action CA15123 and CAS/SAFEA Inter. Partnership Program.

© Springer-Verlag GmbH Germany 2016
M. Amblard et al. (Eds.): LACL 2016, LNCS 10054, pp. 85–98, 2016.
DOI: 10.1007/978-3-662-53826-5_6

of mathematics and software verification, among other things; see, for example, [13] for the proof of the four colour theorem in the proof assistant Coq[1]. The importance and usefulness of proof assistants have also been further proven by some recent research projects, including the very attractive research on Univalent Foundations [28] that aims to develop alternative foundations of mathematics, where the proof assistants Coq and Agda [1] play a crucial role to the whole endeavour (see [34] for an example of formalization of part of the project in Coq).

The use of constructive type theories for the study of NL semantics has also seen a revival in the last decade.[2] A number of approaches that directly employ constructive type theories or are inspired by them have been put forth by various researchers in the recent years and have provided interesting accounts on classic problems of formal semantics (see [2,9,16,23,26,30,33] for examples, although this is not a complete list). In this context, it is worth noting the following:

- Some of the proof assistants, like Coq and Agda, implement constructive type theories;
- The proof assistants are extremely powerful reasoning engines; and
- Constructive type theories have been shown to be a nice alternative to the simple type theory usually in formal semantics.

It seems that the time is right to look at the combination of these three in order to use proof assistants as natural language reasoners and as checkers of the formal validity of formal semantics accounts. Indeed, we have taken the first step in this direction and have used Coq as a natural language reasoner [5,6]. In this paper, we extend this work and create a number of small Coq libraries to show that proof assistant like Coq can provide useful platforms for:

- Formalising NL semantics and, based on it, formally describe various NL phenomena, including co-predication, individuation, common nouns, adjectives and tense, among others. (These libraries are based on earlier theoretical work using Luo's Type Theory with Coercive Subtyping (TTCS for short) [20,21,23].)
- Experimenting with various semantic frameworks: we show how to use Coq to formalise them by implementing some small examples in Rétore's Montagovian Generative Lexicon [30], Cooper's Type Theory with Records (TTR) [9], and neo-Davidsonian event semantics [27].
- Experimenting with new theories: we formalise in Coq a newly developed theory [8] of predicational forms to give semantics to negative sentences and conditionals in constructive type theory. We also look at the issue of individuation and its interaction with copredication from the same perspective.

---

[1] The proof assistant Coq implements a constructive type theory in the tradition of Martin-Löf. The type theory is an impredicative type theory called the Calculus of Inductive Constructions (pCIC) [11], which is similar to the type theory UTT (or TTCS as called in this paper) [18].

[2] The use of constructive type theories has been initiated by the pioneering work of Aarne Ranta [29].

The current paper is structured as follows: in Sect. 2, we provide an introduction to TTCS and the implementation of some of the ideas coated in TTCS with respect to NL semantics in the Coq proof assistant, especially its use in formalising NL semantics in TTCS. In Sect. 3, we present several small libraries: first the one based on our work in type theory, introducing the relevant formal features of TTCS when needed, then several small libraries for other semantic frameworks and, finally, the library for the theory of predicational forms and individuation criteria. In the conclusion, some future work is discussed.

2 Type Theoretical Semantics for NL in Coq

In this section, we shall first introduce formal semantics in a constructive type theory and then how we will discuss the use of Coq to implement the semantics for various features in natural language.

2.1 Formal Semantics in Type Theory with Coercive Subtyping

Type Theory with Coercive Subtyping (TTCS) is a constructive type theory based on Luo's UTT [18] with the addition of an effective subtyping mechanism, that of coercive subtyping [19,26]. TTCS has been effectively used in the study of NL semantics for a range of phenomena including common nouns, adjectives, adverbs and belief intensionality among other things [5,7,20,21,23]. TTCS is a dependent type theory with rich type structures which are exploited for the study of NL semantics. We will refer to this type of semantics in this paper as Modern Type Theoretical (MTT) semantics.[3] In MTT-semantics, some of the major linguistic categories and their interpretation are shown below:

1. A common noun (CN) can be interpreted as a type.
2. A verb (IV) can be interpreted as a predicate over the type D that interprets the domain of the verb (i.e., a function of type $D \to Prop$, where $Prop$ is the type of logical propositions
3. An adjective (ADJ) can be interpreted as a predicate over the type that interprets the domain of the adjective (i.e., a function of type $D \to Prop$).
4. Modified common nouns (MCNs) can be interpreted by means of Σ-types, types of (dependent) pairs.
5. A sentence (S) is interpreted as a proposition of type $Prop$.

See Fig. 1 for a summary with examples.

---

[3] The formal semantics based on Modern Type Theories such as Martin-Löf's type theory or TTCS is usually called MTT-semantics. In the current paper, we shall still talk about MTT-semantics although, if taken seriously, it means formal semantics in TTCS because the Coq implementation of the NL semantics is based on TTCS.

	Example	Montague semantics	Semantics in TTCS
CN	man, human	$[\![man]\!], [\![human]\!] : e \to t$	$[\![man]\!], [\![human]\!] : Type$
IV	talk	$[\![talk]\!] : e \to t$	$[\![talk]\!] : [\![human]\!] \to Prop$
ADJ	handsome	$[\![handsome]\!] : (e \to t) \to (e \to t)$	$[\![handsome]\!] : [\![man]\!] \to Prop$
MCN	handsome man	$[\![handsome]\!]([\![man]\!])$	$\Sigma m : [\![man]\!]. [\![handsome]\!](m) : Type$
S	A man talks	$\exists m : e. [\![man]\!](m) \& [\![talk]\!](m)$	$\exists m : [\![man]\!]. [\![talk]\!](m) : Prop$

Fig. 1. Examples in formal semantics.

2.2 NL Semantics in Coq

Coq [11] implements pCIC, a type theory whose major part is essentially[4] TTCS (UTT with coercive subtyping), based on which the formal semantics briefly described in the previous subsection has been implemented. The encoding of NL semantics based on TTCS is quite straightforward in most of the cases. Let us see some basics of how this can be done.

Starting with the type of logical propositions, nothing needs to be encoded, since Coq already involves a universe of logical propositions, *Prop*. The next step, is to see what the universe of entities would be taken to be. In MG, a coarse-grained type of entities exists, i.e. the type *e* of all entities. In MTT-semantics, the common nouns constitute a universe, denoted as CN; the type CN contains the (interpretations of) CNs, each of which is further interpreted as a type that contains entities belonging to them. CNs are interpreted as types rather than predicates. However, since universe construction (i.e., defining new universes) is not an option in Coq, we equate CN with Coq's predefined universe *Set*.

Σ-types (types of dependent pairs), which are used to give semantics to some modified common nouns among other things, are encoded using Coq's dependent record type mechanism[5] and adjectives and verbs are defined as predicates (objects of type $A \to Prop$). Subsective adjectives like *large* are encoded as polymorphic predicates (see [4]), extending over the universe CN.[6] Subtyping is encoded using Coq's coercion mechanism and the proper names are given suitable domain types: e.g., *John* is assumed to be of type *Man*.

The Coq codes for this basic set up are as follows.

```
Definition CN := Set.
Parameters Man Woman Human Animal Object : CN.
Axiom mh : Man->Human.    Coercion mh : Man >-> Human.
Axiom wh : Woman->Human. Coercion wh : Woman >-> Human.
Axiom ha : Human-> Animal. Coercion ha : Human>->Animal.
Axiom ao : Animal->Object. Coercion ao : Animal>->Object.
Parameter Black : Object->Prop.
```

---

[4] Coq has co-inductive types which are not present in TTCS.

[5] Coq's record types are just Σ-types with global names associated with them.

[6] This is encoded using Π-types as follows: $[\![Adj_{subs}]\!] : \Pi A : \text{CN}. A \to Prop$. The 'forall' part in the code corresponds to Π.

```
Parameter Large : forall A:CN, A->Prop.
Parameter walked: Human->Prop.
Parameter John : Man.
```

Quantifiers can be given polymorphic types as well: a quantifier takes a CN argument A: CN and returns a function of type $(A \to Prop) \to Prop$. Thus, if A is Man the type for the quantified NP will be $(Man \to Prop) \to Prop$ and, if A is $Object$, it is of type $(Object \to Prop) \to Prop$, and so on. As examples, we define the quantifiers *some, all, no* as follows:

```
Definition some := fun A:CN => fun P:A->Prop => exists x:A, P(x).
Definition all  := fun A:CN => fun P:A->Prop => forall x:A, P(x).
Definition no   := fun A:CN => fun P:A->Prop => forall x:A, not(P(x)).
```

Note that the typing is the one we have been describing, taking an A : CN argument, an $A \to Prop$ argument and returning a proposition.

Now, let us see how one can exploit Coq in order to reason with NL sentences based on the implemented semantics. First of all, if one wants to check typing, the command *Check* followed by the element we want to check can be used. Note that Coq is a strongly typed language, so by definition ill-typed constructs cannot be defined, since they will be blocked by Coq. Let us see an NL reasoning example, the one shown below:

(1) John walked $\Rightarrow$ Some man walked

Formalizing this example in Coq, we consider the following 'theorem' whose name is JOHN (to be proved):

Theorem JOHN : walked John -> (some Man) walked.

This will put Coq into proof-mode. We unfold the definition for *some* using cbv and use the tactic *intro*, which will introduce the antecedent as a hypothesis:[7]

```
JOHN < cbv. intro. subgoal
H : walked John
==============================
exists x : Man, walked x
```

What we need to do is substitute John for x and using the tactic assumption, which matches a goal in case there is an identical premise in the context of the proof, the proof is completed and we can save the proof using *Qed*. The whole proof then consists of the steps:

1. cbv (unfolding definitions (in our case the one for *some*))[8]
2. intro (moving the antecedent as a hypothesis)

---

[7] The tactic *cbv* performs all possible reductions.

[8] In general the tactic *cbv* performs all possible reductions. For more information, see [11].

3. exists John (substituting x for John)
4. assumption (matching the goal with a hypothesis)

Remark 1. The MTT-semantics has proved to be a viable alternative to Montague Grammar, with several notable advantages. Here, we think it is worth mentioning one of them: that is, MTT-semantics is both model-theoretic and proof-theoretic, as argued in [24]. It is model-theoretic because, in an MTT-semantics, an MTT is employed as a representational language and it can do so because of its rich representational structures as well as its internal logic. Therefore, it has a wide coverage of linguistic features and can be compared to Montague semantics in this respect. It is also proof-theoretic, in the sense of [14], because MTTs are specified proof-theoretically and the meanings of MTT-judgements, that are used to give semantics to NL sentences, can be understood by means of their inferential roles. Therefore, reasoning with NL can be directly performed in proof assistants like Coq that implement MTTs. This is unique for MTTs and MTT-semantics: such a possibility of having a semantics which is both model-theoretic and proof-theoretic is not available to us until we have the MTT-semantics (for example, if one considers the traditional model-theoretic semantics in set theory, we simple would not have a proof-theoretic representational language: set theory is not proof-theoretic.)

3 Libraries for NL Semantics

We have created a number of small libraries in Coq, encoding NL semantics. They may be classified as follows:

- *MTT-semantics and reasoning*: We have studied various NL phenomena using MTT-semantics and formalised them in Coq.
- *Platform for other semantic frameworks*: We have looked at several semantic frameworks and provided some examples including, for example, Rétore's Montagovian Generative Lexicon [30], Cooper's Type Theory with Records (TTR) [9], and a toy semantic grammar for neo-Davidsonian event semantics [27].
- *Experiments on new semantic theories*: We have done interesting experiments in Coq about some new semantic theories, including that about predicational forms in MTT-semantics [8], as reported here.

The libraries can be found at https://github.com/StergiosCha/CoqLACL.

3.1 MTT Semantics for NL in Coq

The main file for MTT-semantics is MainCoq.v. This includes the Coq implementation of a number of ideas in MTT-semantics. The universe CN includes a number of types (e.g., *Man, Human, Delegate, Woman, Animal, Object*) and subtyping relations between them. Synonym relations are encoded via the *let*-command in Coq. Adjectives are defined in the way specified in the previous section and sometimes some added lexical semantics are inserted. For example, *small* is defined as the opposite of *large*, and both are polymorphically defined as follows:

```
Parameter Large Normalsized: forall A:CN, A->Prop.
Definition Small :=
  fun A:CN => fun a:A => not (Large A a) /\ not (Normalsized A a).
```

Basically the idea here is that small is defined as being not large but furthermore not of normal size. This reflects the idea that something which is not large is not necessarily small.[9] This is needed in order to get the relevant inferences right (see [5]).

In MTT-semantics, there is also a widespread use of Σ-types for factive verbs, adverbs and comparatives. We have not the space here to go in full detail but the idea can be briefly described as follows, taking the case of veridical sentence adverbs as an example. What we need to capture is that the proposition without the adverb is implied by the proposition including the adverb. In order to do this, we first define an auxiliary object:

```
Parameter ADVS : forall (v:Prop), sigT (fun p:Prop => p->v).
```

This basically takes a proposition v and returns a pair whose first component is a proposition p and whose second component is the proposition that p implies v. Then, veridical sentence adverbs (we use *fortunately* as an example) are defined as the first projection of this auxiliary pair:

```
Definition fortunately := fun v:Prop => projT1 (ADVS v).
```

Similar uses of Σ-types can be found for VP adverbs, comparatives as well as factive verbs in the library (see [5] for more details.)

For comparatives, we introduce indexed types for common nouns; for example, humans of type $Human$ may be indexed by a height parameter. Then, a comparative adjective takes two $Human_i$ arguments with $i :: Height$.

```
Inductive HUMAN : nat->Type := HUMAN1:forall n:nat,HUMAN n.
```

A simple model of tense is defined and an attempt to deal with some aspects of tense exists. There is a type $Time$ and a date is defined as triple, taking year, month and day arguments and returning a result in $Time$. A default date is defined which consists of the defaults for year, month and day. Then, verbs are defined with an extra time argument. Present, past and future are then defined using the *precedes* relation with respect to the default time. For example, an adverb like currently is defined as identifying the time argument with the default time:

```
Definition currently := fun P : Time -> Prop => P default_t.
```

---

[9] The level of fine-grainedness with respect to size, i.e. whether sizes between these proposed three will be used, will not bother us here.

The next file is adjectives.v, which involves some more fine-grained issues in adjectival semantics. In particular it deals with multidimensional adjectives and introduces a hack in order to take care of the fact that Coq does not allow subtyping to propagate through constructors (as it is the case in TTCS).[10] *Multidimensional adjectives* do not just involve one dimension (e.g., the dimension of height in the case of *tall*), but more than one. Classical cases are the adjectives like *healthy* and *sick* or even adjectives like *big*. The idea is that an adjective like *healthy* quantifies over a number of dimensions, e.g., blood pressure, cholesterol etc. [32]. Similarly, *big* may involve different dimensions like *height, width* etc. For an adjective like healthy, we define health as an enumerated type including all the relevant dimensions. Then, *Healthy* is defined as taking an argument of type *Human* and assuming that this human is healthy in all dimensions. For sick, the assumption is that the argument is not healthy w.r.t. to at least one dimension. This follows the ideas set out in [32]:

```
Inductive Health:CN:=Heart|Blood|Cholesterol.
Parameter Degree:R. Parameter healthy:Health->Human->Prop.
Definition Sick:=fun y:Human=>~(forall x:Health,healthy x y).
Definition Healthy:=fun y:Human=>forall x:Health,healthy x y.
```

The files FracasCoq.v and test.v are meant to be used in conjunction. Actually FracasCoq loads test.v. FracasCoq.v contains a number of FraCaS test suite examples formalized in Coq along with their proofs. The FraCaS Test Suite [10] arose out of the FraCaS Consortium, a huge collaboration with the aim to develop a range of resources related to computational semantics. The FraCaS test suite is specifically designed to reflect what an adequate theory of NL inference should be able to capture. It comprises NLI examples formulated in the form of a premise (or premises) followed by a question and an answer. Here is a typical example from the suite:

(2) Some Irish delegates finished the survey on time.
    Did any delegate finish the report on time [Yes, FraCaS 055]

The modified CN Irish delegates is defined as a Σ type. Given that π_1 is defined as a coercion, the inference will go through easily. Please see [5] for more details and the code for the actual.

3.2   Other Semantic Frameworks

Proof assistants can be used as platforms to experiment with different semantic frameworks. In this respect, there are three files that have some very small toy

---

[10] Some remark on subtyping propagation in Coq is needed. If $A < B$, then we should have $\Sigma(A, C) < \Sigma(B, C)$ (which follows in TTCS). But this does not follow in Coq. In order to remedy this we have introduced a sort of a hack by overloading the type using unit types (see the actual code and consult [21] for the use of unit types).

semantic grammars of other frameworks that have been used in the study of linguistic semantics. Note that these implementations are shallow implementations in the sense that no deep implementation of the underlying formal systems is done. In other words, we are not doing a faithful implementation of a semantic framework; instead, we emphasize the quick return so that examples can be done. For instance, Retoré's Generative Montagovian Lexicon [30] is based on system F [12,31], but no implementation of system F is done on our part.

In MontagovianLexiconToy.v, we encode some of the ideas in presented in Generative Montagovian Lexicon as presented in [30]. Note that the idea that, representing the interpretation of a common noun, each type has its corresponding predicate cannot be implemented since it is not clear how such correspondence will be formally defined.[11] We, however, encode the idea that a word like *book* has a principal lambda term and then a number of coercions that take care of its dot-type status. This is done by using type overloading via unit types. We further formalize the polymorphic conjunction of [30] and prove that it is equivalent to the semantics of regular conjunction. For example, the definition of polymorphic conjunction is given as follows:

```
Definition PAND := fun a:e => fun b:e => fun P:a->t => fun Q:b->t =>
                   fun x:e => fun y:x => fun f:x->a => fun g:x->b =>
                   and (P(f(y))) (Q(g(y))).
```

Records.v has some very simple experimentations on encoding ideas from Cooper's TTR [9]. For example, the record for *a man owns a donkey* is encoded as:

```
Record amanownsadonkey : Type :=
   mkamanownsadonkey{ x  : Ind;
                      c1 : man x;
                      y  : Ind;
                      c2 : donkey y;
                      c3 : own x y}.
```

From this record type in Coq, one can prove any of the individual fields. For example, one can show that a man exists, that a donkey exists (*man* and *donkey* are defined here as predicates), and that the man owns the donkey.

Lastly, Davidson.v contains a typed neo-Davidsonian toy semantic grammar. It has some simple examples and the welcoming inferential properties of neo-Davidsonian semantics where each modifier adds a conjunct. The grammar presents a typed version of neo-Davidsonian semantics[12]. Similarly, a transitive verb like *stabs* is defined as taking an event argument e and two arguments x

---

[11] For example, one can define both a type book and a predicate book* but linking the two and defining such a process for every common noun is something that we do not know how can be done, without leading to formal difficulties such as undecidability of type-checking [8]. There is not a formal proposal on how to do this in [30] either.

[12] See [25] for a theory of dependent event types which extends Church's simple type theory with dependent event types. This is an initial step towards a theory of events with dependent types.

and y of type Ind and returning a proposition which specifies that there is a stabbing event e1 such that $stabs(x)(y)(e1)$, x is the agent, y is the theme and $e = e1$. This toy semantic grammar can take care of inferences like the following (proofs are in the file):

(3) Brutus stabbed Caesar with a knife in Rome $\Rightarrow$ Brutus stabbed Caesar with a knife

(4) Brutus stabbed Caesar with a knife in Rome $\Rightarrow$ Brutus stabbed Caesar

(5) Brutus stabbed Caesar with a knife in Rome $\Rightarrow$ the agent of the stabbing was Brutus

Remark 2. As we have already mentioned, the above implementations are shallow implementations of fragments of other semantic theories.[13] Coq implements an MTT, which in itself is a very powerful language to represent NL semantics. In a sense, one way of using Coq would be to use this very powerful language in order to embed different semantic theories as kind of modules within Coq's MTT. For example, one might want to define a Natural Logic component (as for example [17] has done), or a neo-Davidsonian fragment as we have very briefly done here. We believe that this is a nice way of looking at how the systems like Coq can be used for NL semantics. Different comparisons can then be performed as regards the different frameworks based e.g. on the predictions they make as regards inference.

3.3   Experiments with New Semantic Theories

Systems like Coq can play a useful role in verifying newly proposed theories in semantics. Here, we consider two cases. The first concerns the theory of predicational forms as studied in [8]. The theory is to deal with negated sentences or conditionals in a type theory where some CNs are interpreted as types in a multi-sorted type system (e.g., the MTT-semantics) and the file predhyp.v contains the experiments done in Coq that formalizes the theory of predicational forms and considered many relevant examples.

Consider the simplest example, where (7) is the (judgemental) interpretation of (6):

(6) John is a man.

(7) $j: Man$

Note that $j: Man$ is a judgment and not a proposition. How do we give semantics to its negation like (8)?

(8) John is not a man.

---

[13] See [15] for an informal explanation of shallow and deep embeddings.

Similarly, a negated sentence like (9) needs to be given semantics, but it would be simply negating the semantics of 'Tables talk' since the latter is meaningless (i.e., ill-typed)[14].

(9) Tables do not talk.

Also, some conditionals correspond to hypothetical judgements and require a treatment as well (we omit the details here).

The theory of predicational forms [8] is a logical theory to deal with the above issues. Based on it, suitable semantic interpretations can be given to negated sentences and conditionals as intended.

The formalisation of the theory (and examples) can be found in predhyp.v.[15] For instance (just showing one example), the following sentences and inferences have been done:

(10) It is not the case that John is not a man.

(11) It is not the case that every human is a logician

(12) Some red tables do not talk $\Rightarrow$ Some tables do not talk

Another theory is to consider how to deal with inferences concerning CNs. Individuation.v contains an account of how individuation criteria should be decided within an MTT. The general idea is that every common noun is associated with its own identity criteria (IC) which can be inherited by other common nouns (see [22] for the theory on this and more detailed discussions on ICs.) For example, one can assume that Man inherits its IC from $Human$. Given this assumption, common nouns are not simple types but setoids whose first component is a type (the domain of the CN), in DomCN (which is the old CN universe) and whose second component is its IC. So under this view, the common noun Human will be represented by the following (we use capitals to denote the new formalization and retain the first letter with uppercase notation to denote the type in DomCN):

(13) $HUMAN = \Sigma(Human, =_H)$

Several IC criteria are defined for different common nouns and dot.types like book are given two different IC criteria depending on whether their physical or informational aspect is individuated. Thus, we have:

---

[14] Note that it is not given false as in MG.

[15] The files FracasCoq.v and test.v are meant to be used in conjunction. Actually FracasCoq loads test.v. FracasCoq.v contains a number of FraCaS test suite examples formalized in Coq along with their proofs.

(14) $BOOK_1 = \Sigma(Book, =_P)$

(15) $BOOK_2 = \Sigma(Book, =_I)$

A number of proofs then follow including, for example, a proof of the following:

(16) John picked up and mastered three books $\Rightarrow$
John picked up three physical objects and mastered three informational objects

Remark 3. One issue that is worth mentioning here, is that of automation. Coq is an interactive theorem prover, which means that the user guides the prover to the proof. However, Coq has a very powerful tactic language that can be used in order to construct composite tactics that can automate part of or whole proofs. We have defined a number of tactics that can automate proofs. The interested reader can check for example the automated tactic AUTO in the files Davidson.v (for example BRUTUS1 to BRUTUS4 are proven using AUTO only) and MontagovianLexicon.v. AUTO can prove all theorems in these two files. A more advanced automatic tactic is needed for the proofs found in the FracasCoq.v file. Such a tactic is AUTOa (this tactic also solves all the goals in the previous files solved by AUTO) [5,6]. All proofs can be automated with this tactic except one that is semiautomated (see FracasCoq.v file).

4   Conclusions and Future Work

In this paper, we have argued for the use of the proof assistant technology for natural language semantics. In particular, we have argued, that the time is mature for such an endeavor given the progress made in both the proof technology itself as well as the use of constructive type theories for natural language semantics. We have prepared a number of small libraries for NL semantics using the proof assistant Coq based on Luo's TTCS and have shown the benefits of such an endeavor by exemplifying the use of proof assistants as natural language reasoners or as checkers of the formal validity of proposals in formal semantics. We have lastly shown how experiments with semantic accounts proposed in several semantic frameworks can also be implemented in Coq.

As future work, we are envisaging the extension of work as regards inference by endorsing a system where a tight correspondence between syntax and semantics exists, in the same way such a correspondence is found in categorial grammar. This builds on theoretical work of second author, where a proposal for extending the Lambek calculus with dependent types can be found. Given such a development one can then define a parser based on this extended Lambek calculus with dependent types, which will automatically give us MTT-semantics as output. These semantics will then be used by Coq to perform reasoning tasks. The ultimate goal is to develop a wide-coverage, robust parser that will then be able to output semantics for larger pieces as well as open text. Similar work using multi-modal categorial grammars or combinatory categorial grammar has been shown to be feasible. If this is the case, this is a great chance of using a more structured semantic framework as well as a specific purpose reasoning device (Coq) in order to deal with NLI.

References

1. Agda proof assistant (2008). http://appserv.cs.chalmers.se/users/ulfn/wiki/agda.php
2. Bekki, D.: Representing anaphora with dependent types. In: Asher, N., Soloviev, S. (eds.) LACL 2014. LNCS, vol. 8535, pp. 14–29. Springer, Heidelberg (2014)
3. de Bruijn, N.: A survey of the project AUTOMATH. In: Hindley, J., Seldin, J., To, H.B. (eds.) Curry: Essays on Combinatory Logic, Lambda Calculus and Formalism. Academic Press, Cambridge (1980)
4. Chatzikyriakidis, S., Luo, Z.: Adjectives in a modern type-theoretical setting. In: Morrill, G., Nederhof, M.-J. (eds.) FG 2012–2013. LNCS, vol. 8036, pp. 159–174. Springer, Heidelberg (2013). doi:10.1007/978-3-642-39998-5_10
5. Chatzikyriakidis, S., Luo, Z.: Natural language inference in Coq. J. Log. Lang. Inf. **23**(4), 441–480 (2014)
6. Chatzikyriakidis, S., Luo, Z.: Natural language reasoning using proof-assistant technology: rich typing and beyond. In: Proceedings of EACL 2014 (2014)
7. Chatzikyriakidis, S., Luo, Z.: Using signatures in type theory to represent situations. In: Logic and Engineering of Natural Language Semantics 11, Tokyo (2014)
8. Chatzikyriakidis, S., Luo, Z.: On the interpretation of common nouns: types v.s. predicates. In: Chatzikyriakidis, S., Luo, Z. (eds.) Modern Perspectives in Type Theoretical Semantics. Studies of Linguistics and Philosophy, Springer, Heidelberg (2016, to appear)
9. Cooper, R.: Records and record types in semantic theory. J. Log. Comput. **15**(2), 99–112 (2005)
10. Cooper, R., Ginzburg, J.: A compositional situation semantics for attitude reports. In: Selignmann, J., Westerstahl, D. (eds.) Logic, Language and Computation, CSLI (1996)
11. The Coq Team: The Coq Proof Assistant Reference Manual (Version 8.1). Inria, Rennes (2007)
12. Girard, J.Y.: Interprétation fonctionelle et élimination des coupures de l'arithmétique d'ordre supérieur. Ph.D. thesis, Université Paris VII (1972)
13. Gonthier, G.: A computer-checked proof of the Four Colour Theorem (2005). http://research.microsoft.com/~gonthier/4colproof.pdf
14. Kahle, R., Schroeder-Heister, P. (eds.): Proof-Theoretic Semantics. Special Issue of Synthese **148**(3), 503–743 (2006)
15. Keller, C., Werner, B.: Importing HOL light into Coq. In: Kaufmann, M., Paulson, L.C. (eds.) ITP 2010. LNCS, vol. 6172, pp. 307–322. Springer, Heidelberg (2010). doi:10.1007/978-3-642-14052-5_22
16. Krahmer, E., Piwek, P.: Presupposition projection as proof construction. In: Bunt, H., Muskens, R. (eds.) Computing Meaning. SLP, vol. 73, pp. 281–300. Springer, Dordrecht (1999)
17. Lungu, G.E., Luo, Z.: Monotonicity reasoning in formal semantics based on modern type theories. In: Asher, N., Soloviev, S. (eds.) LACL 2014. LNCS, vol. 8538, pp. 138–148. Springer, Heidelberg (2014)

18. Luo, Z.: Computation and Reasoning: A Type Theory for Computer Science. Oxford University Press, Oxford (1994)
19. Luo, Z.: Coercive subtyping in type theory. In: Dalen, D., Bezem, M. (eds.) CSL 1996. LNCS, vol. 1258, pp. 275–296. Springer, Heidelberg (1997). doi:10.1007/3-540-63172-0_45
20. Luo, Z.: Type-theoretical semantics with coercive subtyping. In: Semantics and Linguistic Theory 20 (SALT20), Vancouver (2010)
21. Luo, Z.: Contextual analysis of word meanings in type-theoretical semantics. In: Pogodalla, S., Prost, J.-P. (eds.) LACL 2011. LNCS (LNAI), vol. 6736, pp. 159–174. Springer, Heidelberg (2011). doi:10.1007/978-3-642-22221-4_11
22. Luo, Z.: Common nouns as types. In: Béchet, D., Dikovsky, A. (eds.) LACL 2012. LNCS, vol. 7351, pp. 173–185. Springer, Heidelberg (2012). doi:10.1007/978-3-642-31262-5_12
23. Luo, Z.: Formal semantics in modern type theories with coercive subtyping. Linguist. Philos. **35**(6), 491–513 (2012)
24. Luo, Z.: Formal semantics in modern type theories: is it model-theoretic, proof-theoretic, or both? In: Asher, N., Soloviev, S. (eds.) LACL 2014. LNCS, vol. 8535, pp. 177–188. Springer, Heidelberg (2014)
25. Luo, Z., Soloviev, S.: Dependent event types (abstract). In: LACL 2016 (2016)
26. Luo, Z., Soloviev, S., Xue, T.: Coercive subtyping: theory and implementation. Inf. Comput. **223**, 18–42 (2012)
27. Parsons, T.: Events in the Semantics of English. MIT Press, Cambridge (1990)
28. The Univalent Foundations Program: Homotopy type theory: univalent foundations of mathematics. Technical report, Institute for Advanced Study (2013)
29. Ranta, A.: Type-Theoretical Grammar. Oxford University Press, Oxford (1994)
30. Retoré, C.: The montagovian generative lexicon Tyn: a type theoretical framework for natural language semantics. In: Matthes, R., Schubert, A. (eds.) 19th International Conference on Types for Proofs and Programs (TYPES 2013), vol. 26, pp. 202–229 (2013)
31. Reynolds, J.C.: Towards a theory of type structure. In: Robinet, B. (ed.) Programming Symposium. LNCS, vol. 19, pp. 408–425. Springer, Heidelberg (1974). doi:10.1007/3-540-06859-7_148
32. Sassoon, G.: A typology of multidimensional adjectives. J. Semant. **30**(3), 335–380 (2013)
33. Tanaka, R., Mineshima, K., Bekki, D.: Factivity and presupposition in dependent type semantics. In: Type Theories and Lexical Semantics Workshop (2015)
34. Voevodsky, V.: Experimental library of univalent formalization of mathematics. Math. Struct. Comput. Sci. **25**, 1278–1294 (2015)

Compositional Event Semantics in Pregroup Grammars

Gabriel Gaudreault[(✉)]

Concordia University, Montreal, Canada
gabriel.gaudreault@gmail.com

Abstract. A derivational approach to event semantics using pregroup grammars as syntactic framework is defined. This system relies on three crucial components: the explicit introduction of event variables which are linked to the basic types of a lexical item's grammatical type; the unification of event variables following a concatenation of two expressions and the associated type contraction; and the correspondence between pregroup orderings and the change of the available event variables associated to a lexical item, which the meaning predicates take scope over.

Keywords: Pregroup grammars · Formal semantics · Conjunctivism

1 Introduction

This project aims at studying implicit event variables over which meaning predicates take scope and their interaction throughout syntactic derivations. A derivational system will be put in place around the pre-existing pregroup grammar framework to handle these variables compositionally using a unification process, while at the same time providing for a very natural semantics for pregroup grammars.

More concretely, it will be shown how by extending the usual pregroup framework with a semantic layer and by assigning explicit event variables to the syntactic categories of an expression, we can get semantic extraction from pregroup derivations without too many complications. The resulting meaning will be neo-davidsonian and conjunctivist in form, that is, the meaning will be analysed in terms of events, and a single logical operator will be used during the combination of meanings: the conjunction $\wedge$. This is in opposition to the more traditional approach of logical analysis called *Functionalism* that treats semantic composition as function application: a sentence such as (1a) will not have corresponding logical form (1b) but instead have the form (1c) where what are usually treated as arguments to the verb — *John*, *Mary* — are instead related to it by the thematic role they play in the event that the verb characterizes.

(1)   a.   John likes Maria
       b. $[\![John\ likes\ Maria]\!] = like(John)(Maria)$
       c. $[\![John\ likes\ Maria]\!]$
          $= \exists e.Agent(e) = John \wedge like(e) \wedge Theme(e) = Maria$

© Springer-Verlag GmbH Germany 2016
M. Amblard et al. (Eds.): LACL 2016, LNCS 10054, pp. 99–115, 2016.
DOI: 10.1007/978-3-662-53826-5_7

Using conjunctions as sole mean of meaning combination makes it harder at first to analyse certain constructions, but this is a small price to pay for the level of generality and overall derivational simplicity that will be obtained in the end by equating syntactic combination (pregroup contractions) with meaning conjunction.

2 Pregroup Grammars

The syntactic framework that will be used for this project is called Pregroup Grammars and is a recent descendant of the original syntactic calculus which arose from the study of resource sensitive logics [2,8,9]. They are called as such because their syntactic types form a special mathematical structure called a pregroup. The semantic system defined later could be worked out independently of the pregroup framework, though they seem to work well together for multiple reasons.

A pregroup [9] $\mathbb{P} = (P, \rightarrow, {}^r, {}^l, \cdot, 1)$ is a partially ordered monoid on a set of elements P, the set of basic types, in which to every element $a \in P$ corresponds a right and a left adjoint — $a^r \in P$ and $a^l \in P$ respectively — subject to

$$a \cdot a^r \rightarrow 1 \rightarrow a^r \cdot a \qquad\qquad a^l \cdot a \rightarrow 1 \rightarrow a \cdot a^l$$

The left sides of the relations are called contractions and the right sides, expansions. More precisely, the types forming a pregroup satisfy the following properties:

- Existence of an identity element 1: $\qquad\qquad a \cdot 1 = 1 \cdot a = a$, for any $a \in \mathbb{P}$
- Associativity of type concatenation: $\qquad a \cdot (b \cdot c) = (a \cdot b) \cdot c$, for $a, b, c \in \mathbb{P}$
- Reflexivity of the ordering: $\qquad\qquad\qquad\qquad a \rightarrow a$, for any $a \in \mathbb{P}$
- Antisymmetry of the ordering: $\quad$ if $a \rightarrow b$ and $b \rightarrow a$ then $a = b$, for $a, b \in \mathbb{P}$
- Transitivity of the ordering: $\quad$ if $a \rightarrow b$ and $b \rightarrow c$ then $a \rightarrow c$, for $a, b, c \in \mathbb{P}$

The set of types closed under the ^{r} and ^{l} adjoint operations is called the set of simple types.

A pregroup grammar $\mathbb{G} = (\Sigma, P, \rightarrow, {}^r, {}^l, 1, \mathbb{T})$ consists of a lexicon Σ and a typing relation $\mathbb{T} \subseteq \Sigma \times \mathbb{F}$ between the alphabet and the pregroup freely generated by the simple types of P and the ordering relation $\rightarrow$. This simply means that each element of the lexicon is associated with one or more strings of simple types. For instance, $(want, i\phi^l)$ will be used in a sentence like (2a) and $(want, i\bar{j}^l)$ in (2b).

(2)   a. You want for Mark to lead a happy life

      b. You want to eat ice cream

Here are common basic types:

s: declarative sentences	s_2: declarative sentence in the past tense
N: proper nouns	i: infinitives of intransitive verbs
n: common nouns	$\bar{n}$: complete noun phrases
π: subjects/nominative noun phrases	o: objects/accusative noun phrases

and an example derivation:

$$He\ likes\ her$$

(3)
$$\pi_3 \cdot (\pi_3^r \cdot s \cdot o^l) \cdot o$$
$$\rightarrow \pi_3 \cdot \pi_3^r \cdot s \cdot o^l \cdot o \rightarrow 1 \cdot s \cdot 1 \rightarrow s$$

The fact that the structure is partially ordered also allows us to set a specific ordering of grammatical types such as

$$\bar{n} \rightarrow \pi_3 \rightarrow \pi \qquad\qquad s_1 \rightarrow s$$

where $\alpha \rightarrow \beta$ means that α could also be used as β, e.g. a plural noun n_2 such as *cats* could be used as an object o or plural subject π_2, but not as a third person singular subject π_3, i.e. $n_2 \rightarrow \pi_2$, but $n_2 \nrightarrow \pi_3$.

Using the orderings we can now analyse more complex sentences

(4)
$$John\ wants\ for\ the\ cat\ that\ dogs\ fear\ to\ live$$
$$N \quad \pi_3^r s \phi^l \quad \phi \bar{j}^l o^l \quad \bar{n} n^l \quad n \quad n^r n o^{ll} s^l \quad n_2 \quad \pi_3^r s o^l \quad \bar{j} i^l \quad i$$
$$\rightarrow \quad \pi_3 \quad \pi_3^r s \phi^l \quad \phi \bar{j}^l o^l \quad o n^l \quad n \quad n^r n o^{ll} s^l \quad \pi_3 \quad \pi_3^r s o^l \quad \bar{j} i^l \quad i \quad \rightarrow s$$

Note the use of the ordering relations $\bar{n} \rightarrow o$, $n_2 \rightarrow \bar{n} \rightarrow \pi_3$ and $N \rightarrow \pi_3$.

As pregroup types are merely concatenation of types, the order of contractions does not really matter. What really matters in this kind of grammar are the derivation links that tell us how the different lexical items combine with eachother in a given sentence:

$$A \quad man \quad will \quad dance \quad to \quad save \quad humanity$$
$$\bar{n} \quad n^l \quad n \quad \pi^r\ s\ i^l \quad i \quad i^r\ i\ i^l \quad i \quad o^l \quad n$$

(5)

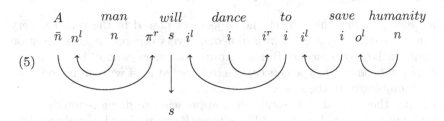

$$s$$

3   Problems with Semantics in Pregroup Grammars

One of the major inconveniences of using pregroup grammars to do semantics is that complex types can often be contracted in multiple ways, as we've seen above. For instance, consider the possible types that could be assigned to the subject position quantifier *every* in different grammatical formalisms:

(6)   a. **Traditional Categorial Grammars:** $(S/(N \setminus S))/N$

b. **Minimalist Grammars[18]:** $=\!\!N\ D\ -\!CASE$

c. **Pregroup Grammars:** $s(\pi^r s)^l n^l$

In the first two cases, the order in which the types or features are used is well-defined and unique:

– **Traditional Categorial Grammars**: Type-elimination follows nestedness, i.e. the quantifier must be joined to a noun phrase, then to a verb phrase

- **Minimalist Grammars**: Feature-checking is from left to right, i.e. the quantifier must be joined to a noun phrase, after which it could be used as a determiner and finally moved by being selected by a higher node with a selectional case feature

On the other hand, pregroup types aren't ordered: any basic type present in a type could theoretically be contracted at any point if it appears on the edge of the type. For instance, in the following sentence, *every* has the possibility to contract with either of its neighbours, *John knows that* or *boy*.

$$
(7) \qquad \begin{array}{ccccc} \textit{John knows that} & \textit{every} & \textit{boy} & \textit{dances} \\ s\bar{s}^l & s(\pi^r s)^l n^l & n & \pi^r s \end{array}
$$

The very liberal type structure of pregroups is essentially the reason why traditional approaches to semantics do not work in that framework. For instance, consider the type of a finite transitive verb

$$
(8) \qquad \begin{array}{c} \textit{kicked} \\ \pi^r s o^l \end{array}
$$

In Montagovian semantics [10, 12], such a verb would correspond to a relation between two entities, and would get assigned meaning:

(9) $\lambda x.\lambda y.kicked(y, x) : e \to e \to t$

The order in which the subject and object get passed to the verb are very important, as a situation where I kick someone is very different from a situation where I get kicked by someone. But pregroup grammars cannot, in this sense, place constraints on which type gets contracted first, at least without introducing unwanted complexity to the system.

Note that there are already established approaches to doing semantics with pregroup grammars, see [4, 6, 14, 15]. The aim with this project is simply to show that other approaches are also possible, that might also be simpler when doing event semantics from a derivational point-of-view than using the λ-calculus as semantic framework.

4   Quick Overview of Event Semantics and Conjunctivism

Conjunctivism [13] is the idea that as smaller expressions concatenate, their meanings simply conjoin. This approach is somewhat controversial, as for the last hundred years formal semanticists have instead related semantic combination with function application. To understand why and how this idea came to be, it is useful to have an understanding of two of the major developments of this branch of semantics: Davidson's characterization of action sentences in terms of events [5] and Parsons' subatomic analysis of events [11].

The traditional way of looking at a verb of action such as *kiss* is as a logical function that takes in two arguments – the subject and the object – and returns a truth value.

(10) $kiss(x, y) = \top \iff x$ kisses $y \iff (x, y) \in kiss_{ext}$

Here $kiss_{ext}$ is the set of pairs of people kissing.

For instance, if the extension of the verb $kiss$ were $kiss_{ext} = \{(J, M)\}$ then we could say that the only kissing happening is between John and Mary, John being the kisser and Mary the kissee.

The values of the following sentences could then be found by translating them into the appropriate logical form and checking if the membership conditions hold:

(11) $[\![John\ kisses\ Mary]\!] = kiss(J, M) = (J, M) \in kiss_{ext} = \top$

but $[\![John\ kisses\ Paola]\!] = \bot$, as $(J, P) \notin kiss_{ext}$.

Another possible analysis would be in terms of *events*.

(12)   John kisses Mary $\iff$ there is an event in which John kisses Mary

i.e. $[\![John\ kisses\ Mary]\!] = \exists e.kiss(e, J, M)$.

Letting the verb take an implicit event argument makes it then much easier to deal with questions such as verb arity and sentential adjuncts, as constructions like temporal, locative and manner adjuncts can now be redefined as independent predicates over events. For instance,

(13)   a. $[\![John\ danced\ at\ the\ ball]\!] = \exists e.danced(e, John) \wedge Location(e, the\ ball)$

   b. $[\![John\ danced\ yesterday]\!] = \exists e.danced(e, John) \wedge yesterday(e)$

This kind of representation makes it also easier to analyse certain cases of entailment relations: it is as simple as using $\wedge$-elimination on the denotation of a sentence with adjunct, to get its meaning without adjunct.

(14) $danced(e, John) \wedge With(e, Michael) \vdash danced(e, John)$

It is also possible to go even further [11] and treat subjects and objects not as arguments of the verb, but in a way similar to how adjuncts are handled using conjunctions. One can do so by introducing predicates standing for thematic relations between events and entities, that share the event with the verb:

(15) $[\![John\ kissed\ Julia]\!] = \exists e.Agent(e, John) \wedge kissed(e) \wedge Theme(e, Julia)$

Having verbs take in event arguments and no other grammatical argument also mirrors the way nouns and adjectives interact; instead of sharing an entity of some sort they are instead sharing an event or a state.

(16)   a. $[\![big\ grey\ cat]\!] = big(x) \wedge grey(x) \wedge cat(x)$

   b. $[\![intense\ dance]\!] = intense(x) \wedge dance(x)$

   c. $[\![dance\ intensely]\!] = dance(e) \wedge intense(e)$

   d. $[\![rain\ violently\ in\ Atlanta]\!] = rain(e) \wedge violent(e) \wedge in\_atlanta(e)$

Cases of embedded sentences can also be treated similarly by letting the event of the embedded sentence be treated as object and letting meaning predicates associated to the lexical items be independent of one another:

(17) ⟦ *John saw that Maria lied* ⟧ $= \exists.e.Agent(e, John) \wedge see(e) \wedge Past(e) \wedge$
      $\exists.e'.\textbf{Theme}(\textbf{e}, \textbf{e}') \wedge lie(e') \wedge Past(e') \wedge Agent(e', Maria)$

The next logical step in our will to generalize even more how the meaning of each lexical item can be seen as a piece of information that simply adds
constraints to the overall meaning of an expression, is to try to only have the
conjunction as mean of combination, and this is what Paul Pietroski brought
with Conjunctivism [13], inspired in part by the work of Schein [16,17] on plurality. Pietroski's proposal is that the semantics of any expression in natural
language consists in a finite conjunction of the meaning of its parts.

Connectives other than the conjunction are often used though in formal
semantics, and it is still not clear how sentences that seem to require them could
be modelled without. The crux of Pietroski's argument relies on using a different
logic for sentential interpretation, namely, plural logic [1]. Plural quantification
is an interpretation of monadic second-order logic, similar to a two-sorted logic,
in which the monadic predicate variable is not interpreted as a set of things, but
instead as taking multiple values.

(18) $x \prec X := x$ is one of the X's

The choice of this logic comes from the need to model plurality in cleaner
ways than with first-order logic. It has also a greater descriptive power and can
model meanings not accessible with first-order logic [1].

For instance, some predicates such as *dance* or *boy* are said to be singular and
require that the input is a singular value for it to be evaluated. The reasoning
behind this is that when someone asks of a group of people if they are boys,
they are not asking whether that group as a whole *is* a boy, but rather, whether
each person of that group is a boy. Therefore, singular predicates evaluate plural
values by evaluating the singular values it is composed of.

This allows us to represent an expression such as *two blue cats* as

(19) $\exists X.two(X) \wedge blue(X) \wedge cat(X) \wedge Plural(X)$

which is then interpreted as:

- There is a (perhaps) plural entity
- This plural entity has two values
- The values of the plural entity are blue
- The values of the plural entity are cats
- The entity has to be plural, i.e. has more than one value

Using this logic is a key component of Conjunctivism, as it also allows one
to model cases that seem to require extra logical tools like implication →, disjunction ∨ or universal quantification ∀. For instance, the quantified sentence
every cat sleep can be represented as:

(20) $\exists E.\exists X.Every\_Ag(E, X) \wedge cat(X) \wedge sleep(E)$

The reason why this works lies in the way each predicate contributes a specific
condition to the global meaning of the expression:

– $Every\_Ag(E, X)$: are the values of X precisely the agents of the events E?
– $cat(X)$: is every value of X a cat?
– $sleep(E)$: is every event of E an event of sleeping?

The goal here is not to defend Conjunctivism as a valuable approach to semantics, but only to show that, although it might not look like much at first, it is powerful enough to handle interesting non-trivial cases.

5   Derivational Event Semantics Using Pregroup Grammars

In this section, it will be shown how one could use the implicit event variables instantiated by a lexical item's corresponding meaning predicate to derive the right neo-davidsonian representation of an expression. The idea is that those variables can be turned into explicit objects that can be unified over as the expressions are combined. To structure the derivations, pregroup grammars will be used as the syntactic framework on top of which will be added the truth-conditional semantic layer and another layer where interaction between event variables take place.

The general process of unifying event variables is not required to take place in the pregroup framework, or even categorial framework, though pregroup grammars do offer some advantages over other syntactic frameworks for this type of analysis, especially since their types are non-functional and some syntactic relations are already defined in the system through type ordering, e.g. $N \to \pi_3$.

It would also be possible to adapt this framework to a montagovian one where denotations are λ-terms and where the syntactic types are functional (see [3, 7, 20] for inspiration), but the end goal is really to show that the full power of the λ-calculus is not needed to get a good compositional semantics. We aim to get something that really highlights how events relate to each other and how simple meaning composition can really be. Simplicity is the key here.

The focal point of the analysis will also be different, in a sense, from what is seen in more traditional approaches to semantics, as instead of focusing on ways to combine expressions' truth conditions and passing around predicates to predicates, the event variables will themselves be the ones moving around the syntactic trees, as they are the main shareable pieces of information in this framework. On the other hand, the semantics predicates' main role will be to constrain the possibility of events to occur by restraining the possible values the event and entity variables can take.

5.1   Motivation

Let's start by looking at the event analysis of a simple sentence.

(21) $[\![John\ dances]\!] = \exists e.Agent(e, John) \wedge dance(e)$

In this case, a single event e is shared by both lexical items. We want to compositionally explain how to get to that denotation so that one could define a derivational system that takes words as input and outputs conjunctivist values.

(22)

$$\text{John dances}$$
$$\exists e. Agent(e, John) \wedge dance(e)$$

$$\overbrace{\text{John} \quad \text{dances}}$$
$$\alpha(e_1) \quad \beta(e_2)$$

In the above case, both α and β stand for the meaning of their respective expressions and take as argument event variables e_1 and e_2. Leaving aside for a moment the question of what the exact values α and β stand for, an important question to answer is: Where does e come from?

The main property of variables is their mutability, they can take any values assigned to them. One does not have to know from the start what value they will be taking at the end. In the following functional example, x does not have any intrinsic value at the beginning, its raison d'être is to take the value of whatever term gets passed to the expression.

(23)

$$dance(John)$$

$$\overbrace{\lambda x. dance(x) \quad John}$$

Now, looking back at the event translation, neither the subject *John* nor *dances* know what they will be taking scope over when the derivation ends. For instance, in a sentence such as

(24)   John knows that Sara dances

they would not have the same event as argument: the subject *John* is related to a first event of *knowing*, while *dances* predicates over a totally different event where Sara is the agent instead of John. The goal is to figure out what they could have started with so that they end up with the right argument assignment. Assuming that they start for instance with values $Agent(e, John)$ and $dance(e)$, over the same event e, does not solve the problem: how did they know that they both take the exact same event as argument? Getting the right variables in the right place will be achieved by using *unification* on the variables.

Here is how the derivation of the logical form $\exists e. Agent(e, John) \wedge dance(e)$ will take place:

1. Distinct variables are instantiated by *john* and *dances*'s semantic predicates, to be taken as arguments: $Agent(e_1, John)$ and $dance(e_2)$
2. Lexical items are concatenated, from which it follows that the variables associated with the syntactic categories that allowed the concatenation to take place are unified. In this case, e_1 and e_2 are unified, i.e. $e_1 = e_2$.

(25)

$$John\ dances$$
$$Agent(e_1, John) \wedge dance(e_2) \wedge e_1 = e_2$$

$$John \qquad dances$$
$$Agent(e_1, John) \quad dance(e_2)$$

3. A final process then takes place that binds the instantiated variables

To improve readability, values of the form $A[e_1, e_2] \wedge e_1 = e_2$ will be automatically replaced by $A[e_1, e_2/e_1]$.

Note that actually keeping the two variables as distinct and binding each one — hence binding twice instead of once — does not actually make any difference in the meaning:

(26) $\exists x. \exists y. A(x) \wedge B(y) \wedge x = y \iff \exists x. A(x) \wedge B(x)$

Now let's have look at the sentence *the cat dances*, whose derivation tree looks something like:

(27)

$$The\ cat\ dances$$
$$the\ cat\ dances$$
$$the\ cat$$

It would be nice to have the derivation process be similar to the one described above, but that poses a problem, as doing so exactly the same way would give us this kind of logical form:

(28) $\exists e. the(e) \wedge cat(e) \wedge Agent(e, e) \wedge dances(e)$

The variable taken by the determiner phrase as argument should be a completely different one from the one taken by the verb. The relation between those two variables seems to be exactly what $Agent(e, x)$ is defining: the variable from the determiner phrase is the agent of the variable taken by the verb phrase.

This problem could be approached through two different angles. The first way (see 29), following Pietroski [13] is to assume that the type of the determiner and noun compound contains a unique implicit variable x it can refer to, which, through a transformation from determiner phrase to subject – or by being assigned case – gets a new semantic constraint $Agent(e, x)$ added to its meaning and a new event variable: the verb phrase can now only access e, which is fresh, and not x anymore. In other words, a new grammatical role is now synonymous with a change of available implicit variable and a closure of the old variable. This solution has the advantage of being more theoretically motivated, but also has its downsides when used in a syntactic framework like PG, which does not restrict compounding of expressions as much as other frameworks.

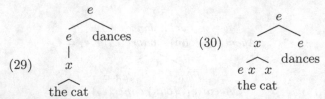

(29)

(30)

The second way (30) this could be dealt with is by assigning different variables to the syntactic categories that form the type of *the*, so that when it first concatenates with *cat*, the syntactic category that allows for the operation to happen will be linked to x, and the one corresponding to the second concatenation will contain a different variable, e.

The variable over which two branches unify is represented in the node.

Looking back at the previous tree, one see that e dominates it and might wonder what it implies. For simplicity, it can be assumed that this node, which is the final one in the tree, is of a basic category, e.g. s or C as opposed to a concatenation of categories, and so that this category corresponds to a single variable, or has a single variable available for further concatenation, e in this case. The reason is that if one were to use this expression within another expression or if one wanted to concatenate extra lexical items to it, what would be shared between the two would be the event variable e, and in no case the entity variable x — assuming the internal structure of that subtree is not modified.

(31)

(32)

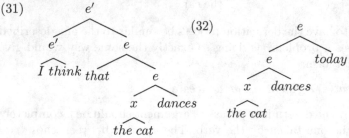

On the left, the sentence is included in the tree as an embedded clause, which semantically would be represented as the variable e now being the theme of e', the event over which *I* and *think* take place. On the right, the expression is concatenated with an adjunct, which shows that the event e is still possibly accessible from the expression *the cat dances*.

In a way, no matter how that constituent — *the cat dances* — is used, the main information that will be shared and that could be quantified over, seems to be the event. This is similar to the way syntactic categories behave, in the sense that no matter how long an expression gets, if its syntactic category is A, it will always be possible to use it anywhere where an expression of category A could be used, no matter what other constituents it might contain. It does not mean that the embedded subtree is completely opaque, but simply that its access is more restricted.

5.2 Semantic Pregroup Types

It will be now shown how to transpose that approach into the pregroup framework.

To stay in the categorial state-of-mind, the system will be as general and require as few rules as possible when it comes to generating the logical form: systematic combination rules will be defined, but constraints as to when and to what lexical items these rules can be applied is mostly left to the lexical items themselves, by carefully specifying their syntactic types. No complexity is added to the syntactic layer from the addition of this new semantic layer, which means that the structure of derivations is no different than that of regular pregroup derivations. This means that the usual parsing algorithms for pregroup grammars can also be used, as long as clauses are added to handle semantics.

Note that the final representation of the meaning one will get at the end of a derivation can be qualified as *raw*, as some aspects of the meaning of a sentence cannot be reached simply by predicate combination, especially since no pre-derivational thematic assignments are assumed to have taken place. Concretely, this means that one might end up with the representation (33a) for the sentence (33b) but then that extra semantic information could be reached through meaning postulates such as (33c) as thematic roles depend on multiple factors such as voice, grammatical functions and the nature of the verb itself.

(33) a. $\exists e.Subject(e, John) \wedge Passive(e) \wedge Past(e) \wedge Kick(e) \wedge Time(e, Monday)$
      b. *John was kicked on Monday*
      c. $Subject(e, A) \wedge Passive(e) \wedge Kick(e) \vdash Patient(e, A)$

The most direct way of building a system to account for the kind of semantics just seen above is to pair lexical items with a syntactic type, a set of available variables and a truth-conditional meaning predicate, which scopes over different values and variables. Those variables will be assumed to be instantiated when the lexical item is first used, and their value will change through the derivation depending on the way types contract.

The full value of a lexical item is then a tuple of the form:

(34) $((a_1, x_1) \cdot (a_2, x_2) \cdot ... \cdot (a_n, x_n), A)$

where a_i is a basic pregroup type, x_i an available implicit variable, and A a logical formula that stands for the expression's meaning.

For instance, the relative pronoun *whom*, will have the form (35a) which will be rewritten as (35b) for clarity, and could be read as: the variable associated with the sentential and subject type is possibly different from the one associated to the noun types. This comes from the fact that *whom* is usually used as *theme* predicate over 2 distinct variables, one corresponding to an event and another corresponding to an entity.

(35) a. $(n^r, x) \cdot (n, x) \cdot (s^l, e) \cdot (\pi_3, e)$
      b. $n_x^r n_x s_e^l \pi_{3,e}$

This is a good example of why having only one available variable per lexical item does not work well with pregroup types: both the event and entity variable have to appear within its meaning predicate at some point of the derivation, but starting with either and trying to introduce the other at a later point brings many complications. Not to say that it is impossible, just much more painful.

(36)   a. the cat whom Caesar stabbed

  b. $\exists x.i(e,x) \land cat(x) \land \exists e.Theme(e,x) \land Agent(e, Caesar) \land stabbed(e)$

In this case, *whom* will have semantic value $Theme(e,x)$, sharing the variable e on the right with *Caesar stabbed* and x, on the left with *cat*.

Note that the *kind* or *type* of the variables is irrelevant in this system. There is no real difference between x, e, or any other variable, and only the constraints put on a variable can say something about it. Using specific characters to represent variables such as x and e only makes reading descriptions easier.

It is tempting to use, for instance, entity types and event types and try to copy what is done in Montagovian semantics, but in the end the types that would end up being required would be very different from the Montagovian ones. For instance, there is not going to be any boolean type passed around.

The reason is simply that the boundaries between entities and events are very blurry: is the *crash* in *The crash was brutal* an event or entity? Similarly, differentiating between activities, accomplishments and other eventualities [19] does not seem necessary: it will be assumed that the kind of eventualities is simply the result of the interplay between features or predicates, e.g.

(37)   a. I built the house for 10 hours : $build(e) \land for(e) \vdash activity(e)$

  b. I built the house in 10 hours : $build(e) \land in(e) \vdash accomplishment(e)$

5.3   Semantic Combinations

This section outlines a method of combining lexical items' meanings given the new tuple types defined above. As previously mentioned, the Conjunctivist approach is followed here and the meaning of an expression takes the form of a conjunction of the meaning of the parts, scoping over given event variables.

To get the right variables at the right place, the variables will have to get unified over the contraction links. What this means is that whenever two syntactic types contract, their contained variables are forced to take the same value. This also affects the distribution of the other variables contained in those types. Here is a simple example to show how it works:

$$
(38) \quad \begin{array}{ccc}
big & cat & \to \\
n_x n_x^l & n_y & \to \\
big(x)\ cat(y) & \to & big(x) \land cat(y) \land x = y
\end{array}
\qquad
\begin{array}{c}
big\ cat \\
n_x
\end{array}
$$

While contracting the types, the constraint that $x = y$ is added to the global meaning. The variables could also be replaced automatically.

What is nice about this is that a derivation can now be represented as semantic predicates linked to each other by the variables they share, which also corresponds to the contraction links. Put another way, the contraction links can be labelled using the contained event and entity variables.

$$(39) \quad
\begin{array}{ccccc}
John & likes & the & big & cat \\
\pi_{3,e} & \pi^r_{3,e}s_3o^l_e & o_e n^l_x & n_x n^l_x & n_x \\
Agent(e, John) & like(e) & Theme(e, x) \wedge the(e, x) & big(x) & cat(x)
\end{array}$$

5.4 Syntactic/Semantic Hierarchy

Let's have a look back at example (27) that was discussed at the very beginning of this chapter.

(40) $[\![The\ cat\ dances]\!] = \exists e.\exists x.the(x) \wedge cat(x) \wedge Agent(e, x) \wedge dances(e)$

Underlyingly, the variables are layered in this kind of way:

$$(41) \quad
\begin{array}{c}
e \\
\overbrace{} \\
x \quad dances \\
\overbrace{the\ cat}
\end{array}$$

The task at hand now is to find a way of going from the variable that is shared between *the* and *cat* to a fresh one that would then get unified with the one coming from *dances*. There is actually a very simple way of relating this to another pregroup operation and that is by extending the pregroup orderings, or syntactic hierarchy, to take into account semantic constructions.

To remind the reader, since pregroups are ordered structures, some grammatical relations can be explicitly defined as orders. For instance,

$\bar{n} \to o$ a determiner phrase can be used as object

$\bar{j} \to \pi_3$ an infinitive verb phrase can be used as subject

A relation such as $\bar{n} \to \pi_3$ could then be rewritten to include information about the variables present in the types and about the extra semantic relation they now play under this new syntactic type:

$$(42) \quad
\begin{array}{cc}
\bar{n}_x & \pi_{3,e} \\
A[x] & \Rightarrow Agent(e, x) \wedge A[x]
\end{array}$$

which is to be interpreted as: using a determiner phrase as a third person subject means having its available entity variable used as the agent of the event specified by the verb it will combine with.

$$(43) \quad
\begin{array}{cccccccc}
He & knows & the & person & whom & John & likes \\
\pi_{e_0} & \pi^r_{e_1}s_{e_1}o^l_{e_1} & \bar{n}_{x_0}n^l_{x_0} & n_{x_1} & n^l_{x_2}n_{x_2}(so^l)^l_{e_2} & N_{john} & \pi^r_{e_3}s_{e_3}o^l_{e_3} \\
Agent(e_0, he) & know(e_1) & the(x_0) & person(x_1) & Theme(e_2, x_2) & \top & likes(e_3)
\end{array}$$

$$\begin{array}{ccccccccc}
\pi_{e_0} & \pi_{e_0}^r s_{e_0} o_{e_0}^l & o_{e_0} n_{x_0}^l & & n_{x_0} & n_{x_0}^l n_{x_0} (so^l)_{e_1}^l & \pi_{e_1} & \pi_{e_1}^r s_{e_1} o_{e_1}^l \\
Agent(e_0, he) & know(e_0) & Theme(e_0, x_0) \wedge the(x_0) & & person(x_0) & Theme(e_0, x_0) & Agent(e_1, John) & likes(e_1)
\end{array}$$

In this case, the rule used on the proper name *John* was:

(44)
$$\begin{array}{cc} N_x & \pi_{3,e} \\ \multicolumn{2}{c}{John(x) \Rightarrow Agent(e,x) \wedge John(x)} \end{array}$$

Note that it is not always necessary to go through a transformation of variable, as potentially distinct variables could be attached to the basic types of an expression, just like for the above case of the relative pronoun *whom*, or in a case like (45) where all pieces naturally combine and the variables over which expressions are unified varies as the concatenation takes place

(45) $\llbracket He\ danced\ at\ school \rrbracket = Agent(e, he) \wedge danced(e) \wedge Loc(e, x) \wedge school(x)$

(46)

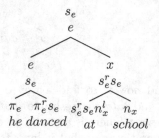

In this case the variable at the top of a branching represents the variable that was unified.

5.5   Existential-Closure

Existential-closure is not as straight-forward to implement in this system and will only be glossed over in this article. The main problem with trying to adapt the kind of semantic system that researchers like Pietroski uses is that syntactic types in pregroup grammars can combine in any order they want as long as they are on the edge of the type. This is problematic, as more mainstream syntactic systems usually does not work this way, and a lot of ordering is constrained by the encoding of the grammatical categories, for instance by using internal/external argument relations.

A simple example is how thematic relation assignment is usually dealt with. A generative way of representing how the phrase *my cat* can play the role of an agent could be as positing a covert agent-node that takes in the clause and changes its domain of predication from an entity to an event

(47) $\llbracket\ Agent\ [my\ cat]\ \rrbracket(e) = \exists x.Agent(e,x) \wedge \llbracket my\ cat \rrbracket(x)$

This way, if something is under the scope of *agent*, it will only have access to x and not e, and on the other hand, x is bound inside the agent-node, hence if one is working outside of it, they will not have access to x, but only to e. By letting the determiner phrase be taken as an argument, the syntactic parsing is also restricted, as the concatenation of *my* and *cat* is something that can only happen lower in the tree, than the assignment of the agent role.

In pregroup grammars, the derivation could take multiple forms.

(48)

On the left, the ordering is used first on $\bar{n} \to \pi$, which corresponds, in a sense, to combining a covert agent node with the determiner, before the latter combines with the noun. On the right, the combination of the determiner and noun takes place before the expression takes on the role of an agent, which is equivalent to the unique generative representation discussed above.

This much more flexible way of combining expressions is the reason it is harder to structure existential closure and is the reason multiple variables per type are needed, as some of those operations take place in parallel, and sometimes multiple variables have to be accessible at the same time. More precisely, in this case, closing the variable as $\bar{n}$ goes to π before contracting the types (left path) blocks the entity variable coming from the noun to be unified with the determiner, which is dramatic. Restricting the order in which pregroup types could combine is also out of the question here.

Two possible alternatives to ∃-closure would be:

- Closing a variable after a contraction of one of the basic types it is present in, only if it does not appear in any other basic type of the complex type it is part of.
- Closing every variable instantiated at the end of a derivation, similarly to the way ∃-introduction is used in logic. This works since the only logical symbols we are working with are the conjunction and the existential quantifier.

Here is a complete table of the correspondence between the syntactic and the semantic structures:

Syntax	Semantics
Concatenation of basic types	Conjunction of logical predicates
Contraction of types	Unification on available event variables
Syntactic type ordering	Conjunction of new semantic predicate
Variable disappears from the types	∃-closure

6   Conclusion

The compositional semantic system defined in this article is an elegant and natural way of defining a semantics for pregroup grammars which relies on light machinery and intuitive operations, which are clearly its key characteristics.

Many questions are left to be answered, especially when it comes to the internal structure of the events and how this kind of system might relate to the typed λ-calculus, which seems to be much more powerful and to have a greater control on how predicates can be moved around and reorganized. The descriptive adequacy of Conjunctivism is also an interesting question to investigate in the future, which was only briefly addressed in this paper.

References

1. Boolos, G.: To be is to be the value of a variable (or to be some values of some variables). J. Philos. **81**, 430–449 (1984)
2. Buszkowski, W.: Lambek calculus and substructural logics. Linguist. Anal. **36**(1), 15–48 (2003)
3. Champollion, L.: Quantification and negation in event semantics. In: Barbara Partee, M.G., Skilters, J. (eds.) Baltic International Yearbook of Cognition, Logic and Communication, vol. 6, pp. 1–23. New Prairie Press, Manhattan (2010)
4. Clark, S., Coecke, B., Sadrzadeh, M.: Mathematical foundations for a compositional distributional model of meaning. Linguist. Anal. **36** (2010)
5. Davidson, D.: The logical form of action sentences. Synthese (1967)
6. Gaudreault, G.: Bidirectional functional semantics for pregroup grammars. In: Kanazawa, M., Moss, L.S., de Paiva, V. (eds.) Third Workshop on Natural Language and Computer Science, NLCS 2015. EPiC Series in Computer Science, vol. 32, pp. 12–28 (2015)
7. de Groote, P., Winter, Y.: A type-logical account of quantification in event semantics. In: Murata, T., Mineshima, K., Bekki, D. (eds.) JSAI-isAI 2014. LNCS (LNAI), vol. 9067, pp. 53–65. Springer, Heidelberg (2015). doi:10.1007/978-3-662-48119-6_5
8. Lambek, J.: Type grammar revisited. In: Lecomte, A., Lamarche, F., Perrier, G. (eds.) LACL 1997. LNCS (LNAI), vol. 1582, pp. 1–27. Springer, Heidelberg (1999). doi:10.1007/3-540-48975-4_1
9. Lambek, J.: From Word to Sentence: A Computational Algebraic Approach to Grammar. Polimetra, Koper (2008)
10. Montague, R.: Formal Philosophy: Papers of Richard Montague. Yale University Press, New Haven (1974). Ed. by R.H. Thomason
11. Parsons, T.: Events in the Semantics of English. The MIT Press, Cambridge (1990)
12. Partee, B. (ed.): Montague Grammar. Academic Press, New York (1976)
13. Pietroski, P.: Events and Semantic Architecture. Oxford University Press, Oxford (2005)
14. Preller, A.: Toward discourse representation via pregroup grammars. J. Logic Lang. Inform. **16**(2), 173–194 (2007)
15. Preller, A., Sadrzadeh, M.: Semantic vector models and functional models for pregroup grammars. J. Logic Lang. Inform. **20**(4), 419–443 (2011)
16. Schein, B.: Plurals and Events. MIT Press, Cambridge (1993)

17. Schein, B.: Events and the semantic content of thematic relations. In: Peter, G.P.G. (ed.) Logical Form and Language, pp. 263–344. Oxford University Press, Oxford (2002)
18. Stabler, E.: Derivational minimalism. In: Retoré, C. (ed.) LACL 1996. LNCS, vol. 1328, pp. 68–95. Springer, Heidelberg (1997). doi:10.1007/BFb0052152
19. Vendler, Z.: Verbs and times. Philos. Rev. **66**, 143–160 (1957)
20. Winter, Y., Zwarts, J.: Event semantics and abstract categorial grammar. In: Kanazawa, M., Kornai, A., Kracht, M., Seki, H. (eds.) MOL 2011. LNCS (LNAI), vol. 6878, pp. 174–191. Springer, Heidelberg (2011). doi:10.1007/978-3-642-23211-4_11

A Compositional Distributional Inclusion Hypothesis

Dimitri Kartsaklis and Mehrnoosh Sadrzadeh[✉]

School of Electronic Engineering and Computer Science,
Queen Mary University of London, London, UK
{d.kartsaklis,mehrnoosh.sadrzadeh}@qmul.ac.uk

Abstract. The distributional inclusion hypothesis provides a pragmatic way of evaluating entailment between word vectors as represented in a distributional model of meaning. In this paper, we extend this hypothesis to the realm of compositional distributional semantics, where meanings of phrases and sentences are computed by composing their word vectors. We present a theoretical analysis for how feature inclusion is interpreted under each composition operator, and propose a measure for evaluating entailment at the phrase/sentence level. We perform experiments on four entailment datasets, showing that intersective composition in conjunction with our proposed measure achieves the highest performance.

Keywords: Computational linguistics · Artificial intelligence · Natural language processing · Textual entailment · Inclusion hypothesis · Compositionality · Distributional models

1 Introduction

Distributional models of meaning, where words are represented by vectors of co-occurrence frequencies gathered from corpora of text, provide a successful model for representing meanings of words and measuring the semantic similarity between them [22]. A pragmatic way for applying these models to entailment tasks is developed via the distributional inclusion hypothesis [8,9,11], which states that a word u entails a word v if whenever u is used so can be v. In distributional semantics terms, this means that contexts of u are included in contexts of v. For example, whenever 'boy' is used, e.g. in the sentence 'a boy runs', so can be 'person'; thus $boy \vdash person$. By projecting this hypothesis onto a truth theoretical model, one may say that u and v stand in an entailment relation if by replacing u with v in a sentence presumed to be true, we produce a new sentence preserving that truth. For example, if the sentence 'a boy runs' is presumed to be true, so is the sentence 'a person runs', obtained by replacing 'boy' by 'person'.

M. Sadrzadeh – Support by EPSRC for Career Acceleration Fellowship EP/J002 607/1 and AFOSR International Scientific Collaboration Grant FA9550-14-1-0079 is gratefully acknowledged.

M. Amblard et al. (Eds.): LACL 2016, LNCS 10054, pp. 116–133, 2016.
DOI: 10.1007/978-3-662-53826-5_8

One problem with distributional models of meaning is that they do not scale up to larger text constituents, such as phrases or sentences. The reason is that these do not frequently occur in corpora of text, thus the process of collecting reliable statistics to represent them as vectors does not witness the distributional hypothesis. This problem is usually addressed with the provision of a composition operator, the purpose of which is to produce vectors for phrases and sentences by combining their word vectors. Compositional distributional models of this form generally fall into three categories: models based on simple element-wise operations between vectors, such as addition and multiplication [19]; tensor-based models in which relational words such as verbs and adjectives are multi-linear maps acting on noun (and noun-phrase) vectors [3,7,10]; and models in which the compositional operator is implemented as part of some neural network architecture [12,21].

The purpose of this paper is to investigate, both theoretically and experimentally, the application of the distributional inclusion hypothesis on phrase and sentence vectors produced in a variety of compositional distributional models. We provide interpretations for the features of these vectors and analyse the effect of each compositional operator on the inclusion properties that hold for them. We further discuss a number of measures that have been used in the past for evaluating entailment at the lexical level. Based on the specificities introduced by the use of a compositional operator on word vectors, we propose an adaptation of the *balAPinc* measure [14]—which is currently considered a state-of-the-art in measuring entailment at the lexical level—for compositional distributional models.

The theoretical discussion is supported by experimental work. We evaluate entailment relationships between simple intransitive sentences, verb phrases, and transitive sentences, on datasets specifically created for the purposes of this work. We also present results on the $AN \vdash N$ task of [2], where the goal was to evaluate the extent to which an adjective-noun compound entails its noun. Our findings suggest that the combination of our newly proposed measure with intersective compositional models achieves the highest discriminating power when evaluating entailment at the phrase/sentence level.

Outline. Sections 2 and 3 provide an introduction to compositional distributional semantics and to distributional inclusion hypothesis, respectively; Sect. 4 studies the inclusion properties of features in a variety of compositional distributional models, while Sect. 5 discusses the adaptation of the *balAPinc* measure to a compositional setting; Sects. 6 and 7 deal with the experimental part; and finally, in Sect. 8 we briefly discuss our findings.

2   Compositional Distributional Semantics

Compositional distributional semantics represents meanings of phrases and sentences by combining the vectors of their words. In the simplest case, this is done by element-wise operations on the vectors of the words [19]. Specifically, the vector representation of a sequence of words $w_1, \ldots, w_n$ is defined to be:

$$\sum_i \vec{w}_i \quad \text{or} \quad \bigodot_i \vec{w}_i \tag{1}$$

where $\odot$ denotes element-wise multiplication.

A second line of research follows a more linguistically motivated approach and treats relational words as linear or multi-linear maps. These are then applied to the vectors of their arguments by following the rules of the grammar [3,7,10]. For example, an adjective is treated as a map $N \to N$, for N a basic noun space of the model. Equivalently, this map can be represented as a matrix living in the space $N \otimes N$. Similarly, a transitive verb is a map $N \times N \to S$, or equivalently, a "cube" or a tensor of order 3 in the space $N \otimes N \otimes S$, for S a basic sentence space of the model. Composition takes place by tensor contraction, which is a generalization of matrix multiplication to higher order tensors. For the case of an adjective-noun compound, this simplifies to matrix multiplication between the adjective matrix and the vector of its noun, while for a transitive sentence it takes the form:

$$\overrightarrow{svo} = (\overline{\text{verb}} \times \overrightarrow{\text{obj}}) \times \overrightarrow{\text{subj}} \tag{2}$$

where $\overline{\text{verb}}$ is a tensor of order 3. Compared to element-wise vector operations, note that tensor-based models adhere to a much stricter notion of composition, where the transition from grammar to semantics takes place via a structure-preserving map [7].

Finally, deep learning architectures have been applied to the production of phrase and sentence vectors, tailored for use in specific tasks. These methods have been very effective and their resulting vectors have shown state-of-the-art performances in many tasks. The main architectures usually employed are that of recursive or recurrent neural networks [5,21] and convolutional neural networks [12]. Neural models are "opaque" for our purposes, in the sense that their non-linear multi-layer nature does not lend itself to be reasoned about in terms of the feature inclusion properties of the distributional inclusion hypothesis, and for this reason we do not deal with them in this paper.

3   The Distributional Inclusion Hypothesis

The distributional inclusion hypothesis (DIH) [8,9,11] is based on the fact that whenever a word u entails a word v, then it makes sense to replace instances of u with v. For example, 'cat' entails 'animal', hence in the sentence 'a cat is asleep', it makes sense to replace 'cat' with 'animal' and obtain 'an animal is asleep'. On the other hand, 'cat' does not entail 'butterfly', and indeed it does not make sense to do a similar substitution and obtain the sentence 'a butterfly is asleep'.

This hypothesis has inherent limitations, the main one being that it only makes sense in contexts that contain no logical words. For instance, the substitution of u for v would not work for sentences that have negations or quantifiers such as 'all' and 'none'. As a result, one cannot replace 'cat' with 'animal' in sentences such as 'all cats are asleep' or 'a cat is not asleep'. Despite this, the

DIH has been subject to a good amount of study in the distributional semantics community and its predictions have been empirically validated to a good extent [9,14].

Formally, if word u entails word v, then the set of features of u are included in the set of features of v. In the context of a distributional model of meaning, the term *feature* refers to a non-zero dimension of the distributional vector of a word. This makes sense since, according to DIH, word v subsumes the meaning of word u. Throughout this paper, we denote the features of a distributional vector $\vec{v}$ by $\mathcal{F}(\vec{v})$, hence we have:

$$u \vdash v \quad \text{whenever} \quad \mathcal{F}(\vec{u}) \subseteq \mathcal{F}(\vec{v}) \tag{3}$$

The research on the DIH can be categorised into two classes. In the first class, the degree of entailment between two words is based on the distance between the vector representations of the words. This distance must be measured by asymmetric means, since entailment is directional. Examples of measures used here are entropy-based measures such as KL-divergence [4]. Abusing the notation and taking $\vec{u}$ and $\vec{v}$ to also denote their underling probability distributions, this is defined as follows:

$$D_{\mathrm{KL}}(\vec{v} \| \vec{u}) = \sum_i v_i (\ln v_i - \ln u_i) \tag{4}$$

KL-divergence is only defined when the support of $\vec{v}$ is included in the support of $\vec{u}$. In order to overcome this restriction, a variant referred to by α-skew [15] has been proposed. This is defined in the following way:

$$s_\alpha(\vec{u}, \vec{v}) = D_{\mathrm{KL}}(\vec{v} \| \alpha \vec{u} + (1 - \alpha) \vec{v}) \tag{5}$$

where $\alpha \in (0, 1]$ serves as a smoothing parameter. *Representativeness* is another way of normalising KL-divergence; it is defined as follows:

$$R_{\mathrm{D}}(\vec{v} \| \vec{u}) = \frac{1}{1 + D_{\mathrm{KL}}(\vec{v} \| \vec{u})} \tag{6}$$

Representativeness turns KL-divergence into a number in the unit interval $[0, 1]$. As a result we obtain $0 \leq R_{\mathrm{D}}(\vec{v} \| \vec{u}) \leq 1$, with $R_{\mathrm{D}}(\vec{v} \| \vec{u}) = 0$ when the support of $\vec{v}$ is not included in the support of $\vec{u}$ and $R_{\mathrm{D}}(\vec{v} \| \vec{u}) = 1$, when $\vec{u}$ and $\vec{v}$ represent the same distribution.

The research done in the second class attempts a more direct measurement of the inclusion of features, with the simplest possible case returning a binary value for inclusion or lack thereof. Measures developed by [6,23] advance this simple methods by arguing that not all features play an equal role in representing words and hence they should not be treated equally when it comes to measuring entailment. Some features are more "pertinent" than others and these features have to be given a higher weight when computing inclusion. For example, 'cat' can have a non-zero coordinate on all of the features 'mammal, miaow, eat, drink, sleep'. But the amount of these coordinates differ, and one can say that, for example, the higher the coordinate the more pertinent the feature. Pertinence is computed by various different measures, the most recent of which is *balAPinc* [14], defined as follows:

$$balAPinc(u,v) = \sqrt{LIN(u,v) \cdot APinc(u,v)} \tag{7}$$

where *LIN* is Lin's similarity [16] and *APinc* is an asymmetric measure defined as below:

$$APinc(u,v) = \frac{\sum_r \left[P(r) \cdot rel'(f_r) \right]}{|\mathcal{F}(\overrightarrow{u})|} \tag{8}$$

APinc applies the DIH via the idea that features with high values in $\mathcal{F}(\overrightarrow{u})$ must also have high values in $\mathcal{F}(\overrightarrow{v})$. In the above formula, f_r is the feature in $\mathcal{F}(\overrightarrow{u})$ with rank r; $P(r)$ is the precision at rank r; and $rel'(f_r)$ is a weight computed as follows:

$$rel'(f) = \begin{cases} 1 - \frac{rank(f, \mathcal{F}(\overrightarrow{v}))}{|\mathcal{F}(\overrightarrow{v})|+1} & f \in \mathcal{F}(\overrightarrow{v}) \\ 0 & o.w. \end{cases} \tag{9}$$

where $rank(f, \mathcal{F}(\overrightarrow{v}))$ shows the rank of feature f within the entailed vector. In general, *APinc* can be seen as a version of average precision that reflects lexical inclusion.

We will return to the topic of entailment measures in Sect. 5, where we propose variations on *APinc* and *balAPinc* that are more appropriate for entailment in compositional distributional models.

4   A Compositional Distributional Inclusion Hypothesis

In the presence of a compositional operator, features of a phrase/sentence adhere to some set-theoretic properties. In what follows, we present these properties for a number of operators in various compositional distributional models.

4.1   Element-Wise Composition

For simple additive and multiplicative models, the set of features of the phrase/sentence are easily derived from the set of features of their words using the set-theoretic operations of union and intersection:

$$\mathcal{F}(\overrightarrow{v_1} + \cdots + \overrightarrow{v_n}) = \mathcal{F}(\overrightarrow{v_1}) \cup \cdots \cup \mathcal{F}(\overrightarrow{v_n}) \tag{10}$$

$$\mathcal{F}(\overrightarrow{v_1} \odot \cdots \odot \overrightarrow{v_n}) = \mathcal{F}(\overrightarrow{v_1}) \cap \cdots \cap \mathcal{F}(\overrightarrow{v_n}) \tag{11}$$

The features of a tensor product of vectors consists of tuples of same-indexed features, taken from their cartesian product:

$$\mathcal{F}(\overrightarrow{v_1} \otimes \cdots \otimes \overrightarrow{v_n}) = \{(v_i^1, \cdots, v_i^n) \mid v_i^j \in \mathcal{F}(\overrightarrow{v_j})\} \tag{12}$$

where v_i^j refers to the ith element of the jth vector. Point-wise minimum and maximum of vectors act inline with intersection and union respectively, providing a feature inclusion behaviour identical to addition and point-wise multiplication.

$$\mathcal{F}(\max(\vec{v_1}, \cdots, \vec{v_n})) = \mathcal{F}(\vec{v_1}) \cup \cdots \cup \mathcal{F}(\vec{v_n}) \qquad (13)$$

$$\mathcal{F}(\min(\vec{v_1}, \cdots, \vec{v_n})) = \mathcal{F}(\vec{v_1}) \cap \cdots \cap \mathcal{F}(\vec{v_n}) \qquad (14)$$

In order to see this, let us consider the max case. In the linear expansion notation, we have:

$$\max(\vec{v_1}, \cdots, \vec{v_n}) = \sum_i \max(v_i^1, v_i^2, \cdots, v_i^n) \, \vec{a}_i$$

where $\{\vec{a_i}\}_i$ is an orthonormal basis of space V where vectors $\vec{v_i}$ live. For any arbitrary dimension $\vec{a_j}$, it is the case that $\vec{a_j} \in \mathcal{F}(\max(\vec{v_1}, \cdots, \vec{v_n}))$ iff $\max(v_j^1, v_j^2, \cdots, v_j^n) \neq 0$. For this to happen, it suffices that one of the v_j^i's is nonzero, that is $v_j^1 \neq 0$ *or* $v_j^2 \neq 0$ *or* $\cdots$ *or* $v_j^n \neq 0$, which is equivalent to saying that $\vec{a_j} \in \mathcal{F}(\vec{v_1}) \cup \cdots \cup \mathcal{F}(\vec{v_n})$. The case for min is similar, with the difference that *or* is replaced with *and*, hence the set theoretic operation $\cup$ with $\cap$.

Element-wise composition has certain desirable properties in relation to the DIH. Firstly, it lifts naturally from the word level to phrase/sentence level; specifically, for two sentences $s_1 = u_1 \ldots u_n$ and $s_2 = v_1 \ldots v_n$ for which $u_i \vdash v_i, \forall i \in [1, n]$, it is always the case that $s_1 \vdash s_2$. This is a special case of a theorem proved in [1] for general tensor-based models. As an example, consider two intransitive sentences "$subj_1 \ verb_1$" and "$subj_2 \ verb_2$", for which we have $\mathcal{F}(\vec{subj_1}) \subseteq \mathcal{F}(\vec{subj_2})$ and $\mathcal{F}(\vec{verb_1}) \subseteq \mathcal{F}(\vec{verb_2})$; then, it is the case that:

$$\mathcal{F}(\vec{subj_1}) \cap \mathcal{F}(\vec{verb_1}) \subseteq \mathcal{F}(\vec{subj_2}) \quad \text{and} \quad \mathcal{F}(\vec{subj_1}) \cap \mathcal{F}(\vec{verb_1}) \subseteq \mathcal{F}(\vec{verb_2})$$

and consequently:

$$\mathcal{F}(\vec{subj_1}) \cap \mathcal{F}(\vec{verb_1}) \subseteq \mathcal{F}(\vec{subj_2}) \cap \mathcal{F}(\vec{verb_2})$$

A similar reasoning holds for the union-based case, since we have:

$$\mathcal{F}(\vec{subj_1}) \subseteq \mathcal{F}(\vec{subj_2}) \cup \mathcal{F}(\vec{verb_2}) \quad \text{and} \quad \mathcal{F}(\vec{verb_1}) \subseteq \mathcal{F}(\vec{subj_2}) \cup \mathcal{F}(\vec{verb_2})$$

thus $\mathcal{F}(\vec{subj_1}) \cup \mathcal{F}(\vec{verb_1}) \subseteq \mathcal{F}(\vec{subj_2}) \cup \mathcal{F}(\vec{verb_2})$. For the case of intersective composition, the above makes clear another DIH property that holds in contexts without logical words; that a phrase can be replaced with each one of its words, i.e. *red car* can be replaced with *car* and with *red*. Note, however, that in this case the same is not true for union-based composition, since the inclusion order becomes reversed, which is clearly unwanted.

4.2   Holistic Phrase/Sentence Vectors

In the ideal (but not so feasible) presence of a text corpus sufficiently large to provide co-occurrence statistics for phrases or even sentences, one could directly create vectors for larger text segments using the same methods as if they were words. This idea has been investigated in the context of entailment by [2], who present promising results for short adjective-noun compounds. *Holistic* vectors of this sort are interesting since they can be seen as representing (at least for short text segments) some form of idealistic distributional behaviour for text segments above the word level. For this reason, we briefly examine the relationship of these models with the compositional models of Sect. 4.1, with regard to their feature inclusion properties.

We consider the case of intersective composition. For a two-word phrase $w_1 w_2$ with a holistic vector $\overrightarrow{w_1 w_2}$, we start by noticing that $\mathcal{F}(\overrightarrow{w_1 w_2})$ is always a subset of $\mathcal{F}(\overrightarrow{w_1}) \cap \mathcal{F}(\overrightarrow{w_2})$ and specifically the subset referring to cases where w_1 and w_2 occur *together* in the same context, that is:

$$\mathcal{F}(\overrightarrow{w_1 w_2}) = [\mathcal{F}(\overrightarrow{w_1}) \cap \mathcal{F}(\overrightarrow{w_2})]_{|w_1, w_2} \subseteq \mathcal{F}(\overrightarrow{w_1}) \cap \mathcal{F}(\overrightarrow{w_2})$$

with the set equality to hold only when w_1 and w_2 occur exclusively in the same contexts, i.e. the presence of w_1 always signifies that w_2 is around and vice versa. The relationship between holistic vectors and intersective composition can be leveraged to the phrase/sentence level. Recall the intransitive sentence example of Sect. 4.1; denoting the holistic vectors of the two sentences as $\overrightarrow{s_1 v_1}$ and $\overrightarrow{s_2 v_2}$, it is the case that:

$$\mathcal{F}(\overrightarrow{s_1 v_1}) \subseteq \mathcal{F}(\overrightarrow{s_2 v_2}) \subseteq \mathcal{F}(\overrightarrow{s_2}) \cap \mathcal{F}(\overrightarrow{v_2})$$

In other words, intersective composition preserves any entailment relation that holds at the holistic vector level, providing a faithful approximation of the holistic distributional behaviour. Note that for the case of union-based composition this approximation will be much more relaxed, and thus less useful in practice.

4.3   Tensor-Based Models

For tensor-based models, one needs a different analysis. These models lie somewhere between intersective and union-based models. Consider the simple case of a matrix multiplication between a $m \times n$ matrix $\mathcal{M}$ and a $n \times 1$ vector $\overrightarrow{v}$, given below:

$$\begin{pmatrix} w_{11} & \cdots & w_{1n} \\ w_{21} & \cdots & w_{2n} \\ \vdots & & \vdots \\ w_{m1} & \cdots & w_{mn} \end{pmatrix} \times \begin{pmatrix} v_1 \\ \vdots \\ v_n \end{pmatrix}$$

The matrix $\mathcal{M}$ can be seen as a list of column vectors $(\overrightarrow{w_1}, \overrightarrow{w_2}, \cdots, \overrightarrow{w_n})$, where $\overrightarrow{w_i} = (w_{1i}, \cdots, w_{mi})^{\mathrm{T}}$. Then the result of the matrix multiplication becomes a combination of scalar multiplications of element v_i of the vector $\overrightarrow{v}$ with its corresponding vectors $\overrightarrow{w_i}$ of the matrix $\mathcal{M}$, as follows:

$$v_1 \overrightarrow{w_1} + v_2 \overrightarrow{w_2} + \cdots + v_n \overrightarrow{w_n}$$

By looking at matrix multiplication $\mathcal{M} \times \overrightarrow{v}$ in this way, we are able to describe the features of $\mathcal{F}(\mathcal{M} \times \overrightarrow{v})$ in terms of the features of $\overrightarrow{v}$ and the features of the $\overrightarrow{w_i}$'s of $\mathcal{M}$. This is as follows:

$$\mathcal{F}(\overline{w} \times \overrightarrow{v}) = \bigcup_{v_i \neq 0} \mathcal{F}(\overrightarrow{w_i}) \tag{15}$$

Generalizing slightly and calling v_i a feature whenever it is non-zero, the above can be written down in the following equivalent form:

$$\bigcup_i \mathcal{F}(\overrightarrow{w_i}) \mid_{\mathcal{F}(v_i)} \tag{16}$$

which means we collect features of each $\overrightarrow{w_i}$ vector but only up to "featureness" of v_i, that is up to v_i being non-zero.

The above procedure can be extended to tensors of higher order; a tensor of order 3, for example, can be seen as a list of matrices, a tensor of order 4 as a list of "cubes" and so on. For the case of this paper, we will not go beyond matrix multiplication and cube contraction. The concrete constructions of these matrices and cubes, presented in the next section, will make the above analysis more clear.

Concrete Tensor-Based Constructions. While the feature inclusion properties of a tensor-based model follow the generic analysis above, their exact form depends on the concrete constructions of their underlying tensors. In this section, we go over a few different methods of tensor construction and derive their feature inclusion properties.

We start by the construction presented in [10], which builds a tensor from the properties of the vectors of its arguments. For example, an intransitive verb gets assigned the vector $\sum_i \overrightarrow{Sbj_i}$, a verb phrase the vector $\sum_i \overrightarrow{Obj_i}$, and a transitive verb the matrix $\sum_i \overrightarrow{Sbj_i} \otimes \overrightarrow{Obj_i}$. Here, Sbj_i/Obj_i are the subjects/objects of the verb across the corpus. The features of the phrases vo and sentences sv, svo (where s/o are the subject/object of the phrase/sentence) are as follows:

$$\mathcal{F}(\overrightarrow{sv}) = \bigcup_i \mathcal{F}(\overrightarrow{Sbj_i}) \cap \mathcal{F}(\overrightarrow{s}) \qquad \mathcal{F}(\overrightarrow{vo}) = \bigcup_i \mathcal{F}(\overrightarrow{Obj_i}) \cap \mathcal{F}(\overrightarrow{o})$$

$$\mathcal{F}(\overrightarrow{svo}) = \bigcup_i \mathcal{F}(\overrightarrow{Sbj_i} \otimes \overrightarrow{Obj_i}) \cap \mathcal{F}(\overrightarrow{s}) \otimes \mathcal{F}(\overrightarrow{o})$$

The disadvantage of this model and a number of other models based on this methodology, e.g. [13,18], is that their resulting representations of verbs have one dimension less than what their types dictate. According to the type assignments, an intransitive verb has to be a matrix and a transitive verb a cube, where as in the above we have a vector and a matrix. We remedy this problem by arguing that the sentence/phrase space should be spanned by the vectors of the arguments of the verb across the corpus. In order to achieve this, we create verb matrices for intransitive sentences and verb phrases by taking the outer product of the argument vectors with themselves, hence obtaining:

$$\overline{v}_{itv} := \sum_i \overrightarrow{Sbj_i} \otimes \overrightarrow{Sbj_i} \qquad \overline{v}_{vp} := \sum_i \overrightarrow{Obj_i} \otimes \overrightarrow{Obj_i} \tag{17}$$

When these verbs are composed with some subject/object to form a phrase/sentence, each vector in the spanning space is weighted by its similarity (assuming normalized vectors) with the vector of that subject/object, that is:

$$\overrightarrow{sv} = \overrightarrow{s} \times \overline{v}_{itv} = \sum_i \langle \overrightarrow{Sbj_i} | \overrightarrow{s} \rangle \overrightarrow{Sbj_i} \tag{18}$$

$$\overrightarrow{vo} = \overline{v}_{vp} \times \overrightarrow{o} = \sum_i \langle \overrightarrow{Obj_i} | \overrightarrow{o} \rangle \overrightarrow{Obj_i} \tag{19}$$

We call this model *projective*. For the case of a transitive verb (a function of two arguments), we define the sentence space to be spanned by the average of the argument vectors, obtaining:

$$\overline{v}_{trv} := \sum_i \overrightarrow{Sbj_i} \otimes \left(\frac{\overrightarrow{Sbj_i} + \overrightarrow{Obj_i}}{2} \right) \otimes \overrightarrow{Obj_i} \tag{20}$$

$$\overrightarrow{svo} = \sum_i \langle \overrightarrow{s} | \overrightarrow{Sbj_i} \rangle \left(\frac{\overrightarrow{Sbj_i} + \overrightarrow{Obj_i}}{2} \right) \langle \overrightarrow{Obj_i} | \overrightarrow{o} \rangle$$

Feature-wise, the above translate to the following:

$$\mathcal{F}(\overrightarrow{sv}) = \bigcup_i \mathcal{F}(\overrightarrow{Sbj_i}) \mid_{\mathcal{F}(\langle \overrightarrow{Sbj_i} | \overrightarrow{o} \rangle)} \qquad \mathcal{F}(\overrightarrow{vo}) = \bigcup_i \mathcal{F}(\overrightarrow{Obj_i}) \mid_{\mathcal{F}(\langle \overrightarrow{Obj_i} | \overrightarrow{o} \rangle)}$$

$$\mathcal{F}(\overrightarrow{svo}) = \bigcup_i \left(\mathcal{F}(\overrightarrow{Sbj_i}) \cup \mathcal{F}(\overrightarrow{Obj_i}) \right) \mid_{\mathcal{F}(\langle \overrightarrow{s} | \overrightarrow{Sbj_i} \rangle) \mathcal{F}(\langle \overrightarrow{Obj_i} | \overrightarrow{o} \rangle)}$$

Informally, we can think of the terms following the | symbol as defining a restriction on feature inclusion based on how well the arguments of the phrase/sentence fit to the arguments of the verb. We close this section by noting that in Sect. 6.2 we briefly present a statistical approach for creating the verb matrices based on holistic phrase vectors, along the lines of [3].

5   Measuring the CDIH

When computing entailment at the lexical level, *balAPinc* (Eq. 7) has been found to be one of the most successful measures [14]. However, the transition from words to phrases or sentences introduces extra complications, which we need to take into account. Firstly, in a compositional distributional model, the practice of considering only non-zero elements of the vectors as features becomes too restrictive and thus suboptimal for evaluating entailment; indeed, depending on the form of the vector space and the applied compositional operator (especially in intersective models), an element can get very low values without however

ever reaching zero. This blurring of the notion of "featureness"— in which zero can be seen as a lower bound in a range of possible values—is in line with the quantitative nature of these models. In this paper we exploit this to the limit by letting $\mathcal{F}(\overrightarrow{w})$ to include all the dimensions of $\overrightarrow{w}$.

Secondly, we further exploit the continuous nature of distributional models by providing a stronger realization of the idea that $u \vdash v$ whenever v occurs in all the contexts of u. Let $f_r^{(u)}$ be a feature in $\mathcal{F}(\overrightarrow{u})$ with rank r and $f_r^{(v)}$ the corresponding feature in $\mathcal{F}(\overrightarrow{v})$, we remind that Kotlerman et al. consider that feature inclusion holds at rank r whenever $f_r^{(u)} > 0$ and $f_r^{(v)} > 0$; we strengthen this assumption by requiring that $f_r^{(u)} \leq f_r^{(v)}$. Incorporating these modifications in the $APinc$ measure, we redefine $P(r)$ and $rel'(f_r)$ in Eq. 8 as:

$$P(r) = \frac{|\{f_r^{(u)}|f_r^{(u)} \leq f_r^{(v)}, 0 < r \leq |\overrightarrow{u}|\}|}{r} \tag{21}$$

$$rel'(f_r) = \begin{cases} 1 & f_r^{(u)} \leq f_r^{(v)} \\ 0 & o.w. \end{cases} \tag{22}$$

Note that the new relevance function essentially subsumes the old one (Eq. 9), since by definition high-valued features in $\mathcal{F}(\overrightarrow{u})$ must be even higher in $\mathcal{F}(\overrightarrow{v})$. We now re-define $APinc$ at the phrase/sentence level to be the following:

$$SAPinc(u,v) = \frac{\sum_r \left[P(r) \cdot rel'(f_r) \right]}{|\overrightarrow{u}|} \tag{23}$$

where $P(r)$ and $rel'(f_r)$ are as defined in Eqs. 21 and 22, respectively, and $|\overrightarrow{u}|$ is the number of dimensions of $\overrightarrow{u}$. We further notice that when using $SAPinc$, a zero vector vacuously entails every other vector in the vector space, and it is entailed only by itself, as is the case for logical entailment.

We now proceed to examine the balanced $APinc$ version, to which Kotlerman et al. refer as $balAPinc$ (Eq. 7). This is the geometric average of an asymmetric measure ($APinc$) with a symmetric one (Lin's similarity). The rationale of including a symmetric measure in the computation was that $APinc$ tends to return unjustifyingly high scores when the entailing word is infrequent, that is, when the feature vector of the entailing word is very short; the purpose of the symmetric measure was to penalize the result, since in this case the similarity of the narrower term with the broader one is usually low. However, now that all feature vectors have the same length, such a balancing action is unnecessary; even more importantly, it introduces a strong element of symmetry in a measure that is intended to be strongly asymmetric. To cope with these issues, we propose to replace Lin's similarity with representativeness on KL-divergence (Eq. 6), and define a sentence-level version of $balAPinc$ between two word vectors $\overrightarrow{u}$ and $\overrightarrow{v}$ as follows:

$$SBalAPinc(u,v) = \sqrt{R_\mathrm{D}(\overrightarrow{u} \| \overrightarrow{v}) \cdot SAPinc(\overrightarrow{u}, \overrightarrow{v})} \tag{24}$$

Recall that $R_\mathrm{D}(p\|q)$ is asymmetric, measuring the extent to which q represents (i.e. is similar to) p. So the term $R_\mathrm{D}(\overrightarrow{u} \| \overrightarrow{v})$ in the above formula measures how well the *broader* term v represents the narrower one u; as an example, we can think that the term 'animal' is representative of 'cat', while the reverse is not true. The new measure aims at: (i) retaining a strongly asymmetric nature; and (ii) providing a more fine-grained element of evaluating entailment.

6   Experimental Setting

We evaluate the compositional models and the entailment measures presented above in four different tasks. Specifically, we measure upward-monotone entailment between (a) intransitive sentences; (b) verb phrases; (c) transitive sentences; and (d) adjective-noun compounds and nouns. The first three evaluations are based on datasets specifically created by us for the purposes of this paper, while for the adjective-noun task we use the dataset of [2]. In all cases, we first apply a compositional model to the phrases/sentences of each pair in order to create vectors representing their meaning, and then we evaluate the entailment relation between the phrases/sentences by using these composite vectors as input to a number of entailment measures. The goal is to see which combination of compositional model/entailment measure is capable of better recognizing strictly directional entailment relationships between phrases and sentences.

In all the experiments, we used a 300-dimensional PPMI vector space trained on the concatenation of UKWAC and Wikipedia corpora. The context was defined as a 5-word window around the target word.

6.1   Datasets

In this section we briefly describe the process we followed in order to create datasets for deciding entailment between subject-verb, verb-object, and subject-verb-object phrases and sentences. Our goal was to produce pairs of phrases/sentences that stand in an upward-monotone entailment relationship to each other. When entailing and entailed phrases have exactly the same structure, as is in our case, one way to achieve that is to ensure that every word in the entailed phrase is a hypernym of the corresponding word in the entailing phrase. We achieved this by using hyponym-hypernym relationships taken by WordNet as follows.

Firstly, we extracted from the concatenation of UKWAC and Wikipedia corpora all verbs occurring at most 2.5 million times and at least 5000 times. Then, each verb was paired with a hypernym of its main synset, creating a list of 4800 pairs of verbs that stand in a hyponym-hypernym relation. Each verb was associated with a list of argument nouns; for the intransitive task this list contained nouns occurring in the corpus as subjects of the verbs, for the verb phrase nouns in an object relationship, and for the transitive task subject/object pairs. Starting from the most frequent cases, each argument of an entailing verb was paired with an argument of the corresponding entailed verb based a number of constraints (for example, each noun could occur at most 3 times as part of an entailing phrase, and a specific phrase can only occur once as entailing phrase).[1] We went through the phrase/sentence pairs manually and discarded any instance where we judged to be nonsensical. This process resulted in 135 subject-verb pairs, 218 verb-object pairs, and 70 subject-verb-object pairs, the phrases/sentences of which stand in a fairly clear entailment relationship. Each dataset was extended with the reverse direction of the entailments as negative examples, creating three strictly directional entailment datasets of 270 (subject-verb), 436 (verb-object) and 140 (subject-verb-object) entries. Table 1 presents a sample of positive entailments from each dataset.[2]

6.2   Compositional Models

We tested the additive and multiplicative compositional operators, as defined in Eq. 1, a point-wise minimum model as discussed in Sect. 4.1, and a variation on the tensor-based model introduced via Eqs. 17–20. In relation to this latter model, informal experimentation showed that by taking into account directly the features of the distributional vector of the verb, the results improve. Let the distributional vector of the verb be $\overrightarrow{v}$ and the verb tensor be $\overline{v}_x$, as computed in Eqs. 17–20, for $x \in \{itv, vp, trv\}$. Then a new tensor is computed via the formula $\widetilde{v}_x := \overrightarrow{v} \odot \overline{v}_x$, the feature inclusion behaviour of which is derivable as follows:

$$\mathcal{F}(\widetilde{v}_x) = \mathcal{F}(\overrightarrow{v}) \cap \mathcal{F}(\overline{v}_x)$$

For the experiments on the intransitive and the verb-phrase datasets, we also use a least-squares fitting model for approximating the distributional behaviour of holistic vectors (see discussion in Sect. 4.2), along the lines of [3]. For each verb, we compute analytically an estimator for predicting the ith element of the resulting vector as follows:

$$\overrightarrow{w_i} = (\mathbf{X}^{\mathrm{T}}\mathbf{X})^{-1}\mathbf{X}^{\mathrm{T}}\overrightarrow{y_i}$$

Here, the rows of matrix $\mathbf{X}$ are the vectors of the subjects (or objects) that occur with our verb, and $\overrightarrow{y_i}$ is a vector containing the ith elements of the holistic phrase vectors across all training instances; the resulting $\overrightarrow{w_i}$'s form the rows of our verb matrix. Finally, a non-compositional baseline, where the phrase is represented by the vector (or tensor) of its head verb, is also evaluated where appropriate.

6.3   Measures and Evaluation

We present results for a variety of entailment measures, including *SAPinc* and *SBalAPinc* as introduced in Sect. 5. KL-divergence is applied on smoothed vectors, as suggested by [4]. For α-skew, we use $\alpha = 0.99$ which in the past has showed the best reporting results [14]. *WeedsPrec* refers to the precision measure introduced by [23], while *ClarkeDE* denotes the degree of entailment measure of [6].

---

[1] These constraints were much more relaxed for the transitive task, because of data sparsity problems.

[2] The datasets will become available at http://compling.eecs.qmul.ac.uk/resources/.

Table 1. Positive entailments from the three tasks at phrase and sentence level.

Subject-verb	Verb-object
Evidence suggest ⊢ information express	Develop skill ⊢ create ability
People believe ⊢ group think	Solve problem ⊢ understand difficulty
Paper present ⊢ material show	Sign contract ⊢ write agreement
Station serve ⊢ facility meet	Reduce number ⊢ decrease amount
Survey reveal ⊢ work show	Publish book ⊢ produce publication
Student develop ⊢ person create	Sing song ⊢ perform music
Company operate ⊢ organization manage	Rejoin army ⊢ join force
Player play ⊢ contestant compete	Gain experience ⊢ obtain education
Study demonstrate ⊢ examination show	Serve purpose ⊢ meet goal
News come ⊢ message travel	Identify area ⊢ determine location
Summer finish ⊢ season end	Promote development ⊢ support event
Report note ⊢ document state	Suffer injury ⊢ experience condition
Book offer ⊢ product supply	Undertake research ⊢ initiate investigation
Tree mature ⊢ plant grow	Drive car ⊢ handle vehicle
Subject-verb-object	
Report describe result ⊢ document explain process	
Report outline progress ⊢ document describe change	
Value suit budget ⊢ number meet standard	
Book present account ⊢ work show evidence	
Woman marry man ⊢ female join male	
Author retain house ⊢ person hold property	
Report highlight lack ⊢ document stress need	
Public trust reference ⊢ people accept message	
Study demonstrate importance ⊢ work show value	
Police fight crime ⊢ force compete activity	
Experiment test hypothesis ⊢ research evaluate proposal	
University publish paper ⊢ body produce research	
Brochure outline feature ⊢ booklet explain concept	
Widow sell estate ⊢ woman exchange property	

We also use strict feature inclusion as a baseline; in this case, entailment holds only when $\mathcal{F}(\overrightarrow{phrase_1}) \subseteq \mathcal{F}(\overrightarrow{phrase_2})$. After composition, all phrase/sentence vectors are normalized to unit length.

Regarding evaluation, since the tasks follow a binary classification objective and our models return a continuous value, we report area under curve (AUC). This reflects the generic discriminating power of a binary classifier by evaluating the task at every possible threshold.

7 Results

7.1 Phrase and Sentence Entailment

Table 2 presents the results for the phrase and sentence entailment experiments. As the numbers show, in all three tasks the highest performance is delivered by a combination of *SBalAPinc* or *SAPinc* with element-wise vector multiplication. Furthermore, it is interesting to note that *SBalAPinc* clearly outperforms *balAPinc* in every compositional model and every task. The ability of the proposed measure to better discriminate between positive and negative entailments is further demonstrated in Fig. 1, where we examine the distributions of the two classes when using *balAPinc* (left) and *SBalAPinc* (right) in conjunction with multiplicative composition for the verb-object task.

Table 2. AUC scores for the three phrase and sentence entailment tasks. *Verb* is a non-compositional baseline based on comparing only the verb vectors of the two phrases, ⊙ is element-wise vector multiplication, + vector addition, ⊗ tensor-based composition, and *LstSqr* a least-square fitting model approximating the holistic distributional behaviour of the phrases.

	Subject-verb						Verb-object						Subject-verb-object					
Measure	Verb	⊙	MIN	+	⊗	LstSqr	Verb	⊙	MIN	+	⊗	LstSqr	Verb	⊙	MIN	+	⊗	LstSqr
Inclusion	0.59	0.54	0.54	0.63	0.59	0.50	0.58	0.52	0.52	0.64	0.58	0.50	0.61	0.55	0.55	0.58	0.64	–
KL-div	0.59	0.66	0.68	0.57	0.59	0.59	0.62	0.64	0.66	0.61	0.60	0.58	0.61	0.65	0.71	0.54	0.60	–
αSkew	0.63	0.75	0.72	0.74	0.65	0.62	0.65	0.74	0.70	0.75	0.66	0.57	0.66	0.74	0.74	0.71	0.70	–
WeedsPrec	0.67	0.75	0.75	0.65	0.67	0.59	0.67	0.70	0.71	0.68	0.67	0.56	0.69	0.79	0.78	0.59	0.69	–
ClarkeDE	0.57	0.66	0.63	0.62	0.59	0.56	0.58	0.67	0.63	0.63	0.60	0.53	0.58	0.67	0.63	0.60	0.61	–
APinc	0.69	0.78	0.78	0.72	0.70	0.60	0.69	0.75	0.75	0.74	0.70	0.56	0.74	0.76	0.77	0.65	0.74	–
balAPinc	0.65	0.72	0.71	0.70	0.67	0.58	0.66	0.70	0.69	0.71	0.67	0.55	0.67	0.71	0.71	0.64	0.70	–
SAPinc	0.65	**0.81**	0.74	0.72	0.71	0.63	0.62	**0.82**	0.74	0.72	0.68	0.58	0.59	**0.80**	0.73	0.67	0.75	–
SBalAPinc	0.65	**0.81**	0.75	0.72	0.69	0.64	0.66	**0.79**	0.74	0.73	0.68	0.59	0.63	**0.80**	0.76	0.67	0.76	–

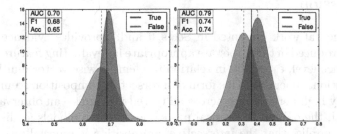

Fig. 1. The distributions of positive and negative entailments when using *balAPinc* (left) and *SBalAPinc* (right) in combination with multiplicative composition on the verb-object task. The dashed red lines indicate the means, while the thick black lines correspond to the thresholds that optimize *informedness*—equivalent to AUC subtended by the highest operating point [20].

7.2   Adjective-Noun Compounds

In this last experiment, we reproduce the $AN \vdash N$ task of [2], the goal of which is to assess the extent to which an adjective-noun compound (such as 'red car') entails the noun of the compound ('car'). The dataset contains 2450 pairs of $AN \vdash N$ entailments, half of which are negative examples that have been created by random permutation of the nouns at the right-hand side. We use this task as a proof of concept for the theory detailed in Sect. 4, since when using element-wise composition this sort of entailment always holds. The results, presented in Table 3, confirm the above in the most definite way. *SBalAPinc* achieves almost perfect classification when combined with multiplicative composition, while *SAPinc* shows top performance for union-based composition.

Table 3. AUC scores for the $AN \vdash N$ task.

Measure	$\odot$	MIN	$+$
Inclusion	1.00	1.00	0.50
KL-divergence	1.00	1.00	0.87
αSkew	0.96	0.97	1.00
WeedsPrec	1.00	1.00	0.85
ClarkeDE	1.00	1.00	0.95
APinc	0.94	0.94	0.84
balAPinc	0.99	0.99	0.84
SAPinc	0.91	0.12	0.97
SBalAPinc	0.99	0.74	0.93

8   Discussion

The experimental work presented in Sects. 6 and 7 provides evidence that the measures introduced in this paper are appropriate for evaluating feature inclusion at the sentence level, especially in relation to element-wise vector multiplication as a compositional operator. This form of intersective composition seems to show a consistently high performance across all tested measures—an observation that is in line with the desired theoretical properties of these models as discussed in Sect. 4. This implies that the intersective composition is especially suitable for sentence entailment evaluation based on the CDIH. The reason may be the feature filtering methods applied by these models. The intersective filtering avoids generation of very dense vectors and thus facilitates entailment judgements based on the CDIH. On the other hand, union-based compositional models, such as vector addition, produce dense vectors for even very short sentences (Fig. 2). In this case, entailment is better handled by information theoretic measures, and

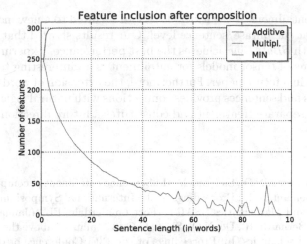

Fig. 2. Feature inclusion on the first million sentences of Wikipedia for three vector-based compositional models (using vectors of 300 dimensions). For sentence lengths greater than 5 words, additive composition produces dense vectors with all elements greater than zero. The feature inclusion behaviour of the two intersective models (vector multiplication and MIN) is identical, showing a polynomial decrease on the number of features for longer sentences.

specifically the α-skew measure (Table 2), without however reaching the performance of intersective models and feature inclusion.

The tensor-based model presented in Sect. 4.3 can be seen as a combination of a union-based model (between the features of the arguments of the verb) and an intersective model (between the features of the distributional vector of the verb and the features of the vector of the verb phrase). While this idea does not seem to work very well in practice—as it returns results lower than those of the vector-based counterparts—the model outperforms the other full tensor model, that is the least-square fitting model. One reason is that tensor-based constructions similar to the ones in Eqs. 17–20 are more robust against data sparsity problems than statistical models based on holistic vectors of phrases and sentences.

In general, while intersective element-wise vector composition seems to be more aligned with a CDIH, tensor-based models, similar to the one presented in Sect. 4.3, provide an abundance of conceptual options, depending on how one creates the verb tensors. At the same time, the tensor-based models preserve the grammatical structure. Hence they can serve as an interesting test-bed for reasoning on entailment relations at the phrase or sentence level.

9   Conclusion and Future Work

In this paper we investigated the application of the distributional inclusion hypothesis on evaluating entailment between phrase and sentence vectors produced by compositional operators. We showed how the popular *balAPinc* measure for evaluating entailment at the lexical level can be lifted to a new measure *SBal-APinc* for use at the phrase/sentence level. Our results showed that intersective composition with *SBalAPinc* achieves the best performance. Experimenting with different versions of tensor models for entailment is an interesting topic that we plan to address in a future paper. Furthermore, the extension of word-level entailment to phrases and sentences provides connections with natural logic [17], a topic that is worth a separate treatment and constitutes a future direction.

References

1. Balkır, E., Kartsaklis, D., Sadrzadeh, M.: Sentence entailment in compositional distributional semantics. In: Proceedings of the International Symposium on Artificial Intelligence and Mathematics (ISAIM). Fort Lauderdale, FL., (January 2016)
2. Baroni, M., Bernardi, R., Do, N.Q., Shan, C.C.: Entailment above the word level in distributional semantics. In: Proceedings of the 13th Conference of the European Chapter of the Association for Computational Linguistics, pp. 23–32. Association for Computational Linguistics, Avignon, France, April 2012. http://www.aclweb.org/anthology/E12-1004
3. Baroni, M., Zamparelli, R.: Nouns are vectors, adjectives are matrices: representing adjective-noun constructions in semantic space. In: Proceedings of the 2010 Conference on Empirical Methods in Natural Language Processing, pp. 1183–1193. Association for Computational Linguistics, Cambridge, MA, October 2010. http://www.aclweb.org/anthology/D10-1115
4. Chen, S.F., Goodman, J.: An empirical study of smoothing techniques for language modeling. In: Proceedings of the 34th Annual Meeting on Association for Computational Linguistics, ACL 1996, pp. 310–318. Association for Computational Linguistics, Stroudsburg, PA, USA (1996). http://dx.doi.org/10.3115/981863.981904
5. Cheng, J., Kartsaklis, D.: Syntax-aware multi-sense word embeddings for deep compositional models of meaning. In: Proceedings of the 2015 Conference on Empirical Methods in Natural Language Processing, pp. 1531–1542. Association for Computational Linguistics, Lisbon, Portugal, September 2015. http://aclweb.org/anthology/D15-1177
6. Clarke, D.: Context-theoretic semantics for natural language: an overview. In: Proceedings of the Workshop on Geometrical Models of Natural Language Semantics, pp. 112–119. Association for Computational Linguistics, Athens, Greece, March 2009. http://www.aclweb.org/anthology/W09-0215
7. Coecke, B., Sadrzadeh, M., Clark, S.: Mathematical foundations for a compositional distributional model of meaning. Linguist. Anal. **36**, 345–384 (2010)
8. Dagan, I., Lee, L., Pereira, F.C.N.: Similarity-based models of word cooccurrence probabilities. Mach. Learn. **34**(1–3), 43–69 (1999). doi:10.1023/A:1007537716579
9. Geffet, M., Dagan, I.: The distributional inclusion hypotheses and lexical entailment. In: Proceedings of the 43rd Annual Meeting of the Association for Computational Linguistics (ACL 2005), pp. 107–114. Association for Computational Linguistics, Ann Arbor, Michigan, June 2005. http://www.aclweb.org/anthology/P05-1014
10. Grefenstette, E., Sadrzadeh, M.: Experimental support for a categorical compositional distributional model of meaning. In: Proceedings of the Conference on Empirical Methods in Natural Language Processing, pp. 1394–1404. Association for Computational Linguistics (2011)

11. Herbelot, A., Ganesalingam, M.: Measuring semantic content in distributional vectors. In: Proceedings of the 51st Annual Meeting of the Association for Computational Linguistics, vol. 2, pp. 440–445. Association for Computational Linguistics (2013)
12. Kalchbrenner, N., Grefenstette, E., Blunsom, P.: A convolutional neural network for modelling sentences. In: Proceedings of the 52nd Annual Meeting of the Association for Computational Linguistics, vol. 1, pp. 655–665. Association for Computational Linguistics, Baltimore, Maryland, June 2014. http://www.aclweb.org/anthology/P14-1062
13. Kartsaklis, D., Sadrzadeh, M., Pulman, S.: A unified sentence space for categorical distributional-compositional semantics: theory and experiments. In: COLING 2012, 24th International Conference on Computational Linguistics, Proceedings of the Conference: Posters, 8–15 December 2012, Mumbai, India, pp. 549–558 (2012)
14. Kotlerman, L., Dagan, I., Szpektor, I., Zhitomirsky-Geffet, M.: Directional distributional similarity for lexical inference. Nat. Lang. Eng. **16**(04), 359–389 (2010)
15. Lee, L.: Measures of distributional similarity. In: Proceedings of the 37th Annual Meeting of the Association for Computational Linguistics on Computational Linguistics, pp. 25–32 (1999)
16. Lin, D.: An information-theoretic definition of similarity. In: Proceedings of the International Conference on Machine Learning, pp. 296–304 (1998)
17. MacCartney, B., Manning, C.D.: Natural logic for textual inference. In: ACL Workshop on Textual Entailment and Paraphrasing. Association for Computational Linguistics (2007)
18. Milajevs, D., Kartsaklis, D., Sadrzadeh, M., Purver, M.: Evaluating neural word representations in tensor-based compositional settings. In: Proceedings of the 2014 Conference on Empirical Methods in Natural Language Processing (EMNLP), pp. 708–719. Association for Computational Linguistics, Doha, Qatar, October 2014. http://www.aclweb.org/anthology/D14-1079
19. Mitchell, J., Lapata, M.: Composition in distributional models of semantics. Cogn. Sci. **34**(8), 1388–1439 (2010)
20. Powers, D.M.: Evaluation: from precision, recall and f-measure to ROC, informedness, markedness and correlation. J. Mach. Learn. Technol. **2**(1), 37–63 (2011)
21. Socher, R., Huval, B., Manning, C.A.N.: Semantic compositionality through recursive matrix-vector spaces. In: 2012 Conference on Empirical Methods in Natural Language Processing (2012)
22. Turney, P.D., Pantel, P.: From frequency to meaning: vector space models of semantics. J. Artif. Intell. Res. **37**(1), 141–188 (2010)
23. Weeds, J., Weir, D., McCarthy, D.: Characterising measures of lexical distributional similarity. In: Proceedings of the 20th International Conference on Computational Linguistics, no. 1015. Association for Computational Linguistics (2004)

Strong and Weak Quantifiers in Focused NL$_{\mathrm{CL}}$

Wen Kokke$^{(\boxtimes)}$

Institute for Logic, Language and Computation, University of Amsterdam,
Amsterdam, Netherlands
wen.kokke@gmail.com

Abstract. We propose an improvement of Barker and Shan's [4] NL$_{\mathrm{CL}}$
for which derivability is decidable, which has a normal-form for proof
search, can analyse scope islands, and distinguish between strong and
weak quantifiers.

Keywords: Categorial grammar · Focusing · Scope Islands ·
Indefinite scope

1 Introduction

In 2014, Kiselyov and Shan [11] published a paper in which they presented an
elegant approach to the anaysis of various scope-related phenomena using, what
they call, the continuation hierarchy. The phenomena they cover are scope ambiguity, scope islands and strong and weak quantifiers. They cover these phenomena using a mechanism which works on the sentence's *semantics*, independent
of whatever form of grammar is used.

At around the same time, Barker and Shan [4] published a book containing
their findings on NL$_\lambda$ and NL$_{\mathrm{CL}}$, a pair of grammar logics, both with the ability
to analyse scope ambiguity using a strictly *syntactic* mechanism. In addition,
these logics can analyse "parasitic scope" [3,4] and a quantifier which change the
result type of the expressions they take scope over. However, neither of these
logics is capable of analysing scope islands or strong and weak quantifiers.

In this paper, we rework NL$_{\mathrm{CL}}$ to a calculus which can analyse both scope
islands and strong and weak quantifiers, without losing the ability to analyse
parasitic scope or changing result types. For this, we base ourselves on work
by Moortgat [13] and Moortgat and Moot [14]. This approach requires a strict
focusing regime. Therefore, as an added bonus, adopting it results in the elimination of spurious ambiguity, and greatly enhances the efficiency of proof search
when compared to Barker and Shan's [4] NL$_{\mathrm{CL}}$.

We will start our discussion by giving several examples of each of the aforementioned phenomena. The following sentences are examples of scope ambiguity, scope islands and weak quantifiers, respectively. They are given together
with their expected semantics, and are based on examples by Szabolcsi [16, p.
608,622].

M. Amblard et al. (Eds.): LACL 2016, LNCS 10054, pp. 134–148, 2016.
https://doi.org/10.1007/978-3-662-53826-5_9

(1) "Someone read every book."
    a. $\exists x.\mathbf{person}(x) \wedge \forall y.\mathbf{book}(y) \supset \mathbf{read}(x, y)$
    b. $\forall y.\mathbf{book}(y) \supset \exists x.\mathbf{person}(x) \wedge \mathbf{read}(x, y)$

(2) "Someone said Kurt wrote every book."
    a. $\exists x.\mathbf{person}(x) \wedge \mathbf{say}(x, \forall y.\mathbf{book}(y) \supset \mathbf{wrote}(\mathbf{kurt}, y))$

(3) "Everyone said [Kurt dedicated a book to Mary]."
    a. $\forall x.\mathbf{person}(x) \supset \mathbf{say}(x, \exists y.\mathbf{book}(y) \wedge \mathbf{dedicate}(\mathbf{kurt}, \mathbf{mary}, y))$
    b. $\forall x.\mathbf{person}(x) \supset \exists y.\mathbf{book}(y) \wedge \mathbf{say}(x, \mathbf{dedicate}(\mathbf{kurt}, \mathbf{mary}, y))$
    c. $\exists y.\mathbf{book}(y) \wedge \forall x.\mathbf{person}(x) \supset \mathbf{say}(x, \mathbf{dedicate}(\mathbf{kurt}, \mathbf{mary}, y))$

The first of these examples is a canonical example of scope ambiguity. Example (2) demonstrates a scope island: there is no reading in which "every book" scopes out of the embedded clause, as this reading would imply that there was potentially a different speaker for each book—"Alex said Kurt wrote Slaughterhouse-Five", "Jules said Kurt wrote Cat's Cradle", "Sam said Kurt wrote..." Example (3) shows that indefinites *can* scope out of scope islands.

We add two more sentences, which are examples of a quantifier which changes the result type, and of parasitic scope, respectively. These examples based on those given by Barker and Shan's [4, p. 208] and Kiselyov [10].

(4) "John read a book [the author of which] feared the ocean."
    a. $\exists x.\mathbf{book}(x) \wedge \mathbf{fear}(\iota(\lambda y.\mathbf{of}(y, \mathbf{author}, x)), \iota(\mathbf{ocean})) \wedge \mathbf{read}(\mathbf{john}, x)$

(5) "Everyone feared the same ocean."
    a. $\exists z.\forall y.\mathbf{fear}(y, \iota(\lambda x.\mathbf{ocean}(x) \wedge x = z))$

These last two examples will play a less important role, as NL$_{\text{CL}}$ is already capable of analysing both. However, in order to demonstrate that we have not lost that capability, we will provide analyses of both near the end of this paper.

2 Background

In this section, we will briefly discuss NL$_{\text{CL}}$ and its sibling, NL$_\lambda$. NL$_{\text{CL}}$ is an extension to the non-associative Lambek calculus [12, NL;]. The history behind NL$_{\text{CL}}$ is somewhat intricate, but helpful to understanding, so we will briefly go over it. The initial idea comes from the practice of encoding quantifier movement as a tree transformation which introduces a binder [9]:

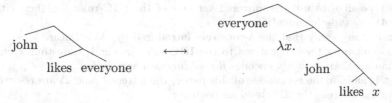

To implement this idea in type-logical grammar, Barker and Shan add a structural λ-construct to NL, and added the following structural postulate:[1]

$$\Sigma[\Gamma] \longleftrightarrow \Gamma \circ \lambda x.\Sigma[x] \qquad (\lambda)$$

As can be seen, the (λ) postulate uses a new connective: the $\circ$ (hollow product). This connective is part of a new residuated family $\{\backslash\!\backslash, \circ, /\!\!/\}$, which starts out as a copy of $\{\backslash, \bullet, /\}$. However, the addition of the (λ) postulate allows you to raise any constituent to the top-left[2] position in the structure, where—if it has the right type—it can be "resolved" against the top-level type as follows:

$$\cfrac{\cfrac{\cfrac{\cfrac{\Sigma[A] \vdash B}{A \circ \lambda x.\Sigma[x] \vdash B}\,(\lambda)}{\lambda x.\Sigma[x] \vdash A \backslash\!\backslash B}\,\mathrm{R}\backslash\!\backslash \qquad C \vdash D}{C /\!\!/ (A \backslash\!\backslash B) \circ \lambda x.\Sigma[x] \vdash D}\,\mathrm{L}/\!\!/}{\Sigma[C /\!\!/ (A \backslash\!\backslash B)] \vdash D}\,(\lambda)$$

Barker and Shan call resulting system NL_λ. While NL_λ fulfils the promise of allowing a syntactic analysis of quantifier raising, scope ambiguity and parasitic scope, it has some problems. Most notably, the system is hard to formalise and to reason about, largely due to the presence of a binding construct in the syntax of structures. While it is not impossible to formalise, the (λ) postulate greatly complicates meta-logical proofs.

To address this issue, and to ease their own investigation of the formal properties of NL_λ, Barker and Shan [4, ch. 17] introduce $\mathrm{NL}_{\mathrm{CL}}$. This system uses the fact that λ-terms can be represented as combinators in combinatory logic, which removes the need for a binding construct. Barker and Shan use a variant of Schönfinkel's mapping to encode the *linear* λ-construct as applications of the combinators **I**, **B** and **C**:[3],[4]

$$\mathbf{I}x = x, \qquad \mathbf{B}xyz = x(yz), \qquad \mathbf{C}xyz = xzy$$

The resulting system is presented in Fig. 1.[5]

Using the system in Fig. 1, we can do quantifier raising in much the same way as we did with the (λ) postulate—although, as we now have to raise the quantifier one step at a time, the proofs are much longer:

---

[1] It is important to note that this construct is *purely structural*, and that it is not accompanied by some implicit form of computation (e.g. β, η-conversions).

[2] It should be noted that the decision to raise quantifiers to the top-*left* position, as opposed to the top-*right*, is a stylistic choice made by Barker and Shan [4]. It is entirely possible to use the mirrored versions of the **IBC**-rules together with the $(B /\!\!/ A) \backslash\!\backslash C$ type for quantifiers.

[3] One can easily verify that the λ-construct introduced by (λ) is linear.

[4] When comparing these equations to the **IBC**-rules in Fig. 1, note that $\bullet$ encodes function application, but $\circ$ encodes *flipped* function application.

[5] In Fig. 1, and for the remainder of this paper, the letters Γ and Δ are reserved for structures, whereas the Σ is used for contexts.

$$\overline{A \vdash A} \; \text{Ax}$$

$$\frac{\Gamma \vdash A \quad \Sigma[B] \vdash C}{\Sigma[\Gamma \bullet A \setminus B] \vdash C} \setminus L \qquad \frac{A \bullet \Gamma \vdash B}{\Gamma \vdash A \setminus B} \setminus R \qquad \frac{\Gamma \vdash A \quad \Sigma[B] \vdash C}{\Sigma[B / A \bullet \Gamma] \vdash C} / L \qquad \frac{\Gamma \bullet A \vdash B}{\Gamma \vdash B / A} / R$$

$$\frac{\Gamma \vdash A \quad \Sigma[B] \vdash C}{\Sigma[\Gamma \circ A \setminus\!\!\setminus B] \vdash C} \setminus\!\!\setminus L \qquad \frac{A \circ \Gamma \vdash B}{\Gamma \vdash A \setminus\!\!\setminus B} \setminus\!\!\setminus R \qquad \frac{\Gamma \vdash A \quad \Sigma[B] \vdash C}{\Sigma[B /\!\!/ A \circ \Gamma] \vdash C} /\!\!/ L \qquad \frac{\Gamma \circ A \vdash B}{\Gamma \vdash B /\!\!/ A} /\!\!/ R$$

$$\frac{\Sigma[A] \vdash B}{\Sigma[A \circ \mathbf{I}] \vdash B} \mathbf{I} \qquad \frac{\Sigma[A \bullet (B \circ C)] \vdash D}{\Sigma[B \circ ((\mathbf{B} \bullet A) \bullet C)] \vdash D} \mathbf{B} \qquad \frac{\Sigma[(A \circ B) \bullet C] \vdash D}{\Sigma[A \circ ((\mathbf{C} \bullet B) \bullet C)] \vdash D} \mathbf{C}$$

Fig. 1. NL_{CL} as presented by Barker and Shan [4]. (When reading this figure, be wary of the difference between the combinators $\mathbf{B}, \mathbf{C}$ and the formulas B, C.)

$$\vdots$$

$$\frac{\dfrac{\dfrac{\dfrac{\dfrac{\dfrac{\dfrac{\dfrac{\dfrac{\text{JOHN} \bullet \text{LIKES} \bullet np \vdash s}{\text{JOHN} \bullet \text{LIKES} \bullet np \circ \mathbf{I} \vdash s} \mathbf{I}}{\text{JOHN} \bullet np \circ (\mathbf{B} \bullet \text{LIKES}) \bullet \mathbf{I} \vdash s} \mathbf{B}}{np \circ (\mathbf{B} \bullet \text{JOHN}) \bullet (\mathbf{B} \bullet \text{LIKES}) \bullet \mathbf{I} \vdash s} \mathbf{B}}{(\mathbf{B} \bullet \text{JOHN}) \bullet (\mathbf{B} \bullet \text{LIKES}) \bullet \mathbf{I} \vdash np \setminus\!\!\setminus s} \setminus\!\!\setminus R \quad s \vdash s}{\text{EVERYONE} \circ (\mathbf{B} \bullet \text{JOHN}) \bullet (\mathbf{B} \bullet \text{LIKES}) \bullet \mathbf{I} \vdash s} /\!\!/ L}{\text{JOHN} \bullet \text{EVERYONE} \circ (\mathbf{B} \bullet \text{LIKES}) \bullet \mathbf{I} \vdash s} \mathbf{B}}{\text{JOHN} \bullet \text{LIKES} \bullet \text{EVERYONE} \circ \mathbf{I} \vdash s} \mathbf{B}}{\text{JOHN} \bullet \text{LIKES} \bullet \text{EVERYONE} \vdash s} \mathbf{I}$$

The labels JOHN, LIKES and EVERYONE abbreviate the types np, $(np \setminus s) / np$ and $s /\!\!/ (np \setminus\!\!\setminus s)$, respectively. For a more detailed account of the relation between NL_λ and NL_{CL}, see Barker and Shan [4]. For a more detailed account of various encodings of combinatorial logic in structural rules, amongst which the encoding of the linear lambda construct used by Barker and Shan, see Finger [8].

3   Scope Islands for NL_{CL}

Our aim for this section is to present an extension to NL_{CL} which will allow us to analyse scope islands, and therefore example (2).

To analyse scope islands, we need some way to block quantifier movement. If you look at the **IBC**-rules in Fig. 1, you will notice that they allow constituents attached to (the left of) a hollow product to move past solid products. This leads us to suggest a fairly simple solution: insert *anything* that is not solid product. For this, we use a residuated pair of unary connectives, $\lozenge$ and $\square$ [13,15]. The relevant rules are presented in Fig. 2.

$$\frac{\Sigma[A] \vdash B}{\Sigma[\langle \square A \rangle] \vdash B} \square L \qquad \frac{\langle \Gamma \rangle \vdash B}{\Gamma \vdash \square B} \square R \qquad \frac{\Sigma[\langle A \rangle] \vdash B}{\Sigma[\lozenge A] \vdash B} \lozenge L \qquad \frac{\Gamma \vdash B}{\langle \Gamma \rangle \vdash \lozenge B} \lozenge R$$

Fig. 2. Scope Islands for NL_{CL}.

Using these connectives, we can assign 'said' the type $(np \backslash s) / \Diamond s$. Instead of taking a sentence-argument from the right, 'said' now takes a *closed-off* sentence—a scope island. Have a look at the derivation for example (2) given below:

$$
\cfrac{
  \cfrac{
    \vdots \\
    \text{KURT} \bullet \text{WROTE} \bullet \text{EVERY} \bullet \text{BOOK} \vdash s
  }{\langle \text{KURT} \bullet \text{WROTE} \bullet \text{EVERY} \bullet \text{BOOK} \rangle \vdash \Diamond s} \Diamond R
  \qquad
  \cfrac{\vdots}{\text{SOMEONE} \bullet (np \backslash s) \vdash s}
}{\text{SOMEONE} \bullet \text{SAID} \bullet \langle \text{KURT} \bullet \text{WROTE} \bullet \text{EVERY} \bullet \text{BOOK} \rangle \vdash s} / L
$$

As long as the scope island (written $\langle \cdot \rangle$) is in place, 'EVERY $\bullet$ BOOK' cannot be raised past it, for there is no rule which allows anything to move past a diamond. But in order to remove the scope island, it has to be eliminated against the $\Diamond s$ argument of 'said', and doing so isolates the embedded clause in its own branch of the proof.[6]

4 Strong and Weak Quantifiers

In the previous section, we presented an extension to NL_{CL} which enabled us to analyse scope islands. This extension blocks all quantifier movement out of scope islands. Example (3) demonstrates that this is too coarse an approach. Specifically, we would like to allow weak quantifiers, such as indefinites, to scope out of scope islands.

We could approach this issue as a syntactic problem, and encode it using structural rules, as we did with quantifier movement and scope islands.[7] However, Szabolcsi [16] writes that "indefinites acquire their existential scope in a manner that does not involve movement and is essentially syntactically unconstrained." Therefore, we feel that a syntactic approach would be out of place.

How do we approach the problem of weak quantifiers as a semantic problem? The solution is to use continuation-passing style (CPS). But how? Early attempts, such as the work by [2], often works by applying a CPS translation directly to the semantic terms. Such approaches, however, face a fundamental dilemma. Because the CPS translation is applied to a solitary semantic term, a deterministic translation cannot introduce scope ambiguity—or any ambiguity, for that matter. However, making the CPS translation sufficiently nondeterministic without causing spurious ambiguity is an arduous task. When Barker makes the translation ambiguous, in order to capture scope ambiguity, this leads to the number of introduced ambigous interpretations growing exponentially with the sentence length. More recent approaches, such as the work by Kiselyov and Shan [11], are much more sophisticated. Their approach allows for the creation of quantifiers of different strengths (e.g. everyone$_1$, everyone$_2$, ...) essentially reducing

---

[6] The presence of the structural diamond in the endsequent may seem problematic, but recall that from the perspective of backward-chaining search we assign semantics to a *known* sentence structure. If we switch to forward-chaining search, i.e. to parsing, the need for a scope island will be inferred from the type of 'said'.

[7] For instance, we could split the family $\backslash\!\backslash, \circ, /\!/$ into two separate families, $\backslash\!\backslash^w, \circ^w, /\!/^w$ and $\backslash\!\backslash^s, \circ^s, /\!/^s$, each with their own copies of the **IBC**-rules, and add a structural rule which selectively allows weak quantifiers to move past scope islands.

scope ambiguity to lexical ambiguity. As a linguistic standpoint, this feels wrong. Furthermore, their framework was engineered to be able to analyse phenomena such as scope islands and weak quantifiers. This makes it too expressive (and intricate) for the task at hand.

Instead, we base our CPS semantics on the approach of Moortgat and Moot [14] and Bastenhof [6], who manage to elegantly integrate CPS semantics into their grammar logic. Moortgat and Moot set up a calculus which enforces one crucial property: every proof in the grammar logic is associated with *unique, normal-form semantics*. In the context of scope ambiguity, this means that each way to interpret a sentence with ambiguous scope corresponds to exactly one proof in the grammar logic.

Focused NL_{CL}. Moortgat and Moot [14, Sect. 3.1] define a normal-form calculus for the Lambek-Grishin calculus (LG). They refer to this calculus as fLG—for *focused* LG, after the technique, pioneered by Andreoli [1], which they use in their calculus. Their version of focusing, however, is more general than that of Andreoli, as they allow for the arbitrary assignment of polarities to atoms. Andreoli's [1] schema can be recovered by assigning all atomic formulas *negative* polarity.

As NL is a fragment of LG, we can trivially extract a normal-form calculus for NL from their work. We will, in their style, refer to this calculus as fNL.

It is important to note that they develop their calculus within the framework of display calculus [7]. One advantage of this framework is that we can freely add structural rules, without fear that we will lose the cut-elimination property. Barker and Shan's [4] extension of NL, NL_{CL}, consists solely of a copy of an existing modality ($\backslash\backslash$, $\circ$, $/\!/$) and a number of structural rules. Therefore, by applying these same changes, we can extend fNL to focsed NL_{CL}—or fNL_{CL}. The result is presented in Fig. 3, together with the focused version of the extension for scope islands from Scct. 3.

Equivalence between fNL_{CL} and NL_{CL} can likely be proven using an intermediate system: display NL_{CL}. One can trivially obtain this system from the focused system in Fig. 3 by dropping the focus marker "☐" and the focusing and unfocusing rules. Equivalence between the display and focused variants of a system was proven for classical NL by Bastenhof [5]. This proof can likely be adapted for NL_{CL}.

However, it is important to realise that, even in the absence of a formal proof of equivalence between NL_{CL} and fNL_{CL}, the second remains a logical system which can analyse all phenomena which Barker and Shan [4] show NL_{CL} can analyse.[8]

Decidable Proof Search. At this point, fNL_{CL} still has a problem, which it shares with NL_{CL}: we do not have a decidable procedure for proof search. Since it is a grammar logic, this means that we do not have a procedure for

---

[8] Throughout the remainder of the paper, whenever we discuss one of the phenomena discussed in Sect. 1, we will give an example proof in fNL_{CL}. In this way, by the end of this paper, this claim will be backed up by evidence.

$$\text{Atom } \alpha \quad ::= s \mid n \mid np \mid \dots$$

$$\text{Type } A, B \quad ::= \alpha \mid A \backslash B \mid B / A \mid A \,\backslash\!\backslash\, B \mid B \,/\!/\, A \mid \Diamond A \mid \Box A$$

$$\text{Struct}^- \Delta \quad ::= \cdot A \cdot \mid \Gamma \backslash \Delta \mid \Delta / \Gamma \mid \Gamma \,\backslash\!\backslash\, \Delta \mid \Delta \,/\!/\, \Gamma \mid [\Delta]$$

$$\text{Struct}^+ \Gamma \quad ::= \cdot A \cdot \mid \Gamma_1 \bullet \Gamma_2 \mid \Gamma_1 \circ \Gamma_2 \mid \mathbf{I} \mid \mathbf{B} \mid \mathbf{C} \mid \langle \Gamma \rangle$$

$$\text{Context } \Sigma \quad ::= \Box \mid \Sigma \bullet \Gamma \mid \Gamma \bullet \Sigma$$

$$\text{Pol}(s) = -, \qquad \text{Pol}(n) = +, \qquad \text{Pol}(np) = +, \qquad \dots$$

$$\text{if Pol}(\alpha) = - \left\{ \quad \frac{}{\boxed{\alpha} \vdash \cdot \alpha \cdot} \, \text{Ax}^L \quad \Bigg| \quad \frac{}{\cdot \alpha \cdot \vdash \boxed{\alpha}} \, \text{Ax}^R \quad \right\} \text{if Pol}(\alpha) = +$$

$$\text{if Pol}(A) = - \left\{ \begin{array}{c} \dfrac{\boxed{A} \vdash \Delta}{\cdot A \cdot \vdash \Delta} \, \text{Foc}^L \quad \Bigg| \quad \dfrac{\Gamma \vdash \boxed{A}}{\Gamma \vdash \cdot A \cdot} \, \text{Foc}^R \\[3ex] \dfrac{\Gamma \vdash \cdot A \cdot}{\Gamma \vdash \boxed{A}} \, \text{Unf}^R \quad \Bigg| \quad \dfrac{\cdot A \cdot \vdash \Delta}{\boxed{A} \vdash \Delta} \, \text{Unf}^L \end{array} \right\} \text{if Pol}(A) = +$$

$$\frac{\Gamma \vdash \boxed{A} \quad \boxed{B} \vdash \Delta}{\boxed{A \backslash B} \vdash \Gamma \backslash \Delta} \backslash L \qquad \frac{\Gamma \vdash \cdot A \backslash \cdot B \cdot}{\Gamma \vdash \cdot A \backslash B \cdot} \backslash R \qquad \frac{\Gamma \vdash \boxed{A} \quad \boxed{B} \vdash \Delta}{\boxed{B / A} \vdash \Delta / \Gamma} / L \qquad \frac{\Gamma \vdash \cdot B \cdot / \cdot A \cdot}{\Gamma \vdash \cdot B / A \cdot} / R$$

$$\frac{\Gamma \vdash \boxed{A} \quad \boxed{B} \vdash \Delta}{\boxed{A \,\backslash\!\backslash\, B} \vdash \Gamma \,\backslash\!\backslash\, \Delta} \,\backslash\!\backslash\, L \qquad \frac{\Gamma \vdash \cdot A \cdot \,\backslash\!\backslash\, \cdot B \cdot}{\Gamma \vdash \cdot A \,\backslash\!\backslash\, B \cdot} \,\backslash\!\backslash\, R \qquad \frac{\Gamma \vdash \boxed{A} \quad \boxed{B} \vdash \Delta}{\boxed{B \,/\!/\, A} \vdash \Delta \,/\!/\, \Gamma} \,/\!/\, L \qquad \frac{\Gamma \vdash \cdot B \cdot \,/\!/\, \cdot A \cdot}{\Gamma \vdash \cdot B \,/\!/\, A \cdot} \,/\!/\, R$$

$$\frac{\Gamma_2 \vdash \Gamma_1 \backslash \Delta}{\Gamma_1 \bullet \Gamma_2 \vdash \Delta} \text{Res}\backslash \bullet \qquad \frac{\Gamma_1 \vdash \Delta / \Gamma_2}{\Gamma_1 \bullet \Gamma_2 \vdash \Delta} \text{Res}/\bullet \qquad \frac{\Gamma_2 \vdash \Gamma_1 \,\backslash\!\backslash\, \Delta}{\Gamma_1 \circ \Gamma_2 \vdash \Delta} \text{Res}\,\backslash\!\backslash\,\circ \qquad \frac{\Gamma_1 \vdash \Delta \,/\!/\, \Gamma_2}{\Gamma_1 \circ \Gamma_2 \vdash \Delta} \text{Res}\,/\!/\,\circ$$

$$\frac{\Gamma \vdash \Delta}{\Gamma \circ \mathbf{I} \vdash \Delta} \mathbf{I} \qquad \frac{\Gamma_1 \bullet (\Gamma_2 \circ \Gamma_3) \vdash \Delta}{\Gamma_2 \circ ((\mathbf{B} \bullet \Gamma_1) \bullet \Gamma_3) \vdash \Delta} \mathbf{B} \qquad \frac{(\Gamma_1 \circ \Gamma_2) \bullet \Gamma_3 \vdash \Delta}{\Gamma_1 \circ ((\mathbf{C} \bullet \Gamma_2) \bullet \Gamma_3) \vdash \Delta} \mathbf{C}$$

$$\frac{\langle \cdot A \cdot \rangle \vdash \Delta}{\cdot \Diamond A \cdot \vdash \Delta} \Diamond L \qquad \frac{\Gamma \vdash \boxed{B}}{\langle \Gamma \rangle \vdash \boxed{\Diamond B}} \Diamond R \qquad \frac{\boxed{A} \vdash \Delta}{\boxed{\Box A} \vdash [\Delta]} \Box L \qquad \frac{\Gamma \vdash [\cdot B \cdot]}{\Gamma \vdash \cdot \Box B \cdot} \Box R \qquad \frac{\Gamma \vdash [\Delta]}{\langle \Gamma \rangle \vdash \Delta} \text{Res}\Box\Diamond$$

Fig. 3. NL$_{\text{CL}}$ reworked as a focused display calculus.

parsing. An easy way to obtain such a procedure is to change the system in such a way that backward-chaining search becomes decidable. The reason this is not decidable in NL$_{\text{CL}}$ is because of the **I**-rule, which does not obey the substructure property.[9]

Admittedly, there are other rules which do not obey the substructure property: the residuation rules and the **B** and **C** rules do not enjoy it. However, the residuation rules still enjoy a weak form this property: they *do not increase* the size of the structure. This means that we can use loop checking to filter out problematic branches of the search. More interestingly, the **B** and **C** rules have the property that "whatever goes up, must come down." At some point, the quantifier will reach the top of the expression, and at that point, there are only two things to do: (1) resolve the quantifier against the top-level type, thereby *eliminating* a connective and breaking out of any loop; or (2) go back down along

---

[9] In the case of NL$_{\text{CL}}$, this is the subformula property.

the same path. Yet when searching for a proof with the **I**-rule, we can always introduce another **I**.

We will address this issue by restricting access to the **I**-rule using a license. This license will be a new unary connective, written **Q**A. Semantically, this logical connective corresponds to a hollow product with a right-hand **I** (i.e. $A \circ \mathbf{I}$). However, as we want neither hollow products, nor the unit **I**, on the logical level, we capture these in a single connective. We remove the **I**-rule, and add the following three rules to the calculus in Fig. 3:

$$\frac{\cdot A \cdot \circ \mathbf{I} \vdash \Delta}{\cdot \mathbf{Q}A \cdot \vdash \Delta}\ IL \qquad \frac{\Gamma \vdash \boxed{B}}{\Gamma \circ \mathbf{I} \vdash \boxed{\mathbf{Q}B}}\ IR \qquad \frac{\Gamma \vdash \Delta}{\Gamma \circ \mathbf{I} \vdash \Delta}\ I^-$$

The first two of these rules are the display calculus rules for right-hand products. The third is the remaining direction of the original **I**-rule. With this change, quantifier raising is restricted to expressions of the form $\mathbf{Q}(C /\!\!/ (A \backslash B))$, and proof search becomes decidable.

One problem which remains is that the **B** and **C** rules cause a huge amount of spurious ambiguity. To see why, note that when raising multiple quantifiers, it is possible to intersperse the various applications of the **B** and **C** rules in many different ways. To solve this, we will take some inspiration from Barker and Shan [4, ch. 17.6], who solve this issue, albeit in a convoluted way. They show that NL$_\lambda$ can be embedded in NL$_{CL}$, using a variant of Schönfinkel's mapping from λ-terms to combinatory logic. Later, they show that a pair of derived rules, $/\!\!/L_\lambda$ and $\backslash R_\lambda$, can serve as a normal-form for the structural rules of NL$_{CL}$. However, these derived rules employ the structural λ which, in the context of NL$_{CL}$, is presumably immediately translated using Schönfinkel's mapping. Instead of employing this two-step process, we exploit the similarities between single-hole contexts and linear λ-terms to derive a variant of the λ-rule which directly uses Schönfinkel's mapping (written $\bar{\ }$) [cf. [4], ch. 17.5]:

$$\overline{\Box} = \mathbf{I}$$
$$\overline{\Sigma \bullet \Gamma} = ((\mathbf{C} \bullet \overline{\Sigma}) \bullet \Gamma) \qquad \frac{\cdot A \cdot \circ \overline{\Sigma} \vdash \Delta}{\Sigma[\cdot \mathbf{Q}A \cdot] \vdash \Delta}\ \uparrow\downarrow$$
$$\overline{\Gamma \bullet \Sigma} = ((\mathbf{B} \bullet \Gamma) \bullet \overline{\Sigma})$$

We can use this mapping in the definition of a derived rule: the $\uparrow\downarrow$-rule, written as $\uparrow$ or $\downarrow$, depending on the direction in which it is applied.[10] We can derive the this rule using three lemmas:

$$\mathbf{Q}/\mathbf{I}: \frac{\Sigma[\cdot A \cdot \circ \mathbf{I}] \vdash \Delta}{\Sigma[\cdot \mathbf{Q}A \cdot] \vdash \Delta}\ \mathbf{Q}/\mathbf{I} \qquad I^{-\prime}: \frac{\Sigma[\Gamma] \vdash \Delta}{\Sigma[\Gamma \circ \mathbf{I}] \vdash \Delta}\ I^{-\prime} \qquad \uparrow\downarrow': \frac{\Gamma \circ \Sigma[\Gamma'] \vdash \Delta}{\Sigma[\Gamma \circ \Gamma'] \vdash \Delta}\ \uparrow\downarrow'$$

Using these lemmas, we can derive the two directions of $\uparrow\downarrow$ as follows:

$$\uparrow: \frac{\dfrac{\cdot A \cdot \circ \overline{\Sigma} \vdash \Delta}{\Sigma[\cdot A \cdot \circ \mathbf{I}] \vdash \Delta}\ \uparrow\downarrow'}{\Sigma[\cdot \mathbf{Q}A \cdot] \vdash \Delta}\ \mathbf{Q}/\mathbf{I} \qquad\qquad \downarrow: \frac{\dfrac{\dfrac{\Sigma[\cdot A \cdot] \vdash \Delta}{\Sigma[\cdot A \cdot \circ \mathbf{I}] \vdash \Delta}\ I^{-\prime}}{\cdot A \cdot \circ \overline{\Sigma} \vdash \Delta}\ \uparrow\downarrow'}{}$$

---

[10] These correspond to Barker and Shan's [4, p. 201] EXPANSION and REDUCTION rules, respectively.

The lemmas themselves can be derived by induction on the structure of the context Σ. The derivation of $\mathbf{Q}/\mathbf{I}$ and $\mathbf{I}^{-\prime}$ is done as follows:

$$\square: \quad \frac{\cdot A \circ \mathbf{I} \vdash \Delta}{\cdot \mathbf{Q} A \cdot \vdash \Delta} \; \mathbf{IL} \qquad \Sigma \bullet \Gamma: \quad \frac{\dfrac{\dfrac{\Sigma[\cdot A \cdot \circ \mathbf{I}] \bullet \Gamma \vdash \Delta}{\Sigma[\cdot A \cdot \circ \mathbf{I}] \vdash \Delta / \Gamma} \; \text{Res}/\bullet}{\Sigma[\cdot \mathbf{Q} A \cdot] \vdash \Delta / \Gamma} \; \mathbf{Q}/\mathbf{I}}{\Sigma[\cdot \mathbf{Q} A \cdot] \bullet \Gamma \vdash \Delta} \; \text{Res}/\bullet \qquad \Gamma \bullet \Sigma: \quad \frac{\dfrac{\dfrac{\Gamma \bullet \Sigma[\cdot A \cdot \circ \mathbf{I}] \vdash \Delta}{\Sigma[\cdot A \cdot \circ \mathbf{I}] \vdash \Gamma \backslash \Delta} \; \text{Res}\backslash \bullet}{\Sigma[\cdot \mathbf{Q} A \cdot] \vdash \Gamma \backslash \Delta} \; \mathbf{Q}/\mathbf{I}}{\Gamma \bullet \Sigma[\cdot \mathbf{Q} A \cdot] \vdash \Delta} \; \text{Res}\backslash \bullet$$

$$\square: \quad \frac{\cdot A \cdot \vdash \Delta}{\cdot A \cdot \circ \mathbf{I} \vdash \Delta} \; \mathbf{I}^{-} \qquad \Sigma \bullet \Gamma: \quad \frac{\dfrac{\dfrac{\Sigma[\cdot A \cdot] \bullet \Gamma \vdash \Delta}{\Sigma[\cdot A \cdot] \vdash \Delta / \Gamma} \; \text{Res}/\bullet}{\Sigma[\cdot A \cdot \circ \mathbf{I}] \vdash \Delta / \Gamma} \; \mathbf{I}^{-\prime}}{\Sigma[\cdot A \cdot \circ \mathbf{I}] \bullet \Gamma \vdash \Delta} \; \text{Res}/\bullet \qquad \Gamma \bullet \Sigma: \quad \frac{\dfrac{\dfrac{\Gamma \bullet \Sigma[\cdot A \cdot] \vdash \Delta}{\Sigma[\cdot A \cdot] \vdash \Gamma \backslash \Delta} \; \text{Res}\backslash \bullet}{\Sigma[\cdot A \cdot \circ \mathbf{I}] \vdash \Gamma \backslash \Delta} \; \mathbf{I}^{-\prime}}{\Gamma \bullet \Sigma[\cdot A \cdot \circ \mathbf{I}] \vdash \Delta} \; \text{Res}\backslash \bullet$$

These rules simply introduce or eliminate the unit $\mathbf{I}$ under some context Σ. The actual movement takes place in the definition of $\uparrow\downarrow'$. In this proof, the base case is simply the identity, as no movement is required to move out of the empty context:

$$\Sigma \bullet \Gamma: \quad \frac{\dfrac{\dfrac{\dfrac{\Gamma \circ ((\mathbf{C} \bullet \Sigma[\Gamma']) \bullet \Gamma'') \vdash \Delta}{(\Gamma \circ \Sigma[\Gamma'] \bullet \Gamma'' \vdash \Delta} \; \mathbf{C}}{\Gamma \circ \Sigma[\Gamma'] \vdash \Delta / \Gamma''} \; \text{Res}/\bullet}{\Sigma[\Gamma \circ \Gamma'] \vdash \Delta / \Gamma''} \; \uparrow\downarrow'}{\Sigma[\Gamma \circ \Gamma'] \bullet \Gamma'' \vdash \Delta} \; \text{Res}/\bullet \qquad \Gamma \bullet \Sigma: \quad \frac{\dfrac{\dfrac{\dfrac{\Gamma \circ ((\mathbf{B} \bullet \Gamma'') \bullet \Sigma[\Gamma']) \vdash \Delta}{\Gamma'' \bullet (\Gamma \circ \Sigma[\Gamma']) \vdash \Delta} \; \mathbf{B}}{\Gamma \circ \overline{\Sigma} \vdash \Gamma'' \backslash \Delta} \; \text{Res}\backslash \bullet}{\Sigma[\Gamma \circ \Gamma'] \vdash \Gamma'' \backslash \Delta} \; \uparrow\downarrow'}{\Gamma'' \bullet \Sigma[\Gamma \circ \Gamma'] \vdash \Delta} \; \text{Res}\backslash \bullet$$

Note that the $\uparrow$-rule eliminates a logical connective—the $\mathbf{Q}$—and therefore has the subformula property. In addition, the $\downarrow$-rule, on the other hand, eliminates the trail of $\mathbf{B}$s and $\mathbf{C}$s, and thus has the substructure property. Because of this, proof search with these rules is decidable.

Furthermore, proof search with the $\uparrow\downarrow$-rule is complete. Briefly, this is true because the **IBC**-rules can do nothing but move a constituent up, or down along an existing path—the $\uparrow\downarrow$-rule mere captures this more succinctly. A formal proof of this can be given by implementing a normalisation function using the commutative conversions for the $\mathbf{B}$ and $\mathbf{C}$ rules: one can move the applications of the $\mathbf{B}$ and $\mathbf{C}$ rules around until they form a continuous sequence (interspersed with residuation rules) starting (or ending) with an application of the $\mathbf{I}$-rule. This sequence of applications can then be replaced by a single application of the $\uparrow\downarrow$-rule. Therefore, proof search using the $\uparrow\downarrow$-rule is complete with repsect to the **IBC**-rules.

We follow Barker and Shan [4], and derive rules corresponding to the $/\!/L_\lambda$- and $\backslash\!\backslash R_\lambda$-rules. These rules combine an application of $\uparrow\downarrow$-rule with an application of $/\!/L$ or $\backslash\!\backslash R$. We name them qL and qR, to signify that they no longer employ a structural λ, and because they can be composed to implement Moortgat's [13] q-connective:

$$qL: \quad \dfrac{\dfrac{\dfrac{\dfrac{\overline{\Sigma} \vdash \boxed{A \setminus B} \qquad \boxed{C} \vdash \Delta}{\boxed{C /\!\!/ (A \setminus B)} \vdash \Delta /\!\!/ \overline{\Sigma}} /\!\!/ L}{\cdot C /\!\!/ (A \setminus B) \cdot \vdash \Delta /\!\!/ \overline{\Sigma}} \mathrm{Foc}^L}{\cdot C /\!\!/ (A \setminus B) \cdot \vdash \circ \overline{\Sigma} \vdash \Delta} \mathrm{Res} /\!\!/ \circ}{\Sigma[\cdot \mathbf{Q}(C /\!\!/ (A \setminus B)) \cdot] \vdash \Delta} \uparrow}$$

$$qR: \quad \dfrac{\dfrac{\dfrac{\dfrac{\Sigma[\cdot A \cdot] \vdash \cdot B \cdot}{\cdot A \cdot \circ \overline{\Sigma} \vdash \cdot B \cdot} \downarrow}{\overline{\Sigma} \vdash \cdot A \cdot \setminus\!\!\setminus \cdot B \cdot} \mathrm{Reso} \setminus\!\!\setminus}{\overline{\Sigma} \vdash \cdot A \setminus\!\!\setminus B \cdot} \setminus\!\!\setminus R}{}$$

As these rules eliminate at least one logical connective each, they still enjoy the subformula property, so proof search with these rules is decidable. In fact, it is *slightly* more efficient than with the $\uparrow\downarrow$-rule. The reason for this is that after raising a quantifier, the only course of action is applying the $/\!\!/L$-rule anyway— and likewise for qR.[11]

Henceforth, if we refer to proof search for fNL$_{\mathrm{CL}}$, we are referring to search using the logical and residuation rules for $\setminus, \bullet, /, \setminus\!\!\setminus, \circ, /\!\!/$ and $\Diamond, \Box$, and the qL and qR rules.[12]

Continuation Semantics for NL$_{\mathrm{CL}}$. A normal-form calculus for proof search is a great improvement, but we were really after Moortgat and Moot's [14, Sect. 3.1] CPS semantics. As with the calculus itself, we can trivially restrict their translation function to fNL, and then extend it to cover fNL$_{\mathrm{CL}}$. In Fig. 4, we present the translation on types and sequents.

We extend the translation on types to a translation on structures as follows: we translate all structural constants ($\mathbf{I}, \mathbf{B}, \mathbf{C}$) as units, forget all unary structural connectives ($\Diamond, \Box$), and translate all binary structural connectives as products. Atomic structures $\cdot A \cdot$ are translated as $[\![A]\!]^-$ or $[\![A]\!]^+$, depending on the polarity of the structure $\cdot A \cdot$.

$$s^* = \mathbf{t}, \qquad n^* = \mathbf{e}\mathbf{t}, \qquad np^* = \mathbf{e}, \qquad \dots$$

$$[\![\alpha]\!]^+ = \begin{cases} \alpha^* & \text{if Pol}(\alpha) = + \\ ((\alpha^*)^R)^R & \text{if Pol}(\alpha) = - \end{cases} \qquad [\![\alpha]\!]^- = (\alpha^*)^R$$

$$[\![A \setminus B]\!]^+ = ([\![A]\!]^+ \times [\![B]\!]^-)^R \qquad\qquad [\![A \setminus B]\!]^- = [\![A]\!]^+ \times [\![B]\!]^-$$

$$[\![B / A]\!]^+ = ([\![B]\!]^- \times [\![A]\!]^+)^R \qquad\qquad [\![B / A]\!]^- = [\![B]\!]^- \times [\![A]\!]^+$$

$$[\![\Diamond A]\!]^+ = [\![A]\!]^+ + \qquad\qquad\qquad\quad [\![\Diamond A]\!]^- = ([\![A]\!]^+ +)^R$$

$$[\![\Box A]\!]^+ = ([\![A]\!]^+ +)^R \qquad\qquad\qquad [\![\Box A]\!]^- = [\![A]\!]^+ +$$

$$[\![\mathbf{Q}A]\!]^+ = [\![A]\!]^+ \qquad\qquad\qquad\qquad [\![\mathbf{Q}A]\!]^- = ([\![A]\!]^+)^R$$

$$[\![\Gamma \vdash \Delta]\!] = [\![\Gamma]\!] \vdash [\![\Delta]\!] \qquad [\![\boxed{A} \vdash \Delta]\!] = [\![\Delta]\!] \vdash [\![A]\!]^- \qquad [\![\Gamma \vdash \boxed{A}]\!] = [\![\Gamma]\!] \vdash [\![A]\!]^+$$

Fig. 4. CPS semantics for focused NL$_{\mathrm{CL}}$.

---

[11] This can be proven using a variant of Barker and Shan's [4, ch. 17.6 and 17.7] proof of equivalence between NL$_\lambda$ and NL$_{\mathrm{CL}}$.

[12] In fact, no proof will ever explicitly use the logical or residuation rules for $\setminus\!\!\setminus, \circ, /\!\!/$, leading to a question of whether it is really necessary for $/\!\!/$ and $\setminus\!\!\setminus$ to be fully residuated logical implications. But this is a matter for another paper.

In this particular CPS translation, all function applications and abstractions are contained within the focusing and unfocusing rules, which are translated as follows:

$$\frac{x : [\![\Delta]\!] \vdash M : [\![A]\!]^-}{k : [\![A]\!]^+ \vdash (\lambda x.k\ M) : [\![\Delta]\!]^R}\ \mathrm{Foc}^L \qquad \frac{x : [\![A]\!]^+ \vdash M : [\![\Delta]\!]^R}{y : [\![\Delta]\!] \vdash (\lambda x.M\ y) : [\![A]\!]^-}\ \mathrm{Unf}^L$$

$$\frac{x : [\![\Gamma]\!] \vdash M : [\![A]\!]^+}{x : [\![\Gamma]\!] \vdash (\lambda k.k\ M) : [\![A]\!]^{-R}}\ \mathrm{Foc}^R \qquad \frac{x : [\![\Gamma]\!] \vdash M : [\![A]\!]^{-R}}{x : [\![\Gamma]\!] \vdash (\lambda y.M\ y) : [\![A]\!]^+}\ \mathrm{Unf}^R$$

The other rules are translated either as axioms (Ax^L, Ax^R), identities ($\backslash R$, $/R$, $\backslash\!\backslash R$, $/\!/R$ and all rules for $\Diamond$, $\Box$) or permutations (the rest). For instance,

$$\frac{x : [\![\Gamma]\!] \vdash M : [\![A]\!]^+ \qquad y : [\![\Delta]\!] \vdash N : [\![B]\!]^-}{z : [\![\Gamma]\!] \times [\![\Delta]\!] \vdash (M[\pi_1 z/x], N[\pi_2 z/y]) : [\![A]\!]^+ \times [\![B]\!]^-}\ \backslash L$$

An exception to this are the **I**-rules. Because we would like to be able to simply forget the **Q**-connective upon translation, so that we do not have to store unnecessary units in our lexicon, we have to insert or remove the units upon using these rules.

Using these semantics, we can assign the indefinite article the type $np\ /\ n$.[13] This will result in *two* interpretations for (2), and *three* interpretations for (3), as required. Let us consider the important steps in the derivation of (3):

1. the quantifier movement and scope taking of "everyone";
2. the collapsing of the scope island, isolating the clause "[$_S$ Kurt .. Mary]" in its own branch of the derivation;
3. the collapsing of "a book", with the indefinite taking scope at the top-level.

If these steps are taken [1,2,3], we obtain interpretation (3a); if they are taken [1,3,2], we obtain (3b); and if they are taken [3,1,2], we obtain (3c).[14]

5   Examples

In this section, we will present a number of analyses of the examples presented in Sect. 1. In the interest of brevity, we will summarise numerous applications of the residuation rules, beginning or ending with focusing or unfocusing rules with 'dp', for *display postulate*. In addition, we will leave out uninteresting subproofs.

First off, we present an analysis of (1), resulting in interpretation (1b). The quantifier EVERY is assigned the type $\mathbf{Q}(s\ /\!/\ (np \backslash\!\backslash s))\ /\ n$, and SOMEONE is assigned the type of a "strong" quantifier—that is to say, $\mathbf{Q}(s\ /\!/\ (np \backslash\!\backslash s))$.

---

[13] Quantifiers such as "someone" should be assigned the type $np\ /\ n \otimes n$, which means we must also extend $\mathrm{NL_{CL}}$ with logical products.

[14] The normal-form requires that 1 occurs before 2, so this list is exhaustive.

$$\vdots$$

$$\cfrac{\cfrac{\cfrac{\cfrac{\cdot np \cdot \bullet \text{READ} \bullet \cdot np \cdot \vdash \cdot s \cdot}{\boxed{\square} \bullet \text{READ} \bullet \cdot np \cdot \vdash \cdot np \backslash\!\backslash s \cdot} \; qR}{\boxed{\square} \bullet \text{READ} \bullet \cdot np \cdot \vdash \boxed{np \backslash\!\backslash s}} \; Unf^R}{\text{SOMEONE} \bullet \text{READ} \bullet \cdot np \cdot \vdash \cdot s \cdot} \quad \cfrac{}{\boxed{s} \vdash \cdot s \cdot} \; Ax^L}{\text{SOMEONE} \bullet \text{READ} \bullet \cdot np \cdot \vdash \cdot s \cdot} \; qL}$$

Below rendered as descriptive proof tree:

Secondly, we present an analysis of (2), resulting in interpretation (2a)—the only interpretation.

As a third example, we show that we can analyse 'a' as a weak quantifier, using the type $np \,/\, n$. We give an analysis of (3), resulting in the interpretation where the indefinite takes wide scope—(3b). The quantifier 'a' takes scope when it is combined with book.

Lastly, we present analyses of examples (4) and (5). We demonstrate changing result types using the word 'which', which we assign the type

$$\mathbf{Q}(((n \backslash n) \,/\, (np \backslash s) \,/\!/\, (np \backslash\!\backslash np)))\text{.}$$

In the second, for parasitic scope, we deviate slightly from Barker's [3] treatment of parasitic scope. We assign 'same' (and 'different') the type

$$\mathbf{Q}(s \,/\!/\, (\mathbf{Q}(np \backslash\!\backslash s \,/\!/\, ((n \,/\!/\, n) \backslash\!\backslash np \backslash\!\backslash s)) \backslash\!\backslash s))\text{.}$$

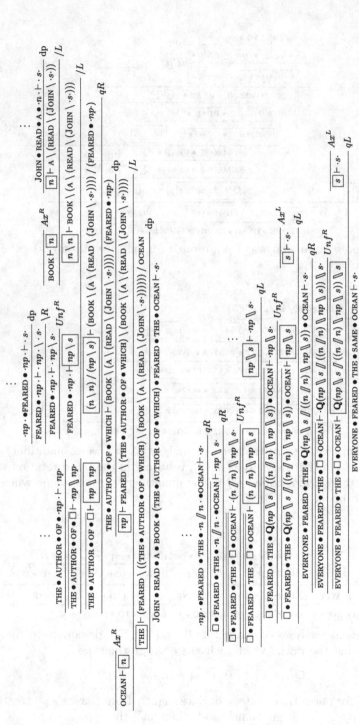

Fig. 5. Analysis of examples (4) and (5).

What this type does is quantify over an expression twice—once normally, to plant its top-level quantifier, and once parasitically. Using this type, we can obtain the semantics advocated by Kiselyov [10]. The proofs for these two examples can be found in Fig. 5.

6   Conclusion

We presented an improvement over Barker and Shan's [4] NL$_{CL}$ for which derivability is decidable, and which has a normal-form for proof search. In addition, it can analyse scope islands, and distinguish between strong and weak quantifiers, shown by the ability to analyse examples (1–5). Of these examples, (1–3) are representative examples of scope islands and strong and weak quantifiers, for which Kiselyov and Shan [11] provides a purely semantic analysis. The remaining examples, (4) and (5), are examples from the work by Barker and Shan [4] which motivated us to start from their syntactic approach.

References

1. Andreoli, J.: Focussing and proof construction. Ann. Pure Appl. Logic **107**(1–3), 131–163 (2001)
2. Barker, C.: Continuations and the nature of quantification. Nat. Lang. Semant. **10**(3), 211–242 (2002)
3. Barker, C.: Parasitic scope. Linguist. Philos. **30**(4), 407–444 (2007)
4. Barker, C., Shan, C.: Continuations and Natural Language. Oxford Studies in Theoretical Linguistics, vol. 53. Oxford University Press, Oxford (2014)
5. Bastenhof, A.: Polarized classical non-associative lambek calculus and formal semantics. In: Pogodalla, S., Prost, J.-P. (eds.) LACL 2011. LNCS (LNAI), vol. 6736, pp. 33–48. Springer, Heidelberg (2011). doi:10.1007/978-3-642-22221-4_3
6. Bastenhof, A.: Categorial symmetry. http://dspace.library.uu.nl/handle/1874/273870
7. Belnap, N.D.: Display logic. J Philos. Logic **11**(4), 375–417 (1982)
8. Finger, M.: Computational solutions for structural constraints. In: Moortgat, M. (ed.) LACL 1998. LNCS (LNAI), vol. 2014, pp. 11–30. Springer, Heidelberg (2001). doi:10.1007/3-540-45738-0_2
9. Heim, I., Kratzer, A.: Semantics in Generative Grammar, vol. 13. Blackwell Oxford, Oxford (1998)
10. Kiselyov, O.: Compositional semantics of same, different, total. In: Proceedings for ESSLLI 2015 Workshop 'Empirical Advances in Categorial Grammar', pp. 71–81 (2015)
11. Kiselyov, O., Shan, C.-C.: Continuation hierarchy and quantifier scope. In: McCready, E., Yabushita, K., Yoshimoto, K. (eds.) Formal Approaches to Semantics and Pragmatics, pp. 105–134. Springer, Heidelberg (2014)
12. Lambek, J.: Observation of strains. Infect Dis. Ther. **3**(1), 35–43 (2011). On the calculus of syntactic types. Structure of language and its mathematical aspects, 166: C178(1961)
13. Moortgat, M.: In situ binding: a modal analysis. In: Dekker, P., Stokhof, M. (eds) Proceedings of the Tenth Amsterdam Colloquium, pp. 539–549. Institute for Logic, Language and Computation (ILLC) (1996)

14. Moortgat, M., Moot, R.: Proof nets for the Lambek-Grishin calculus. In: Compositional Methods in Physics and Linguistics, volume abs/1112.6384 (2011)
15. Morrill, G.: Type Logical Grammar - Categorial Logic of Signs. Kluwer, Alphen aan den Rijn (1994)
16. Szabolcsi, A.: The syntax of scope. In: The Handbook of Contemporary Syntactic Theory, pp. 607–633. Wiley-Blackwell (2000)

Type Reconstruction for λ-DRT Applied to Pronoun Resolution

Hans Leiß[⊠] and Shuqian Wu

Centrum Für Informations- und Sprachverarbeitung,
Ludwig-Maximilians-Universität München, München, Germany
leiss@cis.uni-muenchen.de

Abstract. λ-DRT is a typed theory combining simply typed λ-calculus with discourse representation theory, used for modelling the semantics of natural language. With the aim of type-checking natural language texts in the same vein as is familiar from type-checking programs, we propose *untyped* λ-DRT with automatic type reconstruction. We show a principal types theorem for λ-DRT and how type reconstruction can be used to make pronoun resolution type-correct, i.e. the inferred types of a pronoun occurrence and its antecedent noun phrase have to be compatible, thereby reducing the number of possible antecedents.

Keywords: Pronoun resolution · Discourse representation theory · λ-DRT · Type reconstruction · Principal types

1 Introduction

In order to give a compositional semantics for discourse, [2] have extended the non-compositional and first-order approach of Discourse Representation Theory (DRT, [10]) by adding λ-abstraction and functional application. As is familiar from Montague-semantics, the meaning of an expression can then be defined bottom-up, by abstracting from the meaning contribution of the context; function application is then used to combine this meaning with those of expressions from the context.

While DRT uses *discourse representation structures*, i.e. pairs of variables and quantifier-free formulas, and avoids higher-order logic on its way to translate natural language to first-order logic, Montague-grammar and λ-DRT make heavy use of higher order types and are commonly expressed in a simply typed language.

Our first goal is to have a type-free notation of λ-DRS-terms, such that meanings can be written without types, but checked for typeability by "reconstructing" suitable types from types of built-in constants (polymorphic function words and monomorphic content words in the lexicon) and the context of occurrence. For this, we will show that most general types exist and can be inferred automatically. The second goal is to integrate the type reconstruction into a program for pronoun resolution. We want to be able to type-check when a pronoun

© Springer-Verlag GmbH Germany 2016
M. Amblard et al. (Eds.): LACL 2016, LNCS 10054, pp. 149–174, 2016.
DOI: 10.1007/978-3-662-53826-5_10

resolution (i.e. the unification of the discourse variable of a pronoun with the discourse referent of an antecedent) is type-correct, and moreover, we want to use the type reconstruction for unresolved pronouns to filter possible antecedents by their types and the type of the unresolved pronoun.

2   λ-DRT

Where [18] uses meanings like $a \mapsto \lambda P \lambda Q \exists x(P\,x \wedge Q\,x)$, $man \mapsto \lambda x.man(x)$ and $walks \mapsto \lambda x.walk(x)$ and combines these by application to $a\ man\ walks \mapsto \exists x(man(x) \wedge walk(x))$, in λ-DRT of [2], one uses somewhat different lexical entries

$$\lambda P \lambda Q(\boxed{\begin{array}{c} x \\ \hline \end{array}} \otimes P\,x \otimes Q\,x),\ \lambda x\boxed{\begin{array}{c} \\ \hline man(x) \end{array}},\ \lambda x\boxed{\begin{array}{c} \\ \hline walk(x) \end{array}}$$

and an operation ⊗ of *merging* discourse representation structures as in

$$\boxed{\begin{array}{c} x \\ \hline \end{array}} \otimes \boxed{\begin{array}{c} \\ \hline man(x) \end{array}} \otimes \boxed{\begin{array}{c} \\ \hline walk(x) \end{array}} = \boxed{\begin{array}{c} x \\ \hline man(x),\ walk(x) \end{array}}.$$

In general, two discourse structures are merged by appending their (disjoint) lists of discourse referents (variables) and formulas, respectively:

$$\boxed{\begin{array}{c} x_1,\ldots,x_m \\ \hline \varphi_1,\ldots,\varphi_k \end{array}} \otimes \boxed{\begin{array}{c} y_1,\ldots,y_n \\ \hline \psi_1,\ldots,\psi_p \end{array}} = \boxed{\begin{array}{c} x_1,\ldots,y_1,\ldots \\ \hline \varphi_1,\ldots,\psi_1,\ldots \end{array}}$$

Since a variable in the referent list is seen as a binding, a binder of each merge-factor can bind free variable occurrences in the formulas of *both* merge-factors. In a discourse *A man walks. He talks.*, the meanings of the sentences have to be combined. The pronoun *he* in the second sentence introduces a new discourse referent y with the appropriate property. The combination of the meanings of the sentences is the merging

$$\boxed{\begin{array}{c} x \\ \hline man(x),\ walk(x) \end{array}} \otimes \boxed{\begin{array}{c} y \\ \hline talk(y) \end{array}}$$

of their discourse structures, followed by pronoun resolution: the referent y of the anaphoric pronoun is *resolved* against some previously introduced discourse referent, here x. This can be implemented by adding an equational constraint $x = y$ to the merged DRS, or by unifying the variables.

If one assumes some co-indexing of pronouns and antecedent noun phrases as a result of syntactic analysis, one can use the referent of the antecedent noun phrase as referent of the anaphoric pronoun. Then, the binding is *dynamic*, i.e. the scope extends beyond sentence boundaries as the discourse goes on:

$$\boxed{\begin{array}{c} x \\ \hline man(x),\ walk(x) \end{array}} \otimes \boxed{\begin{array}{c} \\ \hline talk(x) \end{array}}.$$

With type reconstruction for λ-DRT, one could just check the type-soundness of pronoun resolution, i.e. that the semantic type of the pronoun occurrence fits the semantic type of the referent of its antecedent. However, we want to use type reconstruction to help pronoun resolution. To do so, we mark discourse referents as anaphors or possible antecedents, use type reconstruction for λ-DRT to infer types for the discourse referents, and then do pronoun resolution with typed referents. Our typing rules for DRSs and DRT's accessibility relation are closely related.

2.1   Untyped λ-DRS-Terms

We use four kinds of raw expressions: terms, formulas, discourse representation structures, and discourses:

$$
\begin{array}{ll}
\textbf{Term:} & s, t := x \quad (x \in Var) \\
& \mid c \quad (c \in Const)
\end{array}
$$

$$
\begin{array}{ll}
\textbf{λ-DRS:} & D := x \quad (x \in Var) \\
& \mid \lambda x D \\
& \mid (D_1 \cdot D_2) \\
& \mid \langle [], \varphi \rangle \\
& \mid \langle x, D \rangle \\
& \mid (D_1 \otimes D_2)
\end{array}
$$

$$
\begin{array}{ll}
\textbf{Formula:} & \varphi, \psi := \top \\
& \mid R(t_1, \ldots, t_n) \\
& \mid t_1 \dot{=} t_2 \\
& \mid (\varphi \wedge \psi) \\
& \mid \neg D \\
& \mid (D_1 \Rightarrow D_2) \\
& \mid (D_1 \vee D_2)
\end{array}
$$

$$
\begin{array}{ll}
\textbf{Discourse:} & \mathcal{D} := \epsilon \\
& \mid \mathcal{D} \, ; \, D
\end{array}
$$

All terms are atomic. Formulas are built from atomic formulas by conjunction of formulas and (non-conjunctive) Boolean combinations of λ-DRSs.

A *box* or *value*-DRS D is a pair $\langle [x_1, \ldots, x_n], \varphi \rangle$ of a list $[x_1, \ldots, x_n]$ of variables and a formula φ, recursively defined by

$$
\langle [x_1, x_2, \ldots, x_n], \varphi \rangle := \begin{cases} \langle [], \varphi \rangle, & n = 0, \\ \langle x_1, \langle [x_2, \ldots, x_n], \varphi \rangle \rangle, & \text{else.} \end{cases}
$$

Two DRSs D_1 and D_2 may be *merged* to a DRS $(D_1 \otimes D_2)$. So far, the merge-operator $\otimes$ is just a constructor. We will later add reduction rules which provide the intended meaning of the merge of two value-DRSs (with disjoint variable lists) as

$$
\langle [x_1, \ldots, x_n], \varphi \rangle \otimes \langle [y_1, \ldots, y_m], \psi \rangle \ \to^* \ \langle [x_1, \ldots, x_n, y_1, \ldots, y_m], (\varphi \wedge \psi) \rangle.
$$

Finally, we want to have abstraction and application of λ-DRSs. *Note:* We use the pair notation $\langle s, t \rangle$ not for arbitrary terms s, t. Likewise for the types $\sigma \times \tau$: the intention is that σ is an individual type, τ a DRS-type.

The *toplevel referents* and the *free variables* of D are defined by

$$top(x) = \emptyset$$
$$top(\lambda x D) = \emptyset$$
$$top(D_1 \cdot D_2) = \emptyset$$
$$top(\langle [], \varphi \rangle) = \emptyset$$
$$top(\langle x, D \rangle) = \{x\} \cup top(D)$$
$$top(D_1 \otimes D_2) = top(D_1) \cup top(D_2)$$

$$free(x) = \{x\}$$
$$free(\lambda x D) = free(D) \setminus \{x\}$$
$$free(D_1 \cdot D_2) = free(D_1) \cup free(D_2)$$
$$free(\langle [], \varphi \rangle) = free(\varphi)$$
$$free(\langle x, D \rangle) = free(D) \setminus \{x\}$$
$$free(D_1 \otimes D_2) = (free(D_1) \cup free(D_2))$$
$$\setminus top(D_1 \otimes D_2)$$

For formulas built from DRSs, we put

$$free(\neg D) = free(D)$$
$$free((D_1 \Rightarrow D_2)) = free((D_1 \vee D_2))$$
$$= free(D_1) \cup (free(D_2) \setminus top(D_1))$$

This is motivated by considering free variables of D_2 (representing pronouns) as bound by toplevel referents of D_1 (their antecedents). However, these notions are not stable under β-reduction $\rightarrow$: for example, for $D_1 = \lambda y \langle [x], \varphi \rangle \cdot y$ and $D_1' = \langle [x], \varphi \rangle$ we have $D_1 \rightarrow D_1'$, but $top(D_1) = \emptyset \neq top(D_1')$, and so $(D_1 \Rightarrow D_2)$ may bind less variables of D_2 than $(D_1' \Rightarrow D_2)$. Hence these definitions make sense for expressions in β-normal form only.[1]

In Sect. 5 we define the meaning of application $\cdot$ by β-reduction, i.e. by reducing an application $(t \cdot s)$ to the substitution $t[x/s]$ of free occurrences of x in t by s. Some care is needed to avoid variable capture.

We treat toplevel referents of a merge-factor as binders with scope over all factors. Hence, when substituting a free occurrence of x in $(D_1 \otimes D_2)$ by s, we have to α-rename the top-level referents of D_1 and D_2 to avoid capturing free variables of s. But we also have to rename toplevel referents of s when applying $[x/s]$ to $(D_1 \otimes D_2)$, since s might become a merge-factor, as for $D_1 = x$, and then its toplevel referents would capture free variables of D_2. Since D_1, D_2, s might have toplevel referents after some reductions, we define $t[x/s]$ in such a way that *all* bound variables and referents of t and s are renamed to fresh ones before the free occurrences of x are replaced.[2]

---

[1] In Sect. 6.2, the DRSs are computed bottom-up along the syntax tree, and at each syntactic construction, the DRS resulting from a combination of the constituents' DRSs is reduced.

[2] Our implementation actually does the renaming only when applications are involved, so $\lambda P((P \cdot x) \otimes (P \cdot y)) \cdot \lambda z D$ copies $\lambda z D$ to get $(\lambda z D \cdot x) \otimes (\lambda z D \cdot y)$ and then renames referents in D when treating the applications as $(D[z/x] \otimes D[z/x])$. Thus, merge-factors have disjoint reference lists, provided the lexical entries have.

An essential clause in the definition of $D[x/s]$ is:

$$(D_1 \otimes D_2)[x/s] = (D_1'[x/s'] \otimes D_2'[x/s'']),$$

where D_i' is D_i with $top(D_1 \otimes D_2) \cap free(s)$ renamed, and s', s'' are s with $bound(s)$ renamed. Similar clauses are needed to treat $(D_1 \Rightarrow D_2)[x/s]$ and $(D_1 \vee D_2)[x/s]$. For example, if P is not in φ, then in

$$(\langle[x], \varphi\rangle \otimes (P \cdot x))[P/\lambda x D] = (\langle[x'], \varphi[x/x']\rangle \otimes (P \cdot x')[P/\lambda x D'])$$
$$= (\langle[x'], \varphi[x/x']\rangle \otimes D'[x/x']),$$

D' is D with toplevel referents renamed and hence does not bind free variables of φ.

3   Typing Rules

Montague-semantics and λ-DRT usually come with base types e for entities and t for truth values. As boxes are pairs $\langle x, \varphi\rangle$ of a list of individual variables and a formula, it seems natural to give them the pair type $e^* \times t$, where e^* is the type of lists of entities. Instead, all boxes have another base type in [2], and the type $s \to (s \to t)$ of binary relations between situations s (resp. assignments of entities to discourse referents) in [19]).

For the kind of semantic checking of texts we want to do, a more fine-grained typing of DRSs is needed. One should distinguish between entities of different kinds, i.e. replace the base type e by a family $\langle e_i\rangle_{i \in I}$ of base types or *sorts*. The type of a box $\langle x, \varphi\rangle$ then becomes a pair $e \times t$, so that, essentially, a typed DRS $\langle x, \varphi\rangle : e \times t$ is a pair of a *type environment* $x : e$ and a formula $\varphi : t$.

The type $e \times t$ of a merge-DRS $D_1 \otimes D_2$ then ought to be related to the types $e_1 \times t$ and $e_2 \times t$ of the constituents D_1 and D_2 in that e is obtained by *appending* e_1 and e_2, so $e = append(e_1, e_2)$. However, since $\otimes$ is just a DRS-constructor, we will likewise introduce a type constructor $\otimes$ and use a constraint $e = e_1 \otimes e_2$ in the type reconstruction process. Since the length of referent- and type-lists have to match –even if we had only a single sort of entities–, we cannot use the list type constructor *, but build type lists by *consing* a type e_i to a list e of types, $e_i \times e$, beginning with the type $\mathbb{1}$ for the empty list paired with a truth value.

Types:

$\sigma, \tau := \alpha$ (type variables)
    | e_i (atomic types of individuals) | $(\sigma \times \tau)$ (DRSs with non-empty ref-list)
    | t (truth values) | $(\sigma \otimes \tau)$ (merge-DRSs)
    | $\mathbb{1}$ (DRSs with empty ref-list) | $(\sigma \to \tau)$ (functions)

We call a type a *drs-type*, if it is of the forms α, $\mathbb{1}$, $e_i \times \tau$, or $\sigma \otimes \tau$ with drs-types σ and τ. We write $\sigma \times \tau \times \mathbb{1}$ for $(\sigma \times (\tau \times \mathbb{1}))$ and $[\sigma_1, \ldots, \sigma_n]$ for $\sigma_1 \times \ldots \times \sigma_n \times \mathbb{1}$.

Typing rules:

<table>
<tr>
<td>

Typing variables (and constants),
abstractions and applications

</td>
<td>

Using a typed DRS as a type context

</td>
</tr>
</table>

$$\frac{}{x : \sigma, \Gamma \vdash x : \sigma} \; (var_1)$$

$$\frac{x \not\equiv y \qquad \Gamma \vdash x : \sigma}{y : \tau, \Gamma \vdash x : \sigma} \; (var_2)$$

$$\frac{x : \rho, \Gamma \vdash t : \tau}{\Gamma \vdash \lambda xt : (\rho \to \tau)} \; (abs)$$

$$\frac{\Gamma \vdash t : \sigma \to \tau \qquad \Gamma \vdash s : \sigma}{\Gamma \vdash (t \cdot s) : \tau} \; (app)$$

$$\frac{\Gamma \vdash x : \sigma}{\langle [], \varphi \rangle : \mathbb{1}, \Gamma \vdash x : \sigma} \; (var_3)$$

$$\frac{y : \rho, \; D : \tau, \; \Gamma \vdash x : \sigma}{\langle y, D \rangle : \rho \times \tau, \; \Gamma \vdash x : \sigma} \; (var_4)$$

$$\frac{D : \rho, \; E : \sigma, \Gamma \vdash x : \tau}{(D \otimes E) : (\rho \otimes \sigma), \Gamma \vdash x : \tau} \; (var_5)$$

$$\frac{\Gamma \vdash x : \tau}{(D_1 \cdot D_2) : \sigma, \Gamma \vdash x : \tau} \; (var_6)$$

$$\frac{\Gamma \vdash x : \tau}{\lambda y D : \sigma, \Gamma \vdash x : \tau} \; (var_7)$$

An assumption $D : \sigma$ can only be used when D is a variable, a value-DRS, or a merged DRS. The rules (var_3) and (var_4) amount to a typing rule

$$\frac{x_1 : \sigma_1, \ldots, x_n : \sigma_n, \Gamma \vdash x : \sigma}{\langle [x_1, \ldots, x_n], \varphi \rangle : [\sigma_1, \ldots, \sigma_n], \Gamma \vdash x : \sigma} \; (var^+)$$

which says that a typed DRS as assumption is used as a list of typing assumptions of its top-level discourse referents. By (var_5), assuming a typed merged DRS amounts to assuming suitably typed merge-factors. By (var_6) and (var_7), assumptions for typed applications and abstractions can be ignored.

We need typed DRSs as assumptions to type merge-DRSs, disjunctions, implications and discourses (rules $(\otimes), (impl), (disj), (;)$), where part of the DRS to be typed contains top-level referents whose types have to be assumed to type the rest of the DRS.

Typing value DRSs and merged DRSs

$$\frac{\Gamma \vdash \varphi : \mathbf{t}}{\Gamma \vdash \langle [], \varphi \rangle : \mathbb{1}} \; (drs_1^+) \qquad \frac{x : \sigma, \Gamma \vdash D : \tau}{\Gamma \vdash \langle x, D \rangle : (\sigma \times \tau)} \; (drs_2^+)$$

$$\frac{D_2 : \tau_2, \Gamma \vdash D_1 : \tau_1 \qquad D_1 : \tau_1, \Gamma \vdash D_2 : \tau_2}{\Gamma \vdash (D_1 \otimes D_2) : (\tau_1 \otimes \tau_2)} \; (\otimes)$$

Notice that in (drs_2^+) a variable is removed from the context and built into a DRS. Hence, $\langle x, D \rangle$ corresponds to a binding operator, written $\delta x.D$ in Kohlhase e.a. [13] But in $(\otimes)$ a typed DRS is used like a type context to type another DRS, whereby the scope of $\langle x, D \rangle : \sigma$ is extended to terms outside of D. This is what Kohlhase e.a. [13] call "dynamic" binding of variables in D_2 by binding operators of D_1.

Typing formulas

$$\frac{\Gamma \vdash t_1 : \tau_1 \quad \ldots \quad \Gamma \vdash t_n : \tau_n \quad \Gamma \vdash R : \tau_1 \to (\ldots \to (\tau_n \to \mathbf{t})\ldots)}{\Gamma \vdash R(t_1, \ldots, t_n) : \mathbf{t}} \ (rel)$$

$$\frac{\Gamma \vdash D : \sigma}{\Gamma \vdash \neg D : \mathbf{t}} \ (neg)$$

$$\frac{\Gamma \vdash t_1 : e \quad \Gamma \vdash t_2 : e}{\Gamma \vdash t_1 \dot{=} t_2 : \mathbf{t}} \ (eqn)$$

$$\frac{\Gamma \vdash D_1 : \sigma_1 \quad D_1 : \sigma_1, \Gamma \vdash D_2 : \sigma_2}{\Gamma \vdash (D_1 \vee D_2) : \mathbf{t}} \ (disj)$$

$$\frac{\Gamma \vdash \varphi : \mathbf{t} \quad \Gamma \vdash \psi : \mathbf{t}}{\Gamma \vdash (\varphi \wedge \psi) : \mathbf{t}} \ (conj)$$

$$\frac{\Gamma \vdash D_1 : \sigma_1 \quad D_1 : \sigma_1, \Gamma \vdash D_2 : \sigma_2}{\Gamma \vdash (D_1 \Rightarrow D_2) : \mathbf{t}} \ (impl)$$

Discourses are sequences of sentences; to type the sequence of their DRSs, each DRS is typed in the context extended by the previous DRSs. (Thereby we can resolve pronouns anaphoric ally, to referents in the left textual context.)

Typing discourses

$$\frac{\Gamma \vdash D_1 : \tau_1 \quad D_1 : \tau_1, \Gamma \vdash D_2 : \tau_2 \quad \cdots \quad D_n : \tau_n, \ldots, D_1 : \tau_1, \Gamma \vdash D_{n+1} : \tau_{n+1}}{\Gamma \vdash (D_1 ; D_2 ; \ldots ; D_{n+1}) : ((\ldots (\tau_1 \otimes \tau_2)\ldots) \otimes \tau_{n+1})} \ (;\ )$$

In typing a term, a typed assumption $D : \sigma$ can only be used by decomposing it to the typed top-level discourse referents of D, using (var_3) to (var_5). This cannot be done if D is a variable, application, or abstraction. We ignore assumed typed abstractions by (var_7), which is harmless since they cannot evaluate to boxes, but (var_6), ignoring assumed typed applications, is not: they may reduce to a box containing x as a top-level discourse referent and thus block an assumption $x : \tau$ in Γ. We need to restrict (var_6) to have a form of subject-reduction, see Sect. 5.

By induction on the structure of terms, formulas and λ-DRSs t, we obtain:

Lemma 1. *Suppose for all $x \in free(t)$ and all types σ, $\Gamma \vdash x : \sigma$ iff $\Delta \vdash x : \sigma$. Then $\Gamma \vdash t : \tau$ iff $\Delta \vdash t : \tau$.*

Corollary 1. *1. If $\Gamma, \langle [], \varphi \rangle : \mathbb{1}, \Delta \vdash s : \sigma$, then $\Gamma, \Delta \vdash s : \sigma$.*
2. If $x : \rho, E : \tau, \Gamma \vdash s : \sigma$ and x is not a top-level referent of E, then $E : \tau, x : \rho, \Gamma \vdash s : \sigma$.

4 Type Reconstruction

We want to extend Hindley's well-known "principal types"-theorem from (simply typed) λ-calculus to λ-DRT. The theorem says that the set of typings $\Gamma \vdash t : \tau$ of a term t is the set of instances $S\Gamma_0 \vdash t : S\tau_0$ of a single typing $\Gamma_0 \vdash t : \tau_0$, where $S : TyVar \rightarrow Ty$ are the assignments of types to type variables. Then $\Gamma_0 \vdash t : \tau_0$ is a *most general* or *principal typing* of t. A (principal) typing of t *modulo* Γ is a (principal) typing $S\Gamma \vdash t : \sigma$ for some type substitution S and type σ.

It is not hard to see that instances of a DRS-typing are also typings of the DRS.

Lemma 2. *If* $\Gamma \vdash D : \sigma$ *and* $S : TyVar \rightarrow Ty$, *then* $S\Gamma \vdash D : S\sigma$.

More work is needed to show the existence of principal typings.

Theorem 1. *There is an algorithm W that, given a type context Γ and a term t, either returns a pair (U, τ) of a type substitution $U : TyVar \rightarrow Ty$ and a type τ such that $U\Gamma \vdash t : \tau$ is a most general typing of t modulo Γ, or returns 'fail', if there is no (U, τ) such that $U\Gamma \vdash t : \tau$.*

The algorithm W has an easy modification which, on input (Γ, e) where e has a type in some instance of Γ, not only delivers (U, τ) such that $U\Gamma \vdash e : \tau$, but also a variant e' of e where variable bindings are annotated with types.

Proof. The proof is an extension of the proof of [6,9]. We only consider the cases of variables and terms that are new in λ-DRT over the λ-calculus. Define W as follows:

$$
W(\Gamma, x) = \begin{cases} (Id, \tau), & \Gamma = x : \tau, \Gamma' \text{ for some } \Gamma', \\ W((D : \sigma, E : \tau, \Gamma'), x), & \Gamma = (D \otimes E) : (\sigma \otimes \tau), \Gamma', \\ W((z : \sigma, D : \tau, \Gamma'), x) & \Gamma = \langle z, D \rangle : \sigma \times \tau, \Gamma', \\ W(\Gamma', x), & \Gamma = s : \sigma, \Gamma', \text{else}, \\ fail, & \text{else} \end{cases}
$$

$$
W(\Gamma, \langle x, D \rangle) = S\alpha \times S\tau, \quad \text{if } W((x : \alpha, \Gamma), D) = (S, \tau) \text{ for fresh } TyVar \ \alpha
$$

$$
W(\Gamma, (D_1 \otimes D_2)) = \begin{cases} (US_2S_1, (US_2\tau_1 \otimes U\tau_2)), \\ \quad \text{if for some } \tau_1, \tau_2 \text{ and fresh } \alpha_2 \\ \quad W((D_2 : \alpha_2, \Gamma), D_1) = (S_1, \tau_1), \\ \quad W((D_1 : \tau_1, S_1\Gamma), D_2) = (S_2, \tau_2), \\ \quad \text{and } U = mgu(\tau_2, S_2S_1\alpha_2) \neq fail \\ fail, \quad \text{else} \end{cases}
$$

By induction on t, we want to show that for all Γ, S, τ:

(i) $W(\Gamma, t)$ terminates.
(ii) If $W(\Gamma, t) = fail$, then there is no typing of t modulo Γ.
(iii) If $W(\Gamma, t) = (S, \tau)$, then $S\Gamma \vdash t : \tau$ is a principal typing of t modulo Γ.

Case $t = x$: (i) $W(\Gamma, x)$ searches the type context from left to right, unpacking boxes and merge-DRSs to lists of typed referents, and applies (var_1) to the first assumption $x : \tau$ found. Clearly, this terminates. (ii) If $W(\Gamma, x) = fail$, then no assumption $x : \tau$ is found in the (unpacked) context, so x is untypeable, since (var_1) cannot be applied to x. (iii) If $W(\Gamma, x) = (S, \tau)$, then $S = Id$ and $\Gamma = x : \tau, \Gamma'$ for some Γ'. Suppose $R\Gamma \vdash x : \rho$ is a typing of x modulo Γ. Then $R\Gamma = x : R\tau, R\Gamma'$, and hence $\rho = R\tau$ by (var_1). So $R\Gamma \vdash x : \rho$ is obtained from $S\Gamma \vdash x : \tau$ by instantiating with R.

Case $t = (D_1 \otimes D_2)$:

(i) $W(\Gamma, (D_1 \otimes D_2))$ terminates, since by induction, $W((D_2 : \alpha, \Gamma), D_1)$ terminates, for each (S_1, τ_1), $W((D_1 : \tau_1, S_1\Gamma), D_2)$ terminates, and for each (S_2, τ_2), $mgu(\tau_2, S_2S_1\alpha)$ terminates.

(ii) Suppose there is a typing of $(D_1 \otimes D_2)$ modulo Γ. For some S, τ_1, τ_2, the typing derivation ends in

$$\frac{D_2 : \tau_2, S\Gamma \vdash D_1 : \tau_1 \qquad D_1 : \tau_1, S\Gamma \vdash D_2 : \tau_2}{S\Gamma \vdash (D_1 \otimes D_2) : (\tau_1 \otimes \tau_2)} (\otimes).$$

Thus there is a typing of D_1 modulo $D_2 : \alpha_2, \Gamma$, whence, by induction, $W((D_2 : \alpha_2, \Gamma), D_1) \neq fail$, and there is a most general typing $D_2 : S_1\alpha_2, S_1\Gamma \vdash D_1 : \sigma_1$ of D_1 modulo $(D_2 : \alpha_2, \Gamma)$. Since it is most general, there is a type substitution T_1 such that

$$D_2 : \tau_2, S\Gamma \vdash D_1 : \tau_1 \quad \equiv \quad D_2 : T_1S_1\alpha_2, T_1S_1\Gamma \vdash D_1 : T_1\sigma_1.$$

There is also a typing of D_2 modulo

$$(D_1 : \tau_1, S\Gamma) \equiv (D_1 : T_1\sigma_1, T_1S_1\Gamma),$$

hence a typing of D_2 modulo $(D_1 : \sigma_1, S_1\Gamma)$. Therefore, by induction, $W((D_1 : \sigma_1, S_1\Gamma), D_2) \neq fail$, and there is a most general typing $D_1 : S_2\sigma_1, S_2S_1\Gamma \vdash D_2 : \sigma_2$ of D_2 modulo $(D_1 : \sigma_1, S_1\Gamma)$. Since it is most general, there is a type substitution T_2 such that

$$\begin{aligned} D_1 : \tau_1, S\Gamma \vdash D_2 : \tau_2 \quad &\equiv \quad D_1 : T_1\sigma_1, T_1S_1\Gamma \vdash D_2 : \tau_2 \\ &\equiv \quad D_1 : T_2S_2\sigma_1, T_2S_2S_1\Gamma \vdash D_2 : T_2\sigma_2. \end{aligned}$$

So we have $T_2\sigma_2 = \tau_2 = T_1S_1\alpha_2$, and on the type variables of $S_1\Gamma$ and σ_1, $T_1 = T_2S_2$. On type variables β of $S_1\alpha_2$ which are not in $S_1\Gamma$ or σ_1, we have $S_2\beta = \beta$ as S_2 is idempotent. We can assume that β is not in the support of T_2 and put $T_2\beta := T_1\beta$, obtaining $T_1\beta = T_2S_2\beta$. Then from $T_2\sigma_2 = \tau_2 = T_1S_1\alpha_2 = T_2S_2S_1\alpha_2$, we know that σ_2 and $S_2S_1\alpha_2$ unify, so $mgu(\sigma_2, S_2S_1\alpha_2) \neq fail$. By the definition of W, it then follows that $W(\Gamma, (D_1 \otimes D_2)) \neq fail$.

(iii) Suppose $W(\Gamma, (D_1 \otimes D_2)) = (US_2S_1, (US_2\sigma_1 \otimes U\sigma_2))$ with $U, S_1, S_2, \sigma_1, \sigma_2$ as in the definition of W. Then with fresh α_2, $W((D_2 : \alpha_2, \Gamma), D_1) = (S_1, \sigma_1)$, $W((D_1 : \sigma_1, S_1\Gamma), D_2) = (S_2, \sigma_2)$, and $U = mgu(\sigma_2, S_2S_1\alpha_2) \neq fail$. By induction, we know that

(a) $D_2 : S_1\alpha_2, S_1\Gamma \vdash D_1 : \sigma_1$ is a principal typing of D_1 modulo $(D_2 : \alpha_2, \Gamma)$,

(b) $D_1 : S_2\sigma_1, S_2S_1\Gamma \vdash D_2 : \sigma_2$ is a principal typing of D_2 modulo $(D_1 : \sigma_1, S_1\Gamma)$.

By specializing the typing in (a) with US_2 and the one in (b) with U, one obtains

$$D_2 : US_2S_1\alpha_2, US_2S_1\Gamma \vdash D_1 : US_2\sigma_1,$$
$$\text{and} \quad D_1 : US_2\sigma_1, US_2S_1\Gamma \vdash D_2 : U\sigma_2.$$

Since $US_2S_1\alpha_2 = U\sigma_2$, we can apply the rule $(\otimes)$ and obtain a typing

$$US_2S_1\Gamma \vdash (D_1 \otimes D_2) : (US_2\sigma_1 \otimes U\sigma_2)$$

of $(D_1 \otimes D_2)$ modulo Γ. It remains to be shown that this is a most general typing.

So suppose $(D_1 \otimes D_2)$ has a typing modulo Γ. The last step in the typing derivation is

$$\frac{D_2 : \tau_2, S\Gamma \vdash D_1 : \tau_1, \qquad D_1 : \tau_1, S\Gamma \vdash D_2 : \tau_2}{S\Gamma \vdash (D_1 \otimes D_2) : (\tau_1 \otimes \tau_2)} \; (\otimes).$$

For the left subderivation of $D_2 : \tau_2, S\Gamma \vdash D_1 : \tau_1$ we may assume $\tau_2 = S\alpha_2$ for some fresh type variable α_2. So D_1 has a typing (S, τ_1) modulo $D_2 : \alpha_2, \Gamma$. By a) there is a type substitution T_1 such that $(S, \tau_1) = (T_1S_1, T_1\sigma_1)$, whence

$$D_2 : \tau_2, S\Gamma \vdash D_1 : \tau_1 \quad \equiv \quad D_2 : T_1S_1\alpha_2, T_1S_1\Gamma \vdash D_1 : T_1\sigma_1.$$

Now the right subderivation $D_1 : \tau_1, S\Gamma \vdash D_2 : \tau_2$ is a derivation of

$$D_1 : T_1\sigma_1, T_1S_1\Gamma \vdash D_2 : T_1S_1\alpha_2,$$

which is a typing of D_2 modulo $(D_1 : \sigma_1, S_1\Gamma)$. By b), there is a type substitution T_2 with

$$D_1 : \tau_1, S\Gamma \vdash D_2 : \tau_2 \quad \equiv \quad D_1 : T_2S_2\sigma_1, T_2S_2S_1\Gamma \vdash D_2 : T_2\sigma_2.$$

It follows that

$$S\Gamma \vdash (D_1 \otimes D_2) : (\tau_1 \otimes \tau_2) \equiv \& T_2S_2S_1\Gamma \vdash (D_1 \otimes D_2) : (T_2S_2\sigma_1 \otimes T_2\sigma_2).$$

To show that this is an instance of the typing

$$US_2S_1\Gamma \vdash (D_1 \otimes D_2) : (US_2\sigma_1 \otimes U\sigma_2),$$

we need a type substitution R such that $T_2 = RU$ on the type variables of $S_2S_1\Gamma$, $S_2\sigma_1$ and σ_2. We have $T_2\sigma_2 = T_1S_1\alpha_2$. As in (ii), $T_1 = T_2S_2$ on the type variables of $S_1\alpha_2$, so $T_2\tau_2 = T_2S_2S_1\alpha$, and since $U = mgu(\tau_2, S_2S_1\alpha_2)$, $T_2 = RU$ on the type variables of τ_2 and $S_2S_1\alpha_2$. On other type variables β, we have $U\beta = \beta = R\beta$ and can redefine $R\beta := T_2\beta$, to obtain $T_2 = RU$ on all type variables of $S_2S_1\Gamma$, $S_2\sigma_1$ and σ_2.

The remaining cases of t can be treated similarly.

Example 1. The lexicon entry for the indefinite article a in λ-DRT of [13] is

$$\lambda P \lambda Q (\delta x_i \top \otimes P(\hat{} x_i) \otimes Q(\hat{} x_i)) : (d,t), ((d,t),t)$$

where d is the type of individual concepts and t the type of DRSs. Simplifying this to the extensional case and using the DRS-notation from above, type reconstruction yields the principal typing

$$\vdash \lambda P \lambda Q (\langle [x], \top \rangle \otimes Px \otimes Qx) : (\alpha \to \gamma) \to (\alpha \to \delta) \to [\alpha] \otimes \gamma \otimes \delta.$$

Instead of the basic type t for DRSs in [13], we have infinitely many types $[\sigma_1, \ldots, \sigma_n]$. Moreover, we have the principal typing

$$man' : e \to t \ \vdash \ \lambda x \langle [], man' \, x \rangle : e \to \mathbb{1}.$$

The unreduced meaning term for a man therefore is

$$\lambda P \lambda Q (\langle [x], \top \rangle \otimes Px \otimes Qx) \cdot \lambda x \langle [], man' \, x \rangle$$

and has the principal type $(e \to \delta) \to [e] \otimes \mathbb{1} \otimes \delta$.

For the kind of semantic checking of natural language text that we are interested in, we need to distinguish between different sorts of individuals. Lexical entries should assign different base types to the arguments of content words, in particular verbs and nouns. It is then useful, if not imperative, to have a lexicon with polymorphic types for the functional words like the indefinite article above, rather than be forced to put into the lexicon all the instance types needed for a specific application.

The type-checking in texts is slightly different from the one in programs: in programs, we need to check that in applications $f(a_1, \ldots, a_n)$, the type of the arguments equal (or are subtypes of) the argument types of the function, while in texts, in predications $v(np_1, \ldots, np_k)$ the types of the (generally quantified) argument noun phrases have to be related by type-raising to the argument types of the verb.

But in principle, we want to have the same *phase distinction* between type checking and evaluation: we want to build meaning terms according to the syntactic structure, then check if the meaning is typable, and only then perform semantic evaluation. Thus, evaluation only needs to be defined on typed expressions, and type checking would be pointless if evaluation would not preserve the type of expressions.

5   Reduction

We assume the familiar β-reduction and congruence rules of λ-calculus,

$$\frac{}{(\lambda x D \cdot s) \to D[x/s]} \, (\beta) \qquad \frac{D \to D'}{\lambda x D \to \lambda x D'},$$

$$\frac{D \to D'}{(D \cdot E) \to (D' \cdot E)}, \qquad \frac{E \to E'}{(D \cdot E) \to (D \cdot E')}.$$

The intended meaning of the merge $(D_1 \otimes D_2)$ of two value-DRSs with disjoint referent lists, $D_1 = \langle [x_1, \ldots, x_n], \varphi \rangle$ and $D_2 = \langle [y_1, \ldots, y_m], \psi \rangle$, is the value-DRS

$$\langle [x_1, \ldots, x_n, y_1, \ldots, y_m], (\varphi \wedge \psi) \rangle.$$

We therefore define the reduction (resp. evaluation) of DRS-expressions by the following δ-*reduction* rules:

$$\frac{}{\langle [], \varphi \rangle \otimes \langle [], \psi \rangle \rightarrow \langle [], (\varphi \wedge \psi) \rangle} \; (\delta_1), \quad \frac{}{\langle [], \varphi \rangle \otimes \langle y, E \rangle \rightarrow \langle y, \langle [], \varphi \rangle \otimes E \rangle} \; (\delta_2),$$

$$\frac{}{\langle x, D \rangle \otimes E \rightarrow \langle x, D \otimes E \rangle} \; (\delta_3).$$

From these, the intended meaning for the merge of value-DRSs follows:

$$\langle [x_1, \ldots, x_n], \varphi \rangle \otimes \langle [y_1, \ldots, y_m], \psi \rangle \rightarrow^* \langle [x_1, \ldots, x_n, y_1, \ldots, y_m], (\varphi \wedge \psi) \rangle.$$

In order to use (δ_1) - (δ_3), by reductions we must achieve that arguments of $\otimes$ are value-DRSs. Hence we also need congruence rules for $\delta = \langle \cdot, \cdot \rangle$ and $\otimes$:

$$\frac{D \rightarrow E}{\langle x, D \rangle \rightarrow \langle x, E \rangle} \; (\delta_4), \quad \frac{D \rightarrow D'}{(D \otimes E) \rightarrow (D' \otimes E)} \; (\delta_5), \quad \frac{E \rightarrow E'}{(D \otimes E) \rightarrow (D \otimes E')} \; (\delta_6),$$

so that reductions can be performed in subterms of $\langle x, D \rangle$, $(D \otimes E)$ as well as $\lambda x D$ and $(D_1 \cdot D_2)$. Then the following reduction rules are derivable:

$$\frac{E \rightarrow^* E'}{\langle [], \varphi \rangle \otimes \langle y, E \rangle \rightarrow^* \langle y, \langle [], \varphi \rangle \otimes E' \rangle} \; (\delta_2^+), \quad \frac{D \rightarrow^* D', \quad E \rightarrow^* E'}{\langle x, D \rangle \otimes E \rightarrow^* \langle x, D' \otimes E' \rangle} \; (\delta_3^+).$$

Normalization

It is obvious that applications of the δ-reduction rules do not lead to new occurrences of β-redexes. Therefore, expressions can be reduced by first performing β-reductions as long as possible, and only then apply δ-reduction rules. If we start with a typed expression, then from the strong normalization property for simply typed λ-calculus the first will terminate. It is also clear that the δ-reduction rules cannot lead to infinite reduction sequences.

Notice that on value-DRSs with disjoint top-level referents, $\otimes$ is associative, if we consider formula conjunction to be associative, i.e. use list $[\varphi_1, \ldots, \varphi_n]$ of formulas, as we do in Sect. 6.2.

We would like to show that in a derivable typing statement $\Gamma \vdash s : \sigma$, where the "predicate" σ applies to the "subject" s, we may reduce the subject and still the predicate σ applies. However, this is not quite true: when we reduce a merge-DRS, the type constructor $\times$ is interpreted as a *cons* of a referent and a referent list, and $\otimes$ is interpreted as an *append* of referent lists, and since the type of a DRS mirrors its construction, we need to *cons* resp. *append* the lists of types of the referents.

We use the following type reductions, which amount to a recursive definition of *append* ($\otimes$) in terms of the empty list ($\mathbb{1}$) and *cons* ($\times$):

$$\frac{}{\mathbb{1} \otimes \mathbb{1} \rightharpoonup \mathbb{1}}\,(\delta_1')\qquad\frac{}{\mathbb{1} \otimes (\sigma \times \rho) \rightharpoonup \sigma \times (\mathbb{1} \otimes \rho)}\,(\delta_2')$$

$$\frac{}{(\sigma \times \rho) \otimes \tau \rightharpoonup \sigma \times (\rho \otimes \tau)}\,(\delta_3')$$

Moreover, type reduction may operate on embedded type expressions:

$$\frac{\sigma \rightharpoonup \sigma'}{\sigma \times \tau \rightharpoonup \sigma' \times \tau}\,(\times')\qquad\frac{\tau \rightharpoonup \tau'}{\sigma \times \tau \rightharpoonup \sigma \times \tau'}\,(\times'')$$

$$\frac{\sigma \rightharpoonup \sigma'}{\sigma \otimes \tau \rightharpoonup \sigma' \otimes \tau}\,(\otimes')\qquad\frac{\tau \rightharpoonup \tau'}{\sigma \otimes \tau \rightharpoonup \sigma \otimes \tau'}\,(\otimes'')$$

$$\frac{\sigma \rightharpoonup \sigma'}{(\sigma \rightarrow \tau) \rightharpoonup (\sigma' \rightarrow \tau)}\,(\rightarrow')\qquad\frac{\tau \rightharpoonup \tau'}{(\sigma \rightarrow \tau) \rightharpoonup (\sigma \rightarrow \tau')}\,(\rightarrow'')$$

Example 1. (continued) Reducing the above term

$$\lambda P \lambda Q(\langle [x], \top \rangle \otimes Px \otimes Qx) \cdot \lambda x \langle [], man'\, x \rangle$$

by β-reductions gives $\lambda Q(\langle [x], \top \rangle \otimes \langle [], man'\, x \rangle \otimes Qx)$ and reducing further by δ-reductions leads to

$$\lambda Q(\langle [x], \top \wedge man'\, x \rangle \otimes Qx).$$

Its principal type $(e \rightarrow \delta) \rightarrow [e] \otimes \delta$ is obtained from the one of the unreduced term by applications of $(\otimes')$, (δ_3'), and (δ_1') that simplify $[e] \otimes \mathbb{1} \otimes \delta$ to $[e] \otimes \delta$.

Since our types of DRSs closely reflect the construction of their top-level referent lists, in order to have a subject reduction property we need to consider types equivalent when they get equal by interpreting $\otimes$ as *append*, $\times$ as *cons*, and $\mathbb{1}$ as the empty list.

A more serious obstacle to subject-reduction is the typing rule (var_6) which permits us to ignore assumptions $(D_1 \cdot D_2) : \sigma$. In fact, the subject-reduction property does not hold in general.

Example 2. Consider the application of

$$\frac{D_2 : \tau_2, \Gamma \vdash D_1 : \tau_1 \qquad D_1 : \tau_1, \Gamma \vdash D_2 : \tau_2}{\Gamma \vdash (D_1 \otimes D_2) : (\tau_1 \otimes \tau_2)}\,(\otimes)$$

Suppose $(D_1 \otimes D_2) \rightarrow (D_1' \otimes D_2)$ via $D_1 \rightarrow D_1'$. As we have seen above, we may have $x \in top(D_1') \setminus top(D_1)$. In the left subderivation $D_1 : \tau_1, \Gamma \vdash D_2 : \tau_2$, a free occurrence of x in D_2 gets its type from Γ, while in the context $D_1' : \tau_1, \Gamma$, it gets its type from $D_1' : \tau_1$. Hence, it may be impossible to obtain $D_1' : \tau_1, \Gamma \vdash D_2 : \tau_2$. (For example, take $D_1 : \tau_1 = \lambda y \langle x, E \rangle \cdot a : (\sigma \times \tau)$, $D_2 : \tau_2 = \langle [], Px \rangle : \mathbb{1}$.) Thus, $\Gamma \vdash (D_1 \otimes D_2) : (\tau_1 \otimes \tau_2)$ does not imply $\Gamma \vdash (D_1' \otimes D_2) : (\tau_1 \otimes \tau_2)$.

The problem similarly arises for $(D_1 \Rightarrow D_2)$, $(D_1 \vee D_2)$, or $(D_1 \, ; \, D_2)$, where D_1 may β-reduce to a DRS with a new top-level referent occurring free in D_2. This is a defect of λ-DRT terms which admit the binding part D_1 of such expressions to arise from a β-redex like $(\lambda z z \cdot D_1)$.

We will sidestep this problem for the application to pronoun resolution below by assuming

1. all λ-DRS-expressions used as meanings of lexical entries are closed and in normal form,
2. in substitution $t[x/s]$, bound variables (including referents) in t are renamed to make them distinct from free variables of s,
3. in $t[x/s]$, s is in normal form, and referents of s are renamed at each occurrence of x in t (in merge-factors, so that their scope does not extend).[3]
4. all bound variables are pairwise distinct; in particular, no referent is used twice as a binding variable.

In particular, we will use a call-by-value strategy when computing the meaning of phrases: if the meaning of a phrase is an application $\lambda x t \cdot s$, we will have $\lambda x t$ and s in normal form, and deliver a normal form $nf(t[x/s])$ of $t[x/s]$ as value, see the computation rules in Sect. 6.2. We think that the following weak form of the subject reduction property holds under the above assumptions:

Conjecture 1. If t and s are in normal form, and $\Gamma \vdash (\lambda x t \cdot s) : \tau$, then there is τ' with $\tau \rightarrow^* \tau'$ and $\Gamma \vdash nf(t[x/s]) : \tau'$.

However, we do not make use of that in the following; termination of reduction suffices.

6   Application to Pronoun Resolution

There are two possible ways to combine type reconstruction and pronoun resolution. Either one applies a pronoun resolution algorithm and then uses type reconstruction to check if the resolution is type-correct, or one first applies type reconstruction and then does pronoun resolution by exploiting the type information.

6.1   Type Informed Pronoun Resolution

The second way has been implemented [22]. It roughly proceeds as follows:

- Step 1: for each pronoun occurrence, introduce a fresh discourse referent x and extend the DRS by an anaphor-declaration like $anp(x, fem, sg)$. For the discourse referent y of each noun phrase that is not a pronoun, add an antecedent-declaration like $ant(y, masc, sg)$ to the DRS.

---

[3] Notice that $\lambda P(P \otimes P) \cdot \langle [z], [\varphi] \rangle$ then reduces to $(\langle [z_1], [\varphi(z/z_1)] \rangle \otimes \langle [z_2], [\varphi(z/z_2)] \rangle$, and further to $\langle [z_1, z_2], [\varphi(z/z_1), \varphi(z/z_2)] \rangle$, like turning $(\exists z \varphi \wedge \exists z \varphi)$ into prenex form $\exists z_1 \exists z_2 (\varphi_1(z/z_1) \wedge \varphi(z/z_2))$.

- Step 2: apply type reconstruction to get a most general typing for the discourse, including individual types e_i for discourse referents x as inferred from the occurrence context of the pronoun.
- Step 3: "resolve" an anaphoric (or cataphoric) pronoun by unifying its typed discourse referent $x : \alpha$ with some discourse referent $y : \beta$ of a possible antecedent *of the same type*, observing the grammatical properties of gender and number in the corresponding declarations $anp(x, g_x, n_x)$ and $ant(y, g_y, n_y)$.

A more detailed description is best obtained by explaining the relevant parts of the Prolog-program of [22].

A parse tree is represented as a list [Root|Subtrees] where the root is the syntactic category of the parsed expression. A discourse is either empty, with tree [d], or the extension of a discourse by a sentence, and then has tree [d,S,D] where S is the parse tree of the final sentence and D the parse tree of the initial discourse.[4] For each parse tree, sem(+Tree,-DRS) computes a number of meanings. If the tree is a discourse, each meaning is a typed λ-DRS, otherwise an untyped λ-DRS in normal form.

```
% sem(+ParseTree,-DRS); for a discourse, DRS is typed
...
sem([d], drs([],[])) :- !.
sem([d,S,D], Sem) :-
    !, sem(S,SemS), sem(D,SemD), resolve(SemS,SemD,Sem).
```

Having computed a typed meaning SemD for the initial discourse and an untyped meaning SemS for the final sentence, we try to resolve anaphors of SemS, using SemD as accessible DRS for possible antecedents.

```
% resolve(+SemS,+SemD,-Sem)
resolve(SemS,SemD,Sem) :-
    type(⌊⌋,SemS,SemSTy,_TypS),
    resolve_drs([SemSTy,SemD],[DrsS,DrsD]),
    mergeTerm(DrsD + DrsS, Sem).
```

First, type reconstruction type/4 is applied to SemS; as pronouns get fresh discourse referents in SemS, we can use the empty type context to find a principal type TypS for the DRS SemS. Actually, we use a modification of the type reconstruction algorithm that also returns a typed version SemSTy of SemS, which has type annotations at variable bindings (including referents in referent lists). This typed DRS SemSTy is resolved with SemD as accessible DRS, using resolve_drs/2; the modifications DrsS and DrsD are finally merged by appending the referents and formulas of DrsS to those of DrsD.

To resolve a DRS drs(Refs,Fmls) with respect to a stack Ds1 of partially resolved accessible DRSs, we go through the formulas, which may contain unresolved DRSs, resolve these, and construct a resolved form of drs(Refs,Fmls) on top of the stack:

---

[4] To prevent Prolog's top-down parsing strategy from diverging for left-recursive grammar rules d -> d, s., we use a right-recursive rule d --> s, d. for discourses and reverse the sequence of input sentences before parsing.

```
% resolve_drs(+DRSs, -resolvedDRSs)
resolve_drs([drs(Refs,Fmls)|Ds1],RDs):-
    resolve_fml(Fmls,[drs(Refs,[])|Ds1],RDs).
```

If a formula is built from DRSs, like $(D_1 \Rightarrow D_2)$, $(D_1 \vee D_2)$, or $\neg D$, the component DRSs are resolved in term, respecting the accessibility conditions of DRT, and the formula built form the resolved component DRSs is added to the result-DRS under construction, before the remaining formulas are processed:

```
% resolve_fml(+Fmls,[?resultDRS|+accessDRSs],-resolvedDRSs)
resolve_fml([(D1 => D2)|Fmls],Ds,RDs):-
    !, resolve_drs([D1|Ds],[D1r|Dsr]),
    resolve_drs([D2,D1r|Dsr],[D2R,D1R,drs(R,F)|Ds3]),
    resolve_fml(Fmls,[drs(R,[(D1R => D2R)|F])|Ds3],RDs).
```

. . .

If the formula is an anaphor `anp(Ref,Gen,Num)` with typed(!) referent `Ref` and gender and number information, one tries to find a suitable antecedent in the result-DRS under construction (i.e. in the pronoun's left textual context in the current sentence) or the accessible DRSs, or in the remaining formulas of the DRS currently under process:

```
resolve_fml([anp(Ref,Gen,Num)|Fmls], [drs(R,F)|Ds1], RDs) :-
    !, ( ( % in sentence prefix or previous sentences
           resolve_anp(Ref,Gen,Num,[drs(R,F)|Ds1])
         ; % in sentence suffix
           resolve_anp(Ref,Gen,Num,[drs(R,Fmls)])
         ),
         delete_ref(Ref,R,NewR),  % omit duplicates of Ref
         NewD = drs(NewR,F) % omit anp(Ref,..) in the result DRS
       ; NewD = drs(R,[anp(Ref,Gen,Num)|F]) % or: fail, to
       ),                                    % exclude unre-
    resolve_fml(Fmls,[NewD|Ds1],RDs).        % solved anaphors
```

Possessive pronouns are handled by looking for antecedents in their left context only.

To find a suitable antecedent, simply choose some of the accessible DRSs and some antecedent among its formulas that can be unified with the referent:

```
% resolve_anp(+Ref,+Gen,+Num,+DRSs)
resolve_anp(Ref,Gen,Num,Ds):-
    member(drs(_Refs,Fs),Ds),
    member(ant(Ref,Gen,Num),Fs).
```

By using the same variables `Ref`, `Gen`, `Num`, Prolog unifies a typed anaphor `R:Ty` with a typed antecedent `R':Ty'` of the same number and gender features.

Atomic formulas can just be transferred to the result-DRS under construction, and when all formulas of the DRS are processed, the sequence of resolved formulas is reversed to its expected order:

```
resolve_fml([Fml|Fmls],[drs(R,F)|Ds1],RDs):-
    !, resolve_fml(Fmls,[drs(R,[Fml|F])|Ds1],RDs).

resolve_fml([],[drs(R,F)|Ds],[drs(R,Frev)|Ds]):-
    !, reverse(F,Frev).
```

The stack of resolved DRSs with a resolved form of the DRS `drs(Refs,Fmls)` on top is returned.

6.2 Example

We assume that nouns N and relational nouns RN are classified according to gender $g \in \{m, f, n\}$ (masculine, feminine, neuter), and implicitly inflect for number $n \in \{sg, pl\}$ and case c. (We use gender m as in the corresponding German nouns and pronouns to get more possible antecedents below.)

1. Content words are assigned a meaning and a type in the lexicon, for example:

expression	meaning	type
$Galilei : PN$	$galilei$	h
$Jupiter : PN$	$jupiter$	s
$astronomer : N$	$\lambda x \langle [], [ant(x, m, sg), astronomer(x)] \rangle$	$h \rightarrow \mathbb{1}$
$star : N$	$\lambda x \langle [], [ant(x, m, sg), star(x)] \rangle$	$s \rightarrow \mathbb{1}$
$moon : RN$	$\lambda x \lambda y \langle [], [ant(x, m, sg), moon(x, y)] \rangle$	$s \rightarrow (s \rightarrow \mathbb{1})$
$shine : V$	$\lambda x \langle [], [shine(x)] \rangle$	$s \rightarrow \mathbb{1}$
$observe : TV$	$\lambda x \lambda y \langle [], [observe(x, y)] \rangle$	$h \rightarrow (s \rightarrow \mathbb{1})$
$discover : TV$	$\lambda x \lambda y \langle [], [discover(x, y)] \rangle$	$h \rightarrow (s \rightarrow \mathbb{1})$

Pronouns inflect for number, gender, and case, if we consider person fixed to 3rd person. Like determiners, pronouns have polymorphic type; i.e. from their untyped λ-DRS-meaning we reconstruct their most general (schematic) type.

expression	meaning	principal type
$he : Pron$	$\lambda P(\langle [x], [anp(x, m, sg)] \rangle \otimes P\,x)$	$(\alpha \rightarrow \beta) \rightarrow [\alpha] \otimes \beta$
$she : Pron$	$\lambda P(\langle [x], [anp(x, f, sg)] \rangle \otimes P\,x)$	$(\alpha \rightarrow \beta) \rightarrow [\alpha] \otimes \beta$
$his : PossPron$	$\lambda R \lambda P(\langle [x, y], [anposs(y, m, sg)] \rangle$	$(\alpha \rightarrow \beta \rightarrow \gamma)$
	$\otimes (R\,x\,y \otimes P\,x))$	$\rightarrow (\alpha \rightarrow \delta)$
		$\rightarrow [\alpha, \beta] \otimes \gamma \otimes \delta$
$who : RelPron$	$\lambda P \lambda x\,(P\,x)$	$(\alpha \rightarrow \beta) \rightarrow (\alpha \rightarrow \beta)$
$a : Det$	$\lambda N \lambda P(\langle [x], [] \rangle \otimes (N\,x \otimes P\,x))$	$(\alpha \rightarrow \beta) \rightarrow (\alpha \rightarrow \gamma)$
		$\rightarrow [\alpha] \otimes \beta \otimes \gamma$
$every : Det$	$\lambda N \lambda P \langle [], [(\langle [x], [] \rangle \otimes N\,x) \Rightarrow P\,x] \rangle$	$(\alpha \rightarrow \beta) \rightarrow (\alpha \rightarrow \gamma)$
		$\rightarrow \mathbb{1}$
eq	$\lambda x \lambda y.eq(x, y)$	$\alpha \rightarrow (\alpha \rightarrow t)$

Each use of a personal, relative, or possessive pronoun uses a new referent x. Moreover, eq, anp, $anposs$, ant have polymorphic lexical (not reconstructed) type.

2. Compound expressions are built according to grammar rules; each grammar rule is accompanied by one or several meaning computation rules. Some examples are:

$$\frac{p : PN_g}{p : NP} (S\ 1) \qquad \frac{p'}{\lambda P(\langle [x], [\, ant(x, g, sg),\ eq(x, p')]\rangle \otimes P \cdot x)} (C\ 1)$$

$$\frac{p : Pron_{g,n}}{p : NP} (S\ 2) \qquad\qquad \frac{p'}{p'} (C\ 2)$$

$$\frac{d : Det \quad n : N}{d\,n : NP} (S\ 3) \qquad\qquad \frac{d' \quad n'}{nf(d' \cdot n')} (C\ 3)$$

$$\frac{p : PossPron \quad r : RN}{p\,r : NP} (S\ 4) \qquad\qquad \frac{p' \quad r'}{nf(p' \cdot r')} (C\ 4)$$

$$\frac{np_1 : NP \quad v : TV \quad np_2 : NP}{np_1\,v\,np_2 : S} (S\ 5) \qquad \frac{np_1' \quad v' \quad np_2'}{nf(np_1' \cdot \lambda x(np_2' \cdot \lambda y(v' \cdot x \cdot y)))} (C\ 5)$$

$$\frac{}{\epsilon : \mathcal{D}} (S\ 6) \qquad\qquad \frac{}{\langle [], []\rangle} (C\ 6)$$

$$\frac{d : \mathcal{D} \quad s : S}{d\,;\,s : \mathcal{D}} (S\ 7) \qquad\qquad \frac{d' \quad s'}{(d'' \otimes s'')} (C\ 7)$$

An additional computation rule $(C\ 5')$ for sentences $np_1\,v\,np_2 : S$ might give np_2 wide scope. In $(C\ 7)$, d'' and s'' are obtained by pronoun-resolution from most general typings of d' and s' in the empty type context, i.e. $resolve(s', d', d'' \otimes s'')$ by the resolution algorithm explained above.

3. Let us consider the sample discourse *Galilei observed a star. He discovered his moon.* The first sentence is constructed with (S 1), (S 3), and (S 5). We compute the meaning of the subject as

$$np_1' = \lambda P(\langle [x], [\, ant(x, m, sg), eq(x, galilei)]\rangle \otimes P \cdot x),$$

the meaning of the object as

$$np_2' = nf(\lambda N\lambda P(\langle [x], []\rangle \otimes (N\,x \otimes P\,x)) \cdot \lambda x\langle [], [\, ant(x, m, sg), star(x)]\rangle)$$
$$= nf(\lambda P(\langle [x], []\rangle \otimes (\langle [], [\, ant(x, m, sg), star(x)]\rangle \otimes P\,x)))$$
$$= \lambda P(\langle [x], [\, ant(x, m, sg), star(x)]\rangle \otimes P\,x)$$

and from those obtain the sentence meaning by the computation rule for (S 5) as

$$s_1' = nf(np_1' \cdot \lambda x(np_2' \cdot \lambda y(v' \cdot x \cdot y)))$$
$$= nf(np_1' \cdot \lambda x(np_2' \cdot \lambda y(\langle[], [observe(x,y)]\rangle)))$$
$$= nf(np_1' \cdot \lambda x.(\langle[x], [ant(x,m,sg), star(x)]\rangle \otimes P\,x)[P/\lambda y\langle[], [observe(x,y)]\rangle])$$
$$= nf(np_1' \cdot \lambda x(\langle[\tilde{x}], [ant(\tilde{x},m,sg), star(\tilde{x})]\rangle \otimes \langle[], [observe(x,\tilde{x})]\rangle))$$
$$= nf(np_1' \cdot \lambda x\langle[y], [ant(y,m,sg), star(y), observe(x,y)]\rangle)$$
$$= nf(((\langle[x], [ant(x,m,sg), eq(x, galilei)]\rangle \otimes P\,x)[P/\lambda x\langle[y], [ant(\ldots), \ldots]\rangle])$$
$$= nf(((\langle[x], [ant(x,m,sg), eq(x, galilei)]\rangle \otimes \langle[y], [ant(y,m,sg), \ldots]\rangle)))$$
$$= \langle[x,y], [ant(x,m,sg), eq(x, galilei), ant(y,m,sg), star(y), observe(x,y)]\rangle$$

From the type assumptions for nouns and verbs (and eq), type reconstruction can annotate the bound variables of s_1' as

$$\langle[x:h, y:s], [ant(x:h,m,sg), eq(x, galilei), ant(y:s,m,sg), \ldots]\rangle$$

and return a most general type $\langle[h,s], t\rangle$. In the second sentence, the subject *he* has meaning

$$np_1' = \lambda P(\langle[x], [anp(x,m,sg)]\rangle \otimes P\,x),$$

which receives the following annotation and principal type:

$$\lambda P : (\alpha \to \beta)(\langle[x:\alpha], [anp(x:\alpha,m,sg)]\rangle \otimes P\,x) : (\alpha \to \beta) \to [\alpha] \otimes \beta.$$

The object *his moon* gets the meaning[5]

$$np_2' = nf(\lambda R\lambda P(\langle[x,y], [anposs(y,m,sg)]\rangle \otimes (R\,x\,y \otimes P\,x))$$
$$\cdot \lambda x\lambda y\langle[], [ant(x,m,sg), moon(x,y)]\rangle)$$
$$= \lambda P(\langle[x,y], [anposs(y,m,sg), ant(x,m,sg), moon(x,y)]\rangle \otimes P\,x),$$

which type reconstruction annotates to

$$\lambda P : (s \to \alpha)(\langle[x:s, y:s], [\,anposs(y:s,m,sg),$$
$$ant(x:s,m,sg), moon(x,y)]\rangle \otimes P\,x)$$

and to which it assigns a most general type $(s \to \alpha) \to [s,s] \otimes \alpha$. By the computation rule for (S 5), the meaning of the second sentence is

$$s_2' = nf(np_1' \cdot \lambda x(np_2' \cdot \lambda y(v' \cdot x \cdot y)))$$
$$= nf(np_1' \cdot \lambda x(np_2' \cdot \lambda y\langle[], [discover(x,y)]\rangle)))$$
$$= nf(np_1' \cdot \lambda x\langle[\tilde{x},y], [\,anposs(y,m,sg), ant(\tilde{x},m,sg), moon(\tilde{x},y),$$
$$discover(x,\tilde{x})]\rangle)$$
$$= nf(((\langle[x], [anp(x,m,sg)]\rangle \otimes P\,x)[P/\lambda x\langle[\tilde{x},y], [anposs(y,m,sg), \ldots,]\rangle)$$
$$= \langle[x,\tilde{x},y], [\,anp(x,m,sg), anposs(y,m,sg),$$
$$ant(\tilde{x},m,sg), moon(\tilde{x},y), discover(x,\tilde{x})]\rangle.$$

---

[5] By an additional reduction $D_1 \otimes (D_2 \otimes D_3) \to (D_1 \otimes D_2) \otimes D_3$ when D_1, D_2 are value-DRSs.

If several computation rules can be applied, a sentence can get several untyped meanings this way. As normalisation has to return fresh bound variables, we write

$$s_2' = \langle [u, v, z], [\, anp(u, m, sg), anposs(z, m, sg),$$
$$ant(v, m, sg), moon(v, z), discover(u, v)]\rangle.$$

4. Pronoun resolution for the discourse $\epsilon^{\bullet}; s_1 ; s_2$ proceeds as follows.

   (a) The most general typing of the meaning $\langle [], []\rangle$ of ϵ in the empty context is $\vdash \langle [], []\rangle : \mathbb{1}$.

   (b) Type reconstruction is applied to the first sentence, followed by pronoun resolution with $\langle [], []\rangle : \mathbb{1}$. As no pronoun occurred in s_1, the type-annotated version of s_1' is returned:

$$s_1'' = \langle [x : h, y : s], [\, ant(x : h, m, sg), eq(x, galilei),$$
$$ant(y : s, m, sg), star(y), observe(x, y)]\rangle$$
$$= \langle [], []\rangle \otimes s_1''.$$

   (c) Type reconstruction is applied to (each of) the meaning(s) of the next sentence, followed by pronoun resolution with s_1''. Here type reconstructions just returns

$$s_2'' = \langle [u : h, v : s, z : s], [\, anp(u : h, m, sg), anposs(z : s, m, sg),$$
$$ant(v : s, m, sg), moon(v, z), discover(u, v)]\rangle,$$

   where the types of u, v, z are derived from the argument types of nouns and verbs whose argument positions they occupy. The anaphor $u : h$ has no antecedent in the current sentence, as $v : s$ has different type. Assuming that possessives have to be resolved in their left context, the possessive anaphor $z : s$ also cannot be resolved against $v : s$.

   (d) Pronouns of s_2 may also be resolved against antecedents in the type-annotated left context, s_1''. For each typed anaphor, we search for a suitably typed antecedent, unify the referents and remove the anaphor referent in the DRS of the current sentence, s_2''. For the anaphor $anp(u : h, m, sg)$, the only type-compatible antecedent in s_1'' is $ant(x : h, m, sg)$, so we unify u with x (i.e. rename u by x in s_2''), remove $x : h$ from its referent list and $anp(x : h, m, sg)$ from its formulas, getting a partially resolved DRS

$$\langle [v : s, z : s], [\, anposs(z : s, m, sg), ant(v : s, m, sg),$$
$$moon(v, z), discover(x, v)]\rangle.$$

   The next formula is a possessive anaphor $anposs(z : s, m, sg)$. As we want these to be resolved in their left context only, $z : s$ cannot be resolved against $v : s$. But it can be resolved against $ant(y : s, m, sg)$ in s_1'', which leads to

$$r(s_2'') = \langle [v : s], [ant(v : s, m, sg), moon(v, y), discover(x, v)]\rangle$$

   as the resolved"'result"'-DRS of s_2''.

(e) Finally, the resolved version of s_2'' is merged with s_1'', yielding

$$s_1'' \otimes r(s_2'') = \langle [x : h, y : s, v : s],$$
$$[ant(x : h, m, sg), eq(x, galilei), ant(y : s, m, sg), star(y),$$
$$observe(x, y), ant(v : s, m, sg), moon(v, y), discover(x, v)] \rangle$$

as the typed meaning of the discourse $d = \epsilon; s_1; s_2$.

6.3 Type Reconstruction for Bach–Peters-Sentences

One of the motivations for the *symmetric* merge-operator $\otimes$ was hinted at, but not elaborated in [13, p. 480]: the potential to treat Bach-Peters-sentences "in which two phrases are connected by both an anaphor and a cataphor", like [*The boy who deserved it$_y$*]$_x$ *got* [*the prize he$_x$ wanted*]$_y$. We use variants of (S 2), (S 5) and (C 2), (C 5) as syntax and computation rules for relative clauses

$$\frac{p : RelPron}{p : RelNP} \ (S\ 2') \qquad\qquad\qquad \frac{p'}{p'} \ (C\ 2')$$

$$\frac{np_1 : RelNP \quad np_2 : NP \quad v : TV}{np_1 \ np_2 \ v : RelS} \ (S\ 5') \qquad \frac{np_1' \quad v' \quad np_2'}{nf(np_1' \cdot \lambda x(np_2' \cdot \lambda y(v' \cdot x \cdot y)))} \ (C\ 5')$$

$$\frac{d : Det \quad n : N \quad s : RelS}{d \ n \ s : NP} \ (S\ 8) \qquad \frac{d' \quad n' \quad s'}{nf(d' \cdot \lambda x(n' x \land s' x))} \ (C\ 8)$$

Omitting the grammatical features and the uniqueness conditions for the definite article, the untyped meaning of *a boy who deserves it gets the prize he wanted* is obtained via

$$\lambda P(\begin{array}{|l|}\hline x, y \\\hline ant(x) \\ boy(x) \\ anp(y) \\ deserve(x, y) \\\hline\end{array} \otimes P x) \cdot \lambda z \begin{array}{|l|}\hline x', y' \\\hline ant(y') \\ prize(y') \\ anp(x') \\ want(x', y') \\ get(z, y') \\\hline\end{array} \ \to_\beta \ \begin{array}{|l|}\hline x, y \\\hline ant(x) \\ boy(x) \\ anp(y) \\ deserve(x, y) \\\hline\end{array} \otimes \begin{array}{|l|}\hline x', y' \\\hline ant(y') \\ prize(y') \\ anp(x') \\ want(x', y') \\ get(x, y') \\\hline\end{array} .$$

From suitable type assumptions for nouns and verbs in the lexicon, with a type h of humans and e of objects, type reconstruction would infer types $x : h, y : e, x' : h, y' : e$, and hence type-respecting pronoun resolution could only resolve x' against x and y against y', as expected.

The typing rule for $\otimes$-DRSs was designed for merge-DRSs whose factors are linked through resolving cataphors and anaphors by type-independent "coindexing" or referent unification. Type-checking a DRS $\langle [x], \varphi(x, y) \rangle \otimes \langle [y], \psi(x, y) \rangle$ of this kind leads to a typing problem of the form

$$\frac{x : \alpha, y : \beta \vdash \varphi(x, y) : t \quad\quad y : \beta, x : \alpha \vdash \psi(x, y) : t}{\vdots \qquad\qquad\qquad\qquad \vdots}$$
$$\overline{\vdash \langle [x], \varphi(x, y) \rangle \otimes \langle [y], \psi(x, y) \rangle : [\alpha] \otimes [\beta]}$$

The type variables α, β get instantiated when the two typing problems in the premise are solved. As we perform merging of value-DRSs during normalization, we need the typing rule $(\otimes)$ only when a merge-factor is not a value-DRS, not for Bach-Peters-sentences.

6.4   Supporting Pronoun Translation

To translate between natural languages, we need to resolve pronouns in order to translate them correctly: the gender of the translated pronoun is generally not the gender of the source language pronoun, but the gender of the antecedent noun phrase in the target language, which in turn depends on the antecedent of the pronoun in the source sentence. For example, Google translates the English text *The child opened the box. It contained a pen.* into the German *Das Kind öffnete die Schachtel. Es enthielt einen Stift.*, where neuter *es* should be feminine *sie*. A type difference between humans h and things e and the verb type *contain/enthalten* : $e \rightarrow e \rightarrow t$ shows that *it* at position of type e cannot refer to *the child* : $(h \rightarrow t) \rightarrow t$ at position of type h. But only if *it* is resolved to *the box* : $(e \rightarrow t) \rightarrow t$, the gender for the German pronoun *er/sie/es* can be inferred to be the gender of the translation *die Schachtel* of *the box*, i.e. feminine.

6.5   Related Work

On the practical side, discourse representation structures are used as intermediate representation of meaning when translating texts from natural language to first-order logic. This is done for large-scale processing of newspaper texts by the *C&C/Boxer* program[6] [5] and for mathematical texts by the *Naproche* system [4].

The Groningen Meaning Bank [3] (GMB) is a large collection of English texts for which *C&C* computes syntactic analyses in categorial grammar and *Boxer* turns them into DRSs and first-order formulas. By using referents for individuals, events and times and predicates for thematic roles, *Boxer* covers far more of discourse representation theory than we do. In the examples of the GMB, nouns are classified according to animacy (human, non-concrete, etc.), which can be seen as type assignments. But, apparently, these classifications are not related to the meaning of verbs and hence not used in the pronoun resolution process. For example, in *Ein Mann füttert einen Hund; wenn er ihn beißt, schlägt er ihn.*, our system correctly resolves the four pronouns in the only type-compatible way (the first *er* to *Hund*, the second to *Mann* etc.), if we provide types h for humans, a for animals and typings for nouns *Mann* : $h \rightarrow t$, *Hund* : $a \rightarrow t$ and verbs *füttern, schlagen* : $h \rightarrow a \rightarrow t$ and *beißen* : $a \rightarrow h \rightarrow t$. The *C&C/Boxer* program, when we use masculine pronouns in the English input *A man feeds a dog. If he bites him, he beats him.*, resolves both subject

---

[6] Since the link provided in [5] did not work, we were only able to access *C&C/Boxer* via its demo version `gmb.let.rug.nl/webdemo/demo.php` of the Groningen Meaning Bank.

pronouns *he* to the *man* and both object pronouns to the *dog* (as one can infer from the logical formula). Thus, if the argument slots of verbs of the GMB were annotated with animacy, too, its pronoun resolution and meaning translation could be improved by using our type-respecting resolution procedure. As type distinctions are easier to make in mathematics than for natural language, a similar improvement can be expected for the anaphora resolution in systems using DRS-like proof representations like [4,8].

On the theoretical side, there is a growing amount of work (cf. [1,14,17,20]) that uses constructive type theory to develop semantic representations for natural language. In this setting, the notion of type is extended (from simple types, i.e. intuitionistic propositional formulas) to first-order formulas, and proofs of the formulas are the objects of these types. In particular, proofs of existential statements $\exists x \varphi$ consist of pairs (t, p) where t is a term denoting an individual and p a proof of $\varphi[x/t]$. Such terms t may then be used to resolve anaphoric expressions. For example, Mineshima [17] uses constructive type theory enriched by ϵ-terms to treat definite descriptions; the use of an ϵ-term has to be justified by an existential sentence, whose proof object then contains a referent for the description. Instead of ϵ-terms, Bekki [1] has terms $(@ : \gamma \to e)(c)$ of unknown choice functions $@$ applied to contexts c to select suitable referents of type e; by instantiating γ and constructing an object of type $\gamma \to e$ from proof objects in the typing environment Γ, this amounts to "anaphora resolution by proof search and type checking". Clearly, the contexts Γ used in constructive type theory provide a more general domain to search for referents than the typed DRS of the textual left context in our system; for example, one can have background assumptions that do not arise from translation of the textual left context, which is useful to handle bridging anaphora [14]. However, the formulation of background knowledge may often be unfeasible, and proof search in constructive type theory seems more complex that type reconstruction by unification from simple type annotations in the lexicon.

7   Open Problems

Extension to generalized quantifiers and plural pronouns. In [16], we have shown that type reconstruction for Montague grammar with plural noun phrases can be used to resolve some plural ambiguities. The idea is that plural noun phrases in general have several types, for distributive, reciprocal and collective readings, but argument types of predicates only unify with one of those. The type reconstruction program of [16] has been changed in [22] to type reconstruction for λ-DRT and extended to type-respecting pronoun resolution for singular pronouns. So far, type reconstruction for plurals is not adapted to λ-DRT yet. To interpret *She introduced the guests to each other*, for example, we would need discourse referents X for sets of individuals and apply the symmetric predicate distributively to any 2-element subset of X. As our system admits second-order discourse referents X, it seems possible to add type-respecting pronoun resolution for plural pronouns. For this, one should consider if the treatment of plurals and generalised quantifiers via "duplex conditions" [10] can be given a formulation that allows for principal types and type reconstruction.

First-order λ-DRT. In contrast to typed versions of λ-DRT, our untyped version is a kind of "higher-order" DRT: there is no demand that discourse referents have individual type. So we can type some expressions which, from a traditional point of view, should be untypable. For example,

$$P : \sigma \to t \ \vdash \ (\boxed{\begin{array}{c} x \\ \hline Px \end{array}} \otimes x) : [\sigma] \otimes \sigma$$

is a most general typing, using σ both as a referent-type and as a drs-type. To avoid such defects, we could introduce different kinds of types, notice when a type variable must be instantiated by an individual resp. by a drs-type, and forbid to equate types of different kinds. But in realistic cases, conditions of a DRS express properties of referents using predicates with individual argument type, which makes a formal restriction to first-order referents unnecessary.

Principal typings for pronoun resolved discourses. Does type-respecting pronoun resolution as suggested above "preserve principal types"? More precisely, in a merge-DRS $D_1 \otimes D_2$ of two typed DRSs with disjoint toplevel referent lists and principal types, we unify referents $x : \sigma$ of D_1 and $x' : \sigma'$ of D_2 by substituting x for y in D_2 and removing $x' : \sigma'$ from its referent list. Applying the most general unifier U of $x : \sigma$ and $x' : \sigma'$ gives a typed DRS $U D_1 \otimes U D_2$. Can one prove that $U D_1 \otimes U D'_2$ corresponds to the principal typing of $\tilde{D}_1 \otimes \tilde{D}'_2$, where D'_2 is the modification of D_2 by the pronoun resolution, and $\tilde{D}_1$ resp. $\tilde{D}'_2$ are the untyped versions of D_1 and D'_2?

Semantics. A semantics for typed λ-DRT is given in [13,15], with a compositional meaning for the *symmetric* $\otimes$. The relational interpretation of [19] for the unsymmetric merge (;) is not sufficient for our purposes. The *Dynamic lambda calculus* DLC of [11,12] claims to give a typed semantics for a system subsuming typed λ-DRT, but we found their types involving individual variables fairly incomprehensible. In order to show that the typing and reduction rules given here are correct, we ought to interprete typings $\Gamma \vdash t : \tau$ in a suitable domain-model of the *untyped* λ-calculus, like the one in [21], and handle free type variables as universally quantified. We have not yet tried to do so.

8   Conclusion

Our aim was to use semantic type information from the lexicon to reduce the number of possible antecedents of an anaphor to type-compatible ones. For this, a single type e of entities is too crude. Many verbs and nouns in natural language can only be applied to facts/propositions, inanimate physical objects, animals, or humans, respectively. Candidates for pronoun resolution can be reduced with these types quite reasonably in many situations. Of course, in a discourse about humans only, the reduction in candidates may be minimal.

The basic idea is simple: a pronoun gets a type from its occurrence as an argument of a verb, and a noun phrase gets a type from its head noun and the verb argument type of its occurrence; hence, one can filter the set of possible antecedents of a pronoun by comparing their types. To do this efficiently, we prefer a system of simple types with schematic types for function words like determiners, in which complex expressions have principal types that can easily be reconstructed from type assumptions for content words. (A complex expression can have a principal type for each choice of types of its words.)

Using DRSs provides us with DRTs [10] notion of possible "accessible" antecedent noun phrases. Our typing rules for λ-DRT expressions closely reflect the accessibility conditions of DRT; this is to be expected, as the antecedent noun phrase provides a type assumption for its discourse referent, which in turn corresponds to the pronoun occurrences referring to the antecedent. However, the peculiarities of λ-DRT concerning the subject-reduction property might be a good reason to consider a mathematically "cleaner" language for expressing the dynamics of discourse, such as simply typed λ-calculus with continuation semantics [7]. But in contrast to [7], we are not *assuming* pronoun resolution via some oracles, but rather integrate a type reconstruction algorithm into a pronoun resolution algorithm – in a particularly simple way.

Acknowledgement. We thank the referees for a number of critical remarks and questions that helped to improve the presentation.

References

1. Bekki, D.: Representing anaphora with dependent types. In: Asher, N., Soloviev, S. (eds.) LACL 2014. LNCS, vol. 8535, pp. 14–29. Springer, Heidelberg (2014)
2. Bos, J., Mastenbroek, E., McGlashan, S., Millies, S., Pinkal, M.: A compositional DRS-based formalism for NLP-applications. In: Proceedings of International Workshop on Computational Semantics, Tilburg, pp. 21–31 (1994)
3. Bos, J., Basile, V., Evang, K., Venhuizen, N., Bjerva, J.: The Groningen Meaning Bank. In: Ide, N., Pustejovsky, J. (eds.) Handbook of Linguistic Annotation. Springer, Berlin (2017, to appear). http://gmb.let.rug.nl
4. Cramer, M., Fisseni, B., Koepke, P., Kühlwein, D., Schröder, B., Veldman, J.: The Naproche project controlled natural language proof checking of mathematical texts. In: Fuchs, N.E. (ed.) CNL 2009. LNCS (LNAI), vol. 5972, pp. 170–186. Springer, Heidelberg (2010). doi:10.1007/978-3-642-14418-9_11
5. Curran, J.R., Clark, S., Bos, J.: Linguistically motivated large-scale NLP with C&C and boxer. In: Proceedings of ACL, Prague, June 2007, pp. 33–36. Association for Computational Linguistics (2007)
6. Damas, L., Milner, R.: Principal type-schemes for functional programs. In: Proceedings of 9th ACM Symposium on Principles of Programming Languages, pp. 207–212 (1982)
7. de Groote, P.: Towards a Montagovian account of dynamics. In: Gibson, M., Howell, J. (eds.) Proceedings of SALT XVI, vol. 16 (2006)
8. Ganesalingam, M.: The Language of Mathematics. LNCS, vol. 7805. Springer, Heidelberg (2013)
9. Hindley, R.: The principal type-scheme of an object in combinatory logic. Trans. Am. Math. Soc. **146**, 29–60 (1969)

10. Kamp, H., Reyle, U.: From Discourse to Logic. Kluwer, Dordrecht (1993)
11. Kohlhase, M., Kuschert, S.: Towards a dynamic type theory. Technical report, Universität des Saarlands (1996)
12. Kohlhase, M., Kuschert, S.: Dynamic lambda calculus. In: Proceedings of 5th Conference on the Mathematics of Language, Schloß Dagstuhl (1997)
13. Kohlhase, M., Kuschert, S., Pinkal, M.: A type-theoretic semantics for λ-DRT. In: Dekker, P., Strokhof, M. (eds.) Proceedings of 10th Amsterdam Colloquium, pp. 479–498 (1996)
14. Krahmer, E., Piwek, P.: Presupposition projection as proof construction. In: Bunt, H., Muskens, R. (eds.) Computing Meanings: Current Issues in Computational Semantics. Kluwer, Dordrecht (1999)
15. Kuschert, S.: Eine Erweiterung des λ-Kalküls um Diskursrepräsentationsstrukturen. Master's thesis, Universität Saarbrücken (1995)
16. Leiß, H.: Resolving plural ambiguities by type reconstruction. In: Groote, P., Nederhof, M.-J. (eds.) FG 2010-2011. LNCS, vol. 7395, pp. 267–286. Springer, Heidelberg (2012). doi:10.1007/978-3-642-32024-8_18
17. Mineshima, K.: A presuppositional analysis of definite descriptions in proof theory. In: Satoh, K., Inokuchi, A., Nagao, K., Kawamura, T. (eds.) JSAI 2007. LNCS (LNAI), vol. 4914, pp. 214–227. Springer, Heidelberg (2008). doi:10.1007/978-3-540-78197-4_20
18. Montague, R.: The proper treatment of quantification in ordinary English (chapter 8). In: Thomason, R. (ed.) Formal Philosophy, pp. 247–270. Yale University Press, New Haven (1974)
19. Muskens, R.: Combining montague semantics and discourse representation. Linguist. Philos. 19, 143–186 (1996)
20. Ranta, A.: Type-Theoretical Grammar. Clarendon Press, Oxford (1994)
21. Ruhrberg, P.: Simultaneous abstraction and semantic theories. Ph.D. thesis, University of Edinburgh (1996)
22. Wu, S.: Getypte Lambda-Diskursrepräsentationsstrukturen - Typrekonstruktion für die λ-Diskursrepräsentationstheorie. Master's thesis, Centrum für Informations- und Sprachverarbeitung, Universität München (2012)

A Computable Solution to Partee's Temperature Puzzle

Kristina Liefke[1]([✉]) and Sam Sanders[1,2]

[1] Munich Center for Mathematical Philosophy, Ludwig-Maximilians-University
Munich, Geschwister-Scholl-Platz 1, 80539 Munich, Germany
K.Liefke@lmu.de
[2] Department of Mathematics, Ghent University,
Krijgslaan 281 – Building S22, B9000 Ghent, Belgium
sasander@me.com

Abstract. This paper presents a computable solution to Partee's temperature puzzle which uses one of the standard tools of mathematics and the exact sciences: countable approximation. Our solution improves upon the standard Montagovian solution to the puzzle (i) by providing computable natural language interpretations for this solution, (ii) by lowering the complexity of the types in the puzzle's interpretation, and (iii) by acknowledging the role of linguistic and communicative context in this interpretation. These improvements are made possible by interpreting natural language in a model that is inspired by the Kleene-Kreisel model of countable-continuous functionals. In this model, continuous functionals are represented by lower-type objects, called the *associates* of these functionals, which only contain countable information.

Keywords: Temperature puzzle · Individual concepts · Associates · Continuous functionals · Computability

1 Partee's Puzzle and Montague's Solution

Partee's temperature puzzle [33, p. 267] is a touchstone for any formal semantics for natural language. This puzzle regards the incompatibility of our intuitions about the validity of the inference from (1) (i.e. *in*valid) with predictions about the validity of this inference in extensional semantics (cf. [8,32]) (i.e. valid).

a. The temperature is ninety.

b. The temperature rises.   (1)

c. Ninety rises.

We would like to thank three anonymous referees for LACL 2016, Hans Leiss, and Christian Retoré for their valuable comments and suggestions. The research for this paper has been supported by the German Research Foundation (via Kristina Liefke's grant LI 2562/1-1), by the Alexander von Humboldt Foundation (via Sam Sanders' postdoctoral research fellowship), and by LMU Munich's Institutional Strategy LMUexcellent within the framework of the German Excellence Initiative.

© Springer-Verlag GmbH Germany 2016
M. Amblard et al. (Eds.): LACL 2016, LNCS 10054, pp. 175–190, 2016.
DOI: 10.1007/978-3-662-53826-5_11

$$\exists c^{se}\left(\forall c_1^{se}[\text{TEMP}^{(se)t}(c_1)\leftrightarrow c=c_1]\wedge c(@^s)=\text{NINETY}^e\right)$$

$$\frac{\exists c^{se}\left(\forall c_1^{se}[\text{TEMP}^{(se)t}(c_1)\leftrightarrow c=c_1]\wedge \text{RISE}^{(se)t}(c)\right)}{\text{RISE}^{(se)t}(\mathbf{ninety}^{se})} \qquad (2)$$

Montague-style formal semantics (e.g. [13, 17, 29, 33]) solve this puzzle by distinguishing two readings of the DP the temperature: a function-reading (cf. (1b)), on which the DP is interpreted as an *individual concept* (i.e. as a function from indices/world-time pairs to individuals; type[1] se), and a value-reading (cf. (1a)), on which the DP is interpreted as the extension of this concept at the current index, @ (i.e. as an *individual*; type e). The different readings prevent the replacement of the occurrence of the DP the temperature from (1b) by the name ninety (s.t. the conclusion of (1) cannot be derived from the premises) (cf. (2)).[2]

2   Problems with Montague's Solution

Montague's solution to the temperature puzzle is inspired by Carnap's theory of intensions (cf. [7]) and is supported by the fact that Montague semantics already uses indices in the semantic analysis of declarative sentences, which are interpreted as functions from indices to truth-values (cf. also [26]). Because of its ready availability, Montague's solution has been adopted by many contemporary theories of formal semantics.[3] However, there are a number of problems with this solution. These include the non-computability of natural language interpretations in this solution, (ii) the high type-complexity of natural language interpretations in this solution, and (iii) the disregard of relevant contextual parameters in this solution. The latter are described below:

2.1   Problem 1: Non-Computability of NL Interpretations

Intensional (or 'possible world') semantics – which include Montague-style formal semantics – fail to provide computable (or 'effective') interpretations of natural language expressions. This is due to the non-computability of models of possible world semantics and the impossibility of finitely describing the set of possible worlds that provides the meaning of a sentence in the absence of the sentence's translating/intermediate formula (cf. [34]). As a result of these facts, intensional

---

[1] For brevity, we use a short notation for types, where se corresponds to the arrow type $s \to e$ and to Montague's type $\langle s, e\rangle$. We will hereafter indicate types in superscript.

[2] In (2), we assume that **ninety** is s.t. $\forall i^s(\mathbf{ninety}(i) = \text{NINETY})$.

[3] These theories include hyperintensional theories (e.g. [16, 39]), which do not adopt an atomic type for indices, and relational theories (e.g. [35, 48]), which only accept non-atomic types with range Bool. To accommodate the intensionality of DPs like the temperature in (1b), hyperintensional theories introduce an atomic type for individual concepts. Relational theories code individual concepts as binary relations between indices and individuals.

semantics are unable to compute the semantic representation of a given sentence. However, given the need to explain the human ability to form and understand new complex expressions (cf. [9,15,37,44]), such an effective semantics is clearly desirable.

2.2 Problem 2: High-Rank Typing

The interpretation of DPs as individual concepts increases the complexity of the types of natural language interpretations. On Montague's interpretation, proper names and common nouns are expressions of rank 1 (i.e. se) resp. 2 $((se)t)$, rather than of rank 0 (e) resp. 1 (et), as in extensional semantics. Montague semantics even interprets transitive verbs – which have rank 3 (i.e. $((et)t)(et))$ in extensional semantics – in rank 4 (i.e. $(((se)t)t)((se)t))$. But this complicates the type of the interpretations of linguistic expressions analogously to the (much-criticized) treatment of referential DPs as generalized quantifiers (cf. [19,27,38]). Further, while formal semanticists and theoretical computer scientists are used to working with rank-4 (or higher-rank) objects, such objects are highly uncommon in the natural sciences and even in most parts of mathematics.

2.3 Problem 3: Context-Invariance

Montague's solution further neglects the salient role of context in the interpretation of the verb rise (cf. [10]): Intuitively, for different DPs, rise will assert the DP referent's rising *over different-length intervals*. Thus, in (1b), rise will be interpreted with respect to a shorter interval (e.g. minutes, or hours) than in the CP The oil price rises (e.g. weeks, or months). Even when applied to the same DP, rise is often interpreted with respect to different-length intervals. For example, in the context of global climate development, (1b) will be taken to make a claim about a longer interval than in the context of the local weather forecast. Since Montague semantics analyzes intensional intransitive verbs as characteristic functions of sets of individual concepts (which send all occurrences of a DP to the same truth-value), it does not capture this context-sensitivity.

3 Solving the Problems

We solve the above problems by interpreting natural language in a model[4] that is inspired by the *Kleene-Kreisel model of countable-continuous functionals* [21,25] (cf. [30, Ch. 2.3.1]). In this model, continuous functionals are represented by lower-type objects called *associates*.

Following Kleene [21] and Kreisel [25], we hereafter use *finite types* over the natural numbers. The latter are the smallest set of strings that contains the type for natural numbers, 0, and the types for function spaces over natural numbers,

---

[4] To enable a compositional interpretation of the sentences from (1) (cf. Sect. 4), this model extends the Kleene-Kreisel model (which only contains natural numbers and functions over natural numbers) to objects of higher type.

$(\rho \to \tau)$ (with ρ, τ finite types) (cf. [36]). To ease notation, we abbreviate the type for functions over natural numbers, $(0 \to 0)$, as '1', abbreviate the type for *functionals* over sequences of natural numbers, $((0 \to 0) \to 0)$ ($\equiv (1 \to 0)$), as '2', and abbreviate $(n \to 0)$ as '$n+1$'. Our considerations will make special use of *coded* finite sequences of natural numbers (type 0). To distinguish natural numbers which *do* from natural numbers which do *not* code such sequences, we denote the former by '0^*'.

Our solution to the temperature puzzle briefly works as follows: By representing the DP the temperature from (1b) as (a code for) a finite sequence of natural numbers (type 0^*) and by approximating the continuous functional denoted by rise by an associate of type $1 \equiv (0^* \to 0)$, we 'lower' the types of many expressions from (1) (cf. Problem 2). In particular, our solution interprets the DP's occurrence from (1a) as a natural number (type 0) and the DP's occurrence from (1b) as a (coded) sequence of natural numbers (type 0^*). Since distinguishing between types 0 and 0^* is *decidable*, we obtain a *computable* solution to the temperature puzzle (cf. Problem 1). Because associates are introduced through the use of a context-dependent variable, the domain of application of the verb rise is restricted to a specific, contextually salient, temporal interval (cf. Problem 3). As to the computability of our solution, it suffices for now to point out that the Kleene-Kreisel model can be defined inside Martin-Löf type theory and has been implemented in the associated programming language Agda [14, 45–47].

Note the integrative nature of our solution to the above problems: Since associates are *computable*, *lower-type* representations of continuous functionals that approximate these functionals *with regard to a contextually determined parameter*, our solution(s) to the above problems are all sides of the same (three-sided) coin. This contrasts with other solutions to the temperature puzzle (e.g. [3, 20, 27, 41]) which still assume more complex types, are not effective, and/or rely on the use of other methods to render the interpretation of the sentences from (1) context-sensitive.

We describe our solution in some detail below. To this end, we first show how the Montagovian interpretation of the verb rise corresponds to a continuous functional (in Sect. 3.1). Following the informal introduction of associates (in Sect. 3.2), we then outline our *associates*-approach to the temperature puzzle (in Sect. 3.3). This approach receives a compositional implementation in Sect. 4. The empirical domain of our *associates*-approach and the computational properties of associates are discussed in Sects. 5 and 3.4.

3.1 Continuity and the Temperature Puzzle

Our solution to the temperature puzzle starts from the observation that the interpretation of RISE from (2) corresponds to a continuous functional, φ_{rise}, in the space $\mathbb{N}^{\mathbb{N}} \to \mathbb{N}$. The correspondence between RISE and φ_{rise} is based on the possibility of representing individual concepts as sequences over natural numbers (assuming a fixed starting index/world-time pair $\langle w, t \rangle$ and a discrete unit of time measurement; cf. [27]). The latter enables the representation of the individual concept 'the temperature' from (3) as the sequence from (4), and the representation of sets of individual concepts as sets of such sequences.

$$\langle w, t_0 \rangle \mapsto 89, \langle w, t_1 \rangle \mapsto 90, \langle w, t_2 \rangle \mapsto 91, \ldots, \langle w, t_n \rangle \mapsto 89 + n \qquad (3)$$

$$89, 90, 91 \ldots, 89 + n \qquad (4)$$

With this representation in mind, the temperature as given by $\gamma^1 = (T_0, T_1, \ldots)$ (where T_0, T_1, etc. are the values of some temperature measurement) rises, i.e. RISE(γ), iff $\varphi_{\text{rise}}(\gamma) = 1$. The temperature as given by γ does not rise iff $\varphi_{\text{rise}}(\gamma) = 0$.

The *continuity* of the functional φ_{rise} is suggested by (i) the 'finite relevance' of input sequences for φ_{rise} and (ii) the equivalence of sequences which are identical up to some point in time.

Ad (i): Intuitively, after having observed a rise in the values of some temperature measurement *for a certain finite period of time*, even the most ardent skeptic will agree that the values are, in fact, rising. Thus, if the temperature as given by $\gamma = (T_0, T_1, \ldots)$ is rising, i.e. if $\varphi_{\text{rise}}(\gamma) = 1$, we will agree to this fact after having observed the temperature up to some point in time n, i.e. by considering $(T_0, \ldots, T_n)$.

Ad (ii): If the temperature as given by the values of some other measurement $\beta = (T_0', T_1', \ldots)$ is further exactly γ up to the point in time n, we will agree that $\varphi_{\text{rise}}(\beta) = 1$, i.e. that the temperature as given by β is also rising. The functional φ_{rise} is thus continuous in the usual mathematical sense (cf. [30, Ch. 2.3.1]).

Continuity is defined below:

Definition 1 (Continuity of type-2 functionals). *A type-2 functional φ is continuous (on the Baire[5] space) if*

$$\forall \gamma^1 \exists n^0 \forall \beta^1 \left(\overline{\gamma} n = \overline{\beta} n \rightarrow \varphi(\gamma) = \varphi(\beta) \right), \qquad (5)$$

where $\overline{\gamma} n = (T_0, T_1, \ldots, T_n)$ and $\overline{\beta} n = (T_0', T_1', \ldots, T_n')$ (both type 0^*) are the initial segments (up to n) of γ and β.

Above, the point n (for φ_{rise}: a point in time at which everyone agrees that the temperature is rising) is called a *point of continuity* of φ (at γ). Obviously, this point may be different for different sequences. We will use this fact in Sect. 3.2 to explain the dependence of interpretations on the expressions' linguistic context.

The correspondence of the interpretation of RISE to the continuous functional φ_{rise} gives rise to the following 'continuous functional'-version of (2):

$$\exists \gamma^1 \left(\forall \beta^1 [temp^2(\beta) \leftrightarrow \gamma = \beta] \wedge now^2(\gamma) = ninety^0 \right)$$

$$\frac{\exists \gamma^1 \left(\forall \beta^1 [temp^2(\beta) \leftrightarrow \gamma = \beta] \wedge \varphi_{\text{rise}}^2(\gamma) = 1 \right)}{/\,/\,/\,/\,/\,/ \qquad \varphi_{\text{rise}}^2(\boldsymbol{ninety}^1) = 1 \qquad /\,/\,/\,/\,/\,/} \qquad (6)$$

---

[5] The Baire space is usually defined as the set of all infinite sequences of natural numbers with a certain topology. This space has many alternative characterisations (up to isomorphism) as explored in, e.g., [31, Ch. I].

In (6), **ninety** denotes the sequence which is constant *ninety* (s.t. **ninety** serves the function of **ninety**^{se} from (2)). The constant *now* denotes a functional that takes as input non-coded sequences of natural numbers (type 1) and produces as output the value-at-@ in these sequences. The introduction of this constant is made necessary by the absence of indices in (the variant of) our preferred model of countable-continuous functionals (cf. Sect. 4) in which we interpret Partee's temperature puzzle.

We close this section with a remark on the 'coding' of finite sequences as is done in mathematics and computer science (cf. e.g. [6, p. 92]):

Remark 1 (Coding). Finite sequences of natural numbers can be represented (or 'coded') by a single natural number using *pairing functions*. The most widely known of these functions, due to Cantor, is defined as follows:

$$\pi(n, m) := \tfrac{1}{2}(n + m)(n + m + 1) + m$$

Notably, not all natural numbers necessarily code finite sequences (given a certain fixed pairing function).

The coding and the associated decoding of finite sequences has been implemented in most of the common programming languages. In particular, there is a computable function $\mathsf{IsCodeForSeq}(n)$ of comparatively low complexity which outputs '1' if it is indeed the case that the input n codes some finite sequence $(T_0, T_1, \ldots, T_m)$, and '0' otherwise.

As is common in mathematics and computer science, we assume below that a particular coding and decoding function has been fixed (e.g. Gödel numbers as in [6, p. 92]). This assumption allows us to treat finite sequences (type 1) as natural numbers (type 0). We further assume that *ninety* from (6) is a number which does not[6] code a finite sequence. We will see below that this property of pairing functions is essential in our solution to Partee's temperature puzzle (in Sect. 3.3).

This completes our discussion of the interpretation of the verb rise as a continuous functional. We next introduce the notion of *associate* and discuss its role in our solution to the temperature puzzle.

3.2   Associates and the Temperature Puzzle

Intuitively, associates of continuous functionals are countable approximations (or representations) of these functionals which uniquely determine the value of these functionals for every (represented) argument. The Kleene-Kreisel model of countable-continuous functionals is defined in terms of associates (cf. [30, §8.2.1]). Associates are formally defined as follows:

Definition 2 (Associates [21,25]). *An associate, α_φ, of a continuous type-2 functional φ is a sequence of natural numbers (i.e. type $1 \equiv (0^* \to 0)$) such that*

$$\forall \gamma^1 \, \exists n^0 \, \forall N^0 \geq n \big[\alpha_\varphi(\overline{\gamma}N) = \varphi(\gamma) + 1 \wedge (\forall i < n)\, \alpha_\varphi(\overline{\gamma}i) = 0 \big]. \tag{7}$$

---

[6] For the coding from [6, p. 92], there exist numbers which do not code finite sequences.

The associate α_φ thus enumerates[7] the values of φ at all $\overline{\gamma}n$, where n is a point of continuity for γ. In particular, the first conjunct of (7) identifies the value of the associate of φ *for any initial segment of* γ *up to at least* n (here: the value of $\alpha_\varphi(\overline{\gamma}N)$) with the value $+1$ of φ *for* γ. As a result of the identification of $\alpha_\varphi(\overline{\gamma}N)$ and $\varphi(\gamma) + 1$, a continuous functional and its associate contain the same information: Beyond the point of continuity n, φ remains constant, i.e. no new information can be learned.

The '$+1$' in the first conjunct of (7) expresses a kind of partiality: If the input sequence, $\overline{\gamma}k$, of α_φ is 'too short' (i.e. if k is less than the least point of continuity, n, for γ), $\alpha_\varphi(\overline{\gamma}k)$ cannot provide any information about $\varphi(\gamma)$. The second conjunct from (7) captures this possibility by returning the value 0, which is not a possible value for $\varphi(\gamma) + 1$.

The above yields the following intuitive picture for an associate, α_{rise}, of φ_{rise}. Below, γ denotes a temperature-representing sequence (type-1, as in Sect. 3.1); m is a natural number:

$$\alpha_{\text{rise}}(\overline{\gamma}m) = \begin{cases} 0 & \text{if } \overline{\gamma}m \text{ is too short to judge if the temperature is rising;} \\ 1 & \text{if } \varphi_{\text{rise}}(\gamma) = 0 \text{ by (7), i.e. the temperature is not rising;} \\ 2 & \text{if } \varphi_{\text{rise}}(\gamma) = 1 \text{ by (7), i.e. the temperature is rising.} \end{cases}$$

We close this section with an observation about associates and context-dependence:

The variation of the point n in (7) with different input sequences reflects the role of *linguistic* context in the interpretation of verbs like rise and fall: While some occurrences of these verbs only consider comparatively short initial segments of sequences in order to judge whether the sequence rises or falls, others consider longer (or even countably infinite) initial segments of these sequences. Consider the application of the *associates*-interpretation of fall to the type-0^* interpretations of the DPs the water drop and the pitch drop: To confirm that the water drop is, in fact, falling, it suffices to observe its behavior for a short period of time (i.e. for a few (milli-)seconds). In contrast, to confirm that the pitch drop is falling, we need to observe its behavior for a rather long period of time (i.e. for several years).

The (possible) existence of multiple points of continuity for *the same* sequence – and the attendant need to choose a particular point up to which we consider this sequence – further reflects the dependence of the above verbs on the salient *communicative* context. For example, for the sentence The temperature rises (cf. (1b)), we will choose a larger n in the context of global climate development than in the context of the local weather forecast.

---

[7] Note that it is impossible to enumerate the space $\mathbb{N}^{\mathbb{N}}$. Since we can, thus, not enumerate the values of a discontinuous type-2 functional, our approach breaks down for *discontinuous* functionals. We will identify a promising solution to this problem in Sect. 7.

3.3   The *Associates*-Solution to the Temperature Puzzle

We are now ready to present our *associates*-solution to the temperature puzzle. In particular, we can reformulate (6) using the associate, α_{rise}, of φ_{rise} as follows:

$$
\frac{\exists \gamma^1 \big(\forall \beta^1 [temp^2(\beta) \leftrightarrow \beta = \gamma] \wedge now^2(\gamma) = ninety^0 \big)}{\cancel{\frac{\exists \gamma^1 \big(\forall \beta^1 [temp^2(\beta) \leftrightarrow \beta = \gamma] \wedge \exists n^0 [\alpha_{\text{rise}}^1(\overline{\gamma}n) = 2] \big)}{\exists m^0 \big(\alpha_{\text{rise}}^1(\boldsymbol{ninety}\, m) = 2 \big)}}}
\tag{8}
$$

We next show that the inference from (8) indeed does not go through:

Montague semantics solves Partee's temperature puzzle by interpreting the occurrences of the DP the temperature from (1a) and (1b) as an *individual* (cf. the constant NINETY in (2)) resp. as an individual *concept* (cf. the variable c in (2)). Our solution works analogously, but – thanks to the presence of α_{rise} – with lower types. In our solution, the different occurrences of the DP from (1a) and (1b) are interpreted as a natural number which does *not* code a finite sequence of natural numbers (by the assumption following Remark 1) (cf. the constant *ninety* in (8)) and as a natural number, k, which *codes* the finite sequence $\overline{\gamma}n$ from (8). The information whether *ninety* and k do or do not code a sequence of natural numbers is obtained by applying the function IsCodeForSeq(n) from Remark 1. The different types of *ninety* and k (i.e. 0 resp. 0*) – and the subsequent impossibility of replacing the occurrence of $\overline{\gamma}n$ in the second premise of (8) by the constant *ninety* – blocks the temperature puzzle.

In conclusion: the introduction of the associate, α_{rise}, of φ_{rise} allows us to block the inference from (8) while lowering the types of many expressions from (1).

3.4   Computability and the Temperature Puzzle

We have suggested in Sect. 2.1 that our *associates*-solution to the temperature puzzle is computable. To support this claim, we now discuss the computational properties of associates that are relevant for our solution.

An obvious conceptual question about associates is whether every continuous functional has an associate and, if this is the case, whether this associate is computable. We provide three partial answers to this question:

1. Kohlenbach has shown in [24, Sect. 4] that the statement *every continuous functional of type* $(1 \to 1)$ *has an associate* carries no significant logical strength. Thus, as a special case, we may safely assume the existence of an associate for every continuous type-2 functional.
2. In general, there is no *computable* functional which takes as input a continuous type-2 functional and produces as output an associate (cf. [24,25]).
3. However, every primitive recursive functional (in the sense of Gödel's system T) has a *canonical* associate which can be computed via the procedure from [42, p. 139]. Since the class of primitive recursive functionals is rather large, it captures essentially any functional 'occurring in practice'.

A second question about associates regards the computability of the associate's point of continuity n. We here provide two partial answers:

1. There is no *computable* functional which returns a point of continuity on input a continuous type-2 functional and a sequence (cf. [25]).
2. However, the *fan functional* returns a point of (uniform) continuity on input a continuous type-2 functional and a sequence *in a fixed compact space*. The fan functional is present in the Kleene-Kreisel model and has a computable associate (cf. [30, Sect. 8], [47]).

Since temperature measurements come with upper and lower bounds dictated by physics (s.t. they are part of a compact space), a point of continuity of φ_{rise} can always be computed for α_{rise} and a sequence of temperature measurements γ.

This completes our presentation of the *associates*-approach to Partee's temperature puzzle. We next show that this approach can be implemented in a compositional semantics for natural language.

4   Compositional Implementation

To obtain our *associates*-solution to the temperature puzzle, we compositionally interpret natural language in a model, inspired by the Kleene-Kreisel model of countable-continuous functionals, which contains continuous functionals and their associates. This interpretation proceeds via the translation of the relevant subset of the linguistic fragment from [33] into the language of the simply typed lambda logic $\lambda^0_\rightarrow$ ([8]; cf. [36, Ch. 1.1]). This is a logic with a single atomic type, 0, from which all other types are built up through the type constructor $\rightarrow$ (see the definition of *finite types* from Sect. 3). The language and models of $\lambda^0_\rightarrow$ are specified in [2, 8].

To identify the $\lambda^0_\rightarrow$-interpretation of the sentences from (1), we first specify the particular language $\mathcal{L}^{\lambda^0_\rightarrow}$ (abbreviated '$\mathcal{L}$') and frame $\mathcal{F}^{\lambda^0_\rightarrow}$ (abbreviated '$\mathcal{F}$') whose elements translate resp. interpret the syntactic constituents of these sentences. The members of $\mathcal{L}$ are specified in Table 1. Our conventions for the use of $\lambda^0_\rightarrow$ variables are introduced in Table 2.

In the list of non-logical $\lambda^0_\rightarrow$ constants, α_{rise} enables the translation of the verb *rise* as an associate of the continuous functional denoted by *rise* (formerly, φ_{rise}).

Table 1. $\mathcal{L}$ constants.

Constant	$\lambda^0_\rightarrow$ Type
ninety	0
ninety, α_{rise}	1
now, temp, rise	$1 \rightarrow 0$

Table 2. $\mathcal{L}$ variables.

Variable	$\lambda^0_\rightarrow$ Type
m, n, N, x	0
β, γ	1
P, Q	$1 \rightarrow 0$

The interpretation function $\mathcal{I}_\mathcal{F} : \mathcal{L} \to \mathcal{F}$ respects the way in which different content words are conventionally related. Thus, this function identifies the interpretation of the generalized $\lambda_\to^0$-translation, $\lambda P.P(\textbf{\textit{ninety}})$, of the DP ninety as a subset of the interpretation of the $\lambda_\to^0$ translation, $\lambda P\exists\gamma.temp(\gamma) \wedge P(\gamma)$, of the DP a temperature (s.t. ninety is a temperature under this interpretation). To ensure the 'right' interpretation of the syntactic constituents of (1a) to (1c), we demand that the function $\mathcal{I}_\mathcal{F}$ further satisfies a number of semantic constraints.

Definition 3 (Constraints on $\mathcal{L}$ constants). *The function $\mathcal{I}_\mathcal{F}$ satisfies the following semantic constraints:*

(C1) $now(\textbf{\textit{ninety}}) = ninety$;

(C2) $\forall\gamma^1\exists n^0\forall\beta^1\big(\overline{\gamma}n = \overline{\beta}n \to rise(\gamma) = rise(\beta)\big)$;

(C3) $\forall\gamma^1\exists n^0\forall N^0 \geq n\big[\alpha_{\mathrm{rise}}(\overline{\gamma}N) = rise(\gamma) + 1 \wedge (\forall i < n)\alpha_{\mathrm{rise}}(\overline{\gamma}i) = 0\big]$

The constraint (C1) demands that the interpretation of the type-0 constant *ninety* be the output of the functional *now* on input **ninety** (cf. [33, rule T1.(d), MP1]). The constraints (C2) and (C3) demand that the constant *rise* be interpreted as a continuous functional (cf. (C2)) resp. that α_{rise} behaves as an associate of this functional (cf. (C3)).

Admittedly, (C2) and (C3) are additional requirements on our semantic models which are not postulated for the models of Montague's Intensional Logic (cf. [33]). However, since these requirements reflect natural assumptions about the domain of interpretation of the verb rise (cf. Sect. 3.1) – and since continuous functionals can be represented via their associates (cf. Sect. 3.2) –, these requirements are rather innocent.

This completes our specification of the interpretation function $\mathcal{I}_\mathcal{F}$. We next turn to the compositional translation of Partee's temperature puzzle: To enable this translation, we first translate the lexical elements of the sentences from (1). In these translations, $\rightsquigarrow$ is the smallest relation between syntactic trees and $\lambda_\to^0$ terms which conforms to the rules from [22]:

Definition 4 (Basic $\lambda_\to^0$ translations). *The lexical elements of (1a) to (1c) are translated into the following $\lambda_\to^0$ terms:*

$$\text{ninety} \rightsquigarrow \textbf{\textit{ninety}}$$
$$\text{temperature} \rightsquigarrow temp$$
$$\text{rise} \rightsquigarrow \lambda\beta\exists n\big(\alpha_{\mathrm{rise}}(\overline{\beta}n) = 2\big)$$
$$\text{is} \rightsquigarrow \lambda\beta\lambda\gamma\big(now(\gamma) = now(\beta)\big)$$
$$\text{the} \rightsquigarrow \lambda Q\lambda P\exists\gamma\big(\forall\beta[Q(\beta) \leftrightarrow \gamma = \beta] \wedge P(\gamma)\big)$$

As expected, Definition 4 specifies the translation of the verb rise as an associate of the continuous functional denoted by the $\lambda_\to^0$ constant *rise* (cf. (C2), (C3)). The translations of the copula is, of the DP ninety, and of the definite determiner follow the translations of these expressions from [33, cf. rules T1.(b), (d), T2].[8] In particular, our translation of is follows Montague's translation of the copula as the designator of a relation between the *extensions* of (generalized quantifiers over) individual concepts (here: as the designator of a relation between natural numbers, rather than between sequences of numbers).

The above translations enable the compositional $\lambda^0_\rightarrow$ translation of the sentences from (1). We start with the translation of (1a):

1. $[_{\text{VP}}[_{\text{CP}}\text{is}][_{\text{DP}}\text{ninety}]] \rightsquigarrow \lambda\gamma\big(now(\gamma) = now(\boldsymbol{ninety})\big)$ $\hfill (9)$
    $= \lambda\gamma(now(\gamma) = ninety)$

2. $[_{\text{DP}}[_{\text{DET}}\text{the}][_{\text{N}}\text{temperature}]] \rightsquigarrow \lambda P \exists\gamma\big(\forall\beta\,[temp(\beta) \leftrightarrow \gamma = \beta] \wedge P(\gamma)\big)$

3. $[_{\text{S}}[_{\text{DP}}[_{\text{DET}}\text{the}][_{\text{N}}\text{temperature}]][_{\text{VP}}[_{\text{CP}}\text{is}][_{\text{DP}}\text{ninety}]]]$
    $\rightsquigarrow \exists\gamma\big(\forall\beta\,[temp(\beta) \leftrightarrow \gamma = \beta] \wedge (now(\gamma) = ninety)\big)$

Sentences (1b) and (1c) are translated as follows:

$$[_{\text{S}}[_{\text{DP}}[_{\text{DET}}\text{the}][_{\text{N}}\text{temperature}]][_{\text{VP}}[_{\text{IV}}\text{rises}]]] \hfill (10)$$
$$\rightsquigarrow \lambda P \exists\gamma\big(\forall\beta\,[temp(\beta) \leftrightarrow \gamma = \beta] \wedge P(\gamma)\big)\big(\lambda\delta\,\exists n[\alpha_{\text{rise}}(\bar{\delta}n) = 2]\big)$$
$$= \exists\gamma\big(\forall\beta[temp(\beta) \leftrightarrow \gamma = \beta] \wedge \exists n[\alpha_{\text{rise}}(\bar{\gamma}n) = 2]\big)$$

$$[_{\text{S}}[_{\text{DP}}\text{ninety}][_{\text{IV}}\text{rises}]] \rightsquigarrow \exists m\big(\alpha_{\text{rise}}(\overline{\boldsymbol{ninety}}\ m) = 2\big) \hfill (11)$$

The resulting $\lambda^0_\rightarrow$ formulas are exactly the formulas from (8).

We next discuss the empirical scope of our *associates*-approach and the relation of this approach to other solutions to the temperature puzzle.

5 Domain and Scope

Our previous discussion has been restricted to the example of the verb rise. However, the *associates*-approach generalizes to all degree achievement verbs and change-of-state verbs ([28]; cf. [1,5,11]) whose interpretation corresponds to a continuous functional. The latter constitute a sizable[9] class of verbs with the following members:

1. *verbs of continuous calibratable change of state* (cf. [28, pp. 247–248]): decline, drop, grow, increase, plummet, plunge, rocket, rise, soar, surge, ...
2. *verbs of entity-specific continuous change of state* (cf. [28, pp. 246–247]): blush, blossom, burn, ferment, molt, rust, sprout, swell, ...
3. *other verbs of continuous state-change* (cf. [28, pp. 240–246]): abate, advance, age, clog, compress, condense, degrade, distend, mature; in particular:

---

[8] We simplify Montague's translation of the copula to a translation that takes as its first argument the designator of a type-1 object (instead of a *generalized quantifier* over type-1 objects).

[9] For example, Levin [28] lists 369 members of classes 1 to 4.

(a) break-/bend-*verbs:* crack, shatter, split, tear; crumple, fold, wrinkle, ...
(b) *adjective-related verbs:* blunt, clear, cool, dry, empty, narrow, quiet, ...
(c) *change-of-color verbs:* blacken, brown, gray, redden, tan, whiten, ...
(d) -en *verbs:* darken, flatten, harden, ripen, sharpen, strengthen, ...
(e) -ify *verbs:* acidify, humidify, magnify, nitrify, petrify, purify, solidify, ...
(f) -ize *verbs:* crystallize, fossilize, pressurize, pulverize, stabilize, ...
(g) -ate *verbs:* accelerate, coagulate, degenerate, detonate, evaporate, ...

4. *(continuous) directed motion verbs* (cf. [28, pp. 263–264]): arrive, ascend, descend, drop, enter, fall, pass, rise, ...

5. *accomplishment verbs* (cf. [43]): run a mile, draw a circle, build a house, eat a sandwich, play a game of go; grow up, recover from illness, ...

The above-listed verbs all take individual concepts as their arguments (i.e. they are co-classified with the verb rise) (cf. [10]). The intensional interpretation of these verbs is motivated by their particular, non-instantaneous, evaluation procedure: To judge whether John is blushing (cf. class 2), it does not suffice to observe his red face at a particular point in time.[10] Instead, we need to observe John's facial complexion at different neighboring points in time. We can only conclude that John is blushing if he has a normal (non-red) skin color at the earliest observed time-point and an increasingly redder complexion at the later time-points (cf. [27]).

Note that, in contrast to their counterparts from class 1, the 'continuous functional'-interpretations of the verbs from classes 2 to 5 are not restricted to input sequences of *natural numbers* (see *blush*), may describe non-temporal change [10, 18] (see the extent reading of verbs like *narrow* and *darken*)[11], and do not presuppose an *established* scale or unit of measurement (i.e. they describe non-discrete change). For example, in contrast to rising, blushing and narrowing are not properties of sequences of numbers, but of sequences of *temporal states of an individual* (viz. of his/her face) resp. of *spatial states of an object*. Further, there is no established unit of measurement of a person's facial redness (or of a window cracking, a storm arriving, a person recovering from illness, etc.).

The above-described absence of a numerical/measurement structure does not compromise the applicability of our *associates*-approach to the verbs from classes 2 to 5. This is due to the possibility of labelling temporal stages of individuals (or of other physical objects) by natural numbers, of identifying a *contextually salient* unit and scale (here: dominant wavelength or visible change in hue) for the measurement of the relevant property, and of selecting the value of the measurement (under the selected scale and unit of measurement) of the individual's relevant attribute for that property. In particular, the continuous functional-interpretation, *blush*, of the verb blush will return '1' on input a given sequence

---

[10] Maybe John simply suffers from high blood pressure which causes his constant facial redness.

[11] E.g. in The trail narrowed at the summit [10, p. 98] and His skin darkens on his right leg near the femoral artery [10, p. 99]. We thank an anonymous reviewer for reminding us of examples of spatial change.

of temporal 'John'-stages if the values of the measurement (under the contextually presupposed measurement unit) of John's facial complexion at these stages
are increasing, and will return '0' otherwise.

We next discuss the relation of our *associates*-approach to existing work on
the temperature puzzle.

6   Relation to Existing Work

Our *associates*-approach distinguishes itself from existing solutions to the temperature puzzle. This is due to the proximity of our approach to Montague's
original solution from [33] (cf. Sect. 3.3) and to its focus on improving the computational properties of this solution (cf. Sect. 2.2):

Firstly, in contrast to the solutions from [3, 20, 41], and to solutions from event
semantics, our solution is not based on an alternative interpretation of (1a) that
uses a locative interpretation of the copula (i.e. 'is *at* ninety'), a measurement-
explicit interpretation of the DP ninety (i.e. 'is ninety *degrees Fahrenheit*'), or
an event-based interpretation of the verb rise (s.t. 'rise' describes a rising event).

Secondly, in contrast to the solutions from [12, 27, 29, 40], our solution is not
directed at a variant of the temperature puzzle (i.e. *Gupta's problem*; cf. (12))
that arises from the double index-dependence of intensional nouns like temperature; viz. from the dependence of temperature-values on the index-argument of a
particular individual concept [i.e. *inner index-dependence*] and the dependence
of noun-interpretations on the index of evaluation[12] [i.e. *outer index-dependence*]
(cf. [40]). As a result of this double dependence, Montague semantics blocks the
intuitively valid inference from (12):

a. Necessarily, the temperature of the air in my refrigerator is
   the same as the temperature of the air in your refrigerator.

b. The temperature of the air in my refrigerator is rising. (12)
   ──

c. The temperature of the air in your refrigerator is rising.

It should come as no surprise that the different solutions to Gupta's problem
can be integrated into our *associates*-approach to Partee's temperature puzzle.
However, our approach even provides its own solution to the puzzle, which also
involves computability considerations. We will detail this solution in a sequel to
this paper.

7   Conclusion and Outlook

We have presented a computable, low-type, context-sensitive solution to Partee's temperature puzzle which uses the countable approximation of continuous
functionals via their associates. The success of our solution is challenged by the

---

[12] As a result of this dependence, rise may denote a different set of individual concepts
at different indices.

restriction of associates to *continuous* functionals. This restriction prevents the application of our approach to expressions that are traditionally interpreted as *dis*continuous functionals (e.g. mostly above 90).

Its exclusion of discontinuous intensional verbs hampers the generality of the presented approach. However, in natural language, discontinuous expressions are rather rare: of the 369 intensional intransitive verbs listed in [28] (see Sect. 5 for a selection), *only 5* are discontinuous. Their scarcity notwithstanding, discontinuous verbs can be accommodated in Bezem's model $\mathcal{M}$ of strongly majorizable functionals (cf. [23, Ch. 3, 11]). The weak continuity functional ([4, Sect. 5, p. 171]) in this model serves a similar role to the fan functional in the Kleene-Kreisel model: it produces a lower-type correlate of its input functional. However, whereas the associate of a continuous functional is an accurate representation of the continuous functional (in the sense that no information is lost), the output of the weak continuity functional only *partially* represents the input functional in Bezem's model. The detailed development of this account is a project for future work.

References

1. Abusch, D.: On Verbs and Time. Doctoral dissertation, University of Massachusetts, Amherst (1985)
2. Barendregt, H., Dekkers, W., Statman, R.: Lambda Calculus with Types. Perspectives in Logic. Cambridge University Press and ASL, Cambridge (2010)
3. Bennett, M.R.: Some Extensions of a Montague Fragment of English. Indiana University Linguistics Club (1975)
4. Berger, U., Oliva, P.: Modified bar recursion. Math. Struct. Comput. Sci. **16**(2), 163–183 (2006)
5. Bertinetto, P.M., Squartini, M.: An attempt at defining the class of 'gradual completion verbs'. In: Bianchi, V., Higginbotham, J., Squartini, M. (eds.) Temporal Reference: Aspect and Actionality. Semantic and Syntactic Perspectives, vol. 1. Rosenberg and Sellier (1995)
6. Buss, S.R.: An Introduction to Proof Theory. In: Handbook of Proof Theory. Studies in Logic and the Foundations of Mathematics, vol. 137, pp. 1–78. North-Holland Publishing Co. (1998)
7. Carnap, R.: Meaning and Necessity: A Study in Semantics and Modal Logic. University of Chicago Press, Chicago (1988)
8. Church, A.: A formulation of the simple theory of types. J. Symb. Logic **5**(2), 56–68 (1940)
9. Davidson, D.: Truth and meaning. Synthese **17**, 304–323 (1967)
10. Deo, A., Francez, I., Koontz-Garboden, A.: From change to value difference in degree achievements. In: Snider, T. (ed.) Proceedings of SALT 23. University of California, Santa Cruz (2013)
11. Dowty, D.R.: Word Meaning and Montague Grammar: The Semantics of Verbs and Times in Generative Semantics and in Montague's PTQ. Synthese Language Library, vol. 7. D. Reidel Publishing Company (1979)
12. Dowty, D.R., Wall, R.E., Peters, S.: Introduction to Montague Semantics. Studies in Linguistics and Philosophy, vol. 11. Kluwer Academic Publishers, Berlin (1981)

13. van Eijck, J., Unger, C.: Computational Semantics with Functional Programming. Cambridge University Press, Cambridge (2010)
14. Escardó, M.H., Xu, C.: A constructive manifestation of the Kleene-Kreisel continuous functionals. Ann. Pure Appl. Logic (to appear)
15. Fodor, J.A.: Language, thought, and compositionality. Mind Lang. 16(1), 1–15 (2001)
16. Fox, C., Lappin, S., Pollard, C.: A higher-order fine-grained logic for intensional semantics. In: Proceedings of the 7th International Symposium on Logic and Language (2002)
17. Gamut, L.T.F.: Intensional Logic and Logical Grammar. Logic, Language, and Meaning, vol. 2. University of Chicago Press, Chicago (1991)
18. Gawron, M.: The lexical semantics of extent verbs. San Diego State University (2009)
19. Hendriks, H.: Flexible Montague Grammar. ITLI Prepublication Series for Logic, Semantics and Philosophy of Language, vol. 08 (1990)
20. Jackendoff, R.: How to keep ninety from rising. Linguist. Inq. 10(1), 172–177 (1979)
21. Kleene, S.C.: Countable functionals. In: Heyting, A. (ed.) Constructivity in Mathematics. North-Holland, Amsterdam (1959)
22. Klein, E., Sag, I.: Type-driven translation. Linguist. Philos. 8(2), 163–201 (1985)
23. Kohlenbach, U.: Applied Proof Theory. Springer, Heidelberg (2008)
24. Kohlenbach, U.: Foundational and mathematical uses of higher types. In: Reflections on the Foundations of Mathematics. LNCS, vol. 15, pp. 92–116. Association for Symbolic Logic, Natick (2002)
25. Kreisel, G.: Interpretation of analysis by means of constructive functionals of finite types. In: Heyting, A. (ed.) Constructivity in Mathematics. North-Holland, Amsterdam (1959)
26. Kripke, S.A.: Semantical considerations on modal logic. Acta Philos. Fennica 16, 83–94 (1963)
27. Lasersohn, P.: The temperature paradox as evidence for a presuppositional analysis of definite descriptions. Linguist. Inq. 36(1), 127–134 (2005)
28. Levin, B.: English Verb Classes and Alternations: A Preliminary Investigation. The University of Chicago Press, Chicago (1993)
29. Löbner, S.: Intensional verbs and functional concepts: more on the "rising temperature" problem. Linguist. Inq. 12(3), 471–477 (1981)
30. Longley, J., Normann, D.: Higher-Order Computability. Springer, Heidelberg (2015)
31. Kechris, A.S.: Classical Descriptive Set Theory. Graduate Texts in Mathematics, vol. 156. Springer, Heidelberg (1995)
32. Montague, R.: English as a formal language. In: Thomason, R.H. (ed.) Formal Philosophy: Selected Papers of Richard Montague. Yale University Press (1976)
33. Montague, R.: The proper treatment of quantification in ordinary English. In: Formal Philosophy: Selected Papers of Richard Montague. Yale University Press (1976)
34. Moot, R., Retoré, C.: Natural language semantics and computability. Manuscript (2016)
35. Muskens, R.: A relational formulation of the theory of types. Linguist. Philos. 12(3), 325–346 (1989)
36. Normann, D.: Recursion on the Countable Functionals. Lecture Notes in Mathematics, vol. 811. Springer, Heidelberg (1980)

37. Partee, B.: Compositionality. In: Landman, F., Veltman, F. (eds.) Varieties of Formal Semantics: Proceedings of the 4th Amsterdam Colloquium. Groningen-Amsterdam Studies in Semantics, vol. 3 (1984)
38. Partee, B.: Noun phrase interpretation and type-shifting principles. In: Groenendijk, J., de Jong, D., Stokhof, M. (eds.) Studies in Discourse Representation Theory and the Theory of Generalized Quantifiers. Foris Publications (1987)
39. Pollard, C.: Hyperintensions. J. Logic Comput. **18**(2), 257–282 (2008)
40. Schwager, M.: Bodyguards under cover: the status of individual concepts. In: Friedman, T., Gibson, M. (eds.) Proceedings of SALT XVII (2007)
41. Thomason, R.H.: Home is where the heart is. In: French, P.A., Uehling, T.E., Wettstein, H.K. (eds.) Contemporary Perspectives in the Philosophy of Language (1979)
42. Troelstra, A.S.: Metamathematical Investigation of Intuitionistic Arithmetic and Analysis. Lecture Notes in Mathematics, vol. 344. Springer, Heidelberg (1973)
43. Vendler, Z.: Verbs and times. Philos. Rev. **66**(2), 143–160 (1957)
44. Werning, M.: Right and wrong reasons for compositionality. In: Werning, M., Machery, E., Schurz, G. (eds.) The Compositionality of Meaning and Content: Volume I: Foundational issues. Ontos Verlag (2005)
45. Xu, C.: A continuous computational interpretation of type theories. Ph.D. thesis, University of Birmingham (2015)
46. Xu, C.: A continuous computational interpretation of type theories, developed in Agda (2015). http://cj-xu.github.io/ContinuityType/
47. Xu, C., Escardó, M.: A constructive model of uniform continuity. In: Hasegawa, M. (ed.) TLCA 2013. LNCS, vol. 7941, pp. 236–249. Springer, Heidelberg (2013). doi:10.1007/978-3-642-38946-7_18
48. Zalta, E.N.: A comparison of two intensional logics. Linguist. Philos. **11**(1), 59–89 (1988)

Actuality Entailments: When the Modality is in the Presupposition

Alda Mari$^{(\boxtimes)}$

Institut Jean Nicod, CNRS/ENS/EHESS, Paris, France
alda.mari@ens.fr

Abstract. In natural language, modals are not implicative. However, when the modality is combined with the perfective, it shows an implicative (or factive) behavior. This phenomenon is called 'actuality entailment'. We show that actuality entailments arise with goal-oriented modality only and endorse Belnap's view of that goal-oriented modals use historical accessibility with a fixed past and an open future. This modal-theoretic assumption allows us to spell out the precise modal-temporal configuration in which the actuality entailment arises and our predictions are borne out by the data, cross-linguistically. We also show that, when any assumption about the identity of worlds at branching point is leveled - which appears to be the case with generic deontic and opportunity modals, the actuality entailments disappear. We also predict that the entailment disappears with prospectivity. Finally, we argue that modal sentences giving rise to actuality entailments are informative, insofar as the contribution of the modality survives as a presupposition that the modal base is non-homogeneous.

Keywords: Modality · Presupposition · Actuality entailments · Goal · Intentionality · Implicative verbs

1 Introduction

Modals in natural language are not implicative.[1] This is observed for existential (e.g. 'might') and universal (e.g. 'must') modals, both epistemic ((1-b), (1-d)) and deontic ((1-a), (1-c)).

(1) a. He is allowed to go to school. $\nrightarrow$ He goes to school.
     b. He might be sick. $\nrightarrow$ He is sick.
     c. He must go to school. $\nrightarrow$ He goes to school.
     d. It must be raining. $\nrightarrow$ It rains.

---

[1] Special thanks to Anastasia Giannakidou for the long discussions on several aspects of this work. I am also grateful to Chris Kennedy, Itamar Francez, Malte Willer, Guillaume Thomas and the three anonymous reviewers for comments and suggestions. This research was funded by ANR-10- LABX-0087 IEC and ANR-10-IDEX-0001-02 PSL. This paper was written during my stay at the University of Chicago 2014–2016. We also gratefully thank the CNRS-SMI 2015.

© Springer-Verlag GmbH Germany 2016
M. Amblard et al. (Eds.): LACL 2016, LNCS 10054, pp. 191–210, 2016.
DOI: 10.1007/978-3-662-53826-5_12

However, as observed by [4], in some modal-temporal combinations the modality is implicative. In the specific context of the study of modality in interaction with time this phenomenon has been called *actuality entailment* [4] and maintain here this terminology.

Actuality entailments were immediately observed as arising when the modal is in the perfective. Bhatt's observation has been replicated across a variety of languages (see, e.g. for French, [19, 21, 29]; for Italian, [28]; for Greek [17] a.o.).[2] In French, the language studied in this paper, the actuality entailment arises with the *passé composé*[3].

(2)   Jean a   pu   prendre le   train, #mais il   ne   l'a   pas pris.
       John has can.pp take   the train, #but   he   not that-has   taken.
       Intended: 'John managed to move the table, #but he did not do it.'

The *imparfait* cancels the actuality entailment in French (*a contrario*, see [9, 17]).

(3)   John pouvait   prendre le   train, mais il   ne   l'a   pas pris.
       John can.impf take   the train, but   he   not that-has   taken.
       'John could have taken the train, but he did not take it.'

Bhatt (*ibid.*) proposes that the modal is ambiguous and that in addition to a non-implicative *can$_1$*, there is an implicative *can$_2$* that behaves just like the implicative *manage to*. Bhatt also argues that the imperfective conveys generic information, which prevents the actuality entailment from arising. [29] observe that imperfectivity *cannot* cancel the implication with implicative verbs like 'arriver à' (*manage to*) and thus that the modality cannot be implicative to begin with.

With the aim to provide a unified theory for modals, theoreticians have built on the assumption that modals in natural language are non-implicative. The debate has been very active since [4] and, most prominently, [19], and various proposals have sought to maintain the non-implicativity of the modals.[4]

The major challenge faced by any theory of actuality entailments is distinguishing between modal statements giving rise to the entailment ((4-a) and (4-b)) and non-modal statements (4-c). In this paper, we focus on existential modals, since the entailment of actuality is unexpected under any approach of possibility modals.

(4)   a.   Jean a   pu   prendre le   train.
             John has can.pp take   the train.
             'John managed to take the train.

---

[2] Several authors do not subscribe to an aspectual analysis, though, and some of them argue that aspect does not play a role at all (see e.g., [17]).

[3] In the glosses pp is for 'past participle', and impf for 'imperfective'.

[4] For a discussion of available accounts, see a draft version of this paper at http://ling.auf.net/lingbuzz/002634.

b. Jean a   dû   prendre le   train.
    John has must.pp take   the train.
    'John had to take the train (and he took it).'

c. Jean a   pris le   train.
    John has taken the train.
    'John took the train.'

All existing approaches ([19, 21, 29]) derive the entailment via complex calculi, begging the question of why the speaker would choose such a complex interpretation to ultimately entail p, rather than asserting a non modal statement to begin with.

Likewise, since $\Diamond p$ is asymmetrically entailed by p, the question should be posed of how the Gricean Maxim of Quality would be respected in the case where the modal is implicative.

On the assumption that truth of modal statements is evaluated with respect to a set of possible worlds, the modal base, we propose that the following axiom (informally, for now), holds for all modals in natural language (see [8, 12, 15, 28])[5].

(5) *Non-Homogeneity Axiom of modals* - [15]
     Modal bases triggered by a modal are non-homogeneous, i.e. they contain
     p and non-p worlds.

In order to disentangle modal from non-modal statements in the passé composé, we need to show how the non-homogeneity conditions of the modals is fulfilled when the actuality entailment arises.

Our claim is that this condition survives as a presuppositions of those sentences in which the modal gives rise to the entailment.

The paper is structured as follows. We discuss new data in Sect. 2, present the analysis in Sect. 3 and discuss remaining questions in Sect. 4.

2   Goals and Expectations: New Facts

With [5, 19], we observe that the actuality entailment arises with abilitative (6), teleological (7) and non-generic[6] deontic (8) modality in the passé composé.

(6) Jean a   pu   déplacer la   table, #mais il ne l'a   pas
     John has can.pp move   the table, #but he not that-has
     déplacée.
     move.pp.fem.
     'John managed to move the table, #but he did not move it.'

(7) Jean a   pu   prendre le train, #mais il ne l'a   pas pris.
     John has can.pp take   the train, #but he not that-has   taken.
     'John managed to take the train, #but he did not take it.'

---

[5] For more discussion on the notion of non-homogeneity, see also [13, 16].
[6] For generic deontic modality and the distinction between generic and goal-oriented
    deontic modality, see [28].

(8)   Jean a   pu   rentrer à la piscine   grâce au   nouveau
       John has can.pp enter   to the swimming-pool thanks to-the new
       règlement, #mais il n'est pas rentré.
       rules,   #but he not-is   enter.pp.masc.
       'John could enter (and did enter) to the swimming-pool thanks to the
       new rules, #but he did not enter.'

Most importantly, common to the cases in which the entailment arises is the fact
that the entity denoted by the subject[7] pursues a goal. None of the sentences
above can be continued by 'but he did not want it.' (note that 'manage to' does
not trigger this intentionality component). For space limitations, we observe this
generalization only for (6).

(9)   Jean a pu déplacer la table, #mais il ne voulait pas la déplacer.
       John managed to move the table, but he did not want to move it.

Note also, that (10) is felicitous only if John has the intention of being liked (see
[25, 26]).

(10)   John a   pu   plaîre.
        John has can.pp be   liked.
        'John managed to being liked.'

This intentionality feature is absent from the meaning of the implicative 'arriver
à' ('manage to'). When we contrast past modals triggering the actuality entailment with the implicative verb *arriver à* ('manage to'), we see that there is
intentionality with the modal sentence but not with *arriver à*. The English sentence 'He managed to be dumped' can be translated in two different ways (11-a)
and (11-b).

(11)   a.   Jean   est arrivé   à se faire quitter. (no intentionality)
              Gianni is   arrive.pp to refl make dump.
        b.   Jean   a   pu   se faire quitter. (intentionality)
              Gianni has can.pp refl make dump.
              'He managed to be dumped'.

In (11-a), John plays the role of the victim who has been dumped by his girlfriend. In (11-b), his girlfriend is the victim, as the sentence conveys that John
had the goal of being dumped.

   This leads us consider the abilitative, teleological and deontic modalities in
(6), (7) and (8), as instances of goal-oriented modality. Portner [33] uses the term
'dynamic modality' to subsume these three types of goal-oriented modals and
dedicates the term 'goal-oriented modal' for one subtype of dynamic modality.
The term goal-oriented modality which we maintain here will help us recall that
across the instances of goal-oriented modality, agents and entities have goals.

---

[7] It can also be a contextually relevant entity, like the captain of a boat in 'Le navire
   a pu rentrer au port' (The boat managed to enter into the harbor).

Another key factor enhancing the emergence of the entailment (gone unnoticed in the literature – see [25] though) is that the modal giving rise to the entailment can only be used only if the participants in the conversation expect that the goal cannot be fulfilled. This expectation, as we now show, is a presupposition.

Consider the following scenario. As is well-known, Usain Bolt is the fastest runner in the world, who can run 100 meters in 9.58 seconds.

(12)   Usain Bolt a   pu   battre le   record du   monde des   100
        Usain Bolt has can.pp break the record of-the world of-the 100
        mètres grâce   à son entranement.
        meters thanks to his training.
        'Usain bolt managed to break the 100-meter world record thanks to his training.'

Breaking the world record is never granted, and the possibility that even Usain Bolt does not break it is open at a time prior to the race. The sentence is felicitous. Sentence (14), instead, is infelicitous in Context 1 and felicitous in Context 2 described in (13).

(13)   a.   *Context 1*: Usain Bolt is in his best shape and at the climax of his career.
        b.   *Context 2*: Usain Bolt is recovering from a long cold and is far from his highest standards.

(14)   (#)Usain Bolt a   pu   courir 100 mètres en 15 secondes
        Usain   Bolt has can.pp run   100 meters in 15 seconds
        aujourd'hui.
        today.
        'Usain Bolt managed to run 100 meters in 15 seconds today.'

Consider context (13-a), in which sentence (14) is infelicitous. Since Usain Bolt can run 100 meters in 9.58 seconds, it is taken for granted that, in his best shape, he can run 100 meters in fifteen seconds, and the possibility that he does not run 100 meters in fifteen seconds was not even considered.

Sentence (14) is instead felicitous in context 2 (13-b), where Usain Bolt is recovering from a very bad cold. In this context, running 100 meters in fifteen seconds is not granted; the possibility of $\neg p$ was expected to be realized.

The un-modalized sentence (15) is felicitous in both contexts (13-a) and (13-b), instead. It does not require that $\neg p$ was expected.

(15)   Usain Bolt a   couru   100 mètres en 15 secondes.
        Usain Bolt has run.pp 100 meters in 15 seconds.
        'Usain Bolt has run 100 meters in 15 seconds.'

Importantly, such expectation triggered by past goal-oriented modals must be part of the utterance context prior to utterance, and encodes what the participants take for granted (on this property of presuppositions, see e.g. [35,37]), as the 'wait a minute' test (designed to detect presuppositions – [10]) shows. Consider the following scenario. My mother has to take the train to her home in the south of the country. She generally goes there every weekend, and she phones my husband or me to tell us that she has arrived. She generally phones me on Saturday. My husband comes home and asks whether she has arrived (see (16)).

(16)   Est-ce que ta   mère   est arrivée?
              your mother is   arrived?
        'Did your mother arrive?

If I reply (17), and my husband is not aware that it was not granted that my mother would take the train, he would be entitled to ask (18).

(17)   Oui, elle a   pu   prendre le   train.
        Yes, she has can.pp take   the train.
        'Yes, she managed to take the train.'

(18)   Attends, il y   avait   un problème?
        Wait,   it there have.3sg.impf a   problem?
        'Wait a minute, there was a problem?'

This shows that both participants must know that prior to the time at which p is realized, there was a time t'' such that $\neg p$ was expected to be realized. If this presupposition is not met, the sentence is infelicitous.

    The following family of sentences also reveals that we are dealing with a pre-suppositions. Again, (19)-(20)-(21) are felicitous only in contexts implying that not running 100 meters in fifteen seconds is expected (Usain Bolt is recovering from a cold – see Context 2 in (13-b)).

(19)   (#)Est-ce qu' il   a   pu   courir 100 mètres en 15 secondes,
                      He has can.pp run   100 meters in 15 seconds,
        aujourd'hui?
        today?
        'Did he manage to run 100 meters in 15 seconds?'

(20)   (#)Il est possible qu'il   ait   pu   courir 100 mètres en 15
        It   is   possible that-he has.3sg.subj can.pp run   100 meters in 15
        secondes.
        seconds.
        'It is possible that he managed to run 100 meters in 15 seconds'.

(21)   (#)S'il a   pu   courir 100 mètres en 15 secondes, alors il
        If-he   has can.pp run   100 meters in 15 seconds,   then he
        va   bientôt se   remettre.
        go.3sg.pres soon refl be-fine.
        'If he managed to run 100 meters in 15 seconds, then he is going to be fine soon.'

In view of these data, we can conclude that the modal contributes meaning by introducing a meaning component conveying 'expectation.'

3 Analysis

3.1 Representing Goal-Oriented Modality

[2] is the first to propose an analysis of goal-oriented modality within a branching time framework [36].[8] We endorse this model theoretical framework for goal-oriented modality in French as well. As we show, this choice will allows us to derive a variety of predictions, cross-linguistically.

The Modal-Temporal Skeleton. Thomason's world-time model uses $W \times T$ frames. A branching structure is generated. Each branching point determines an equivalence class of worlds with a unique past and present and an open future. A three-place relation $\simeq$ on $T \times W \times W$ is defined such that (i) for all $t \in T$, $\simeq_t$ is an equivalence relation; (ii) for any $w, w' \in W$ and $t, t' \in T$, if $w' \simeq_{t'} w$ and t precedes t', then $w' \simeq_t w$ (we use the symbols $\prec$ and $\succ$ for temporal precedence and succession, respectively). In words, w and w' are historical alternatives at least up to t' and thus differ only, if at all, in what is future to t'.

Figure 1 depicts two equivalence classes of worlds, determined at t_1 and t_2.

(22)   a. $w_0 \simeq_{t_1} w_1 \simeq_{t_1} w_2 \simeq_{t_1} w_3 \simeq_{t_1} w_4$ (historical alternatives at t_1).
         b. $w_0 \simeq_{t_2} w_2 \simeq_{t_2} w_3$ (historical alternatives at t_2).

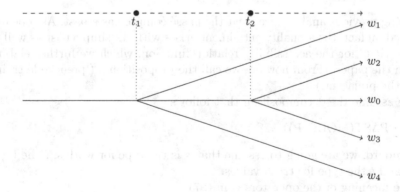

Fig. 1. Equivalence classes of worlds

For any time $t \in T$, on can define the **historical alternatives** $\mathcal{I}$ as the set of worlds that are **identical** to the actual world w_0 at least up to and including t.

---

[8] See [8,22,27] for discussion of this framework in the linguistic literature.

(23) $\mathcal{I}(w_0)(t) := \{w \mid w \simeq_t w_0\}$

Figure 2 depicts the historical alternatives, determined at time t.

(24) $\mathcal{I}(w_0)(t) = \{w_1, w_2, w_0, w_3, w_4\}$

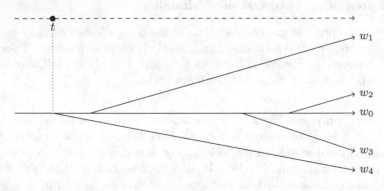

Fig. 2. $\mathcal{I}(t)$

3.2   Calculating the Asserted Meaning

Let us work through the compositional semantics of the sentence in (25).

(25)   John a pu [prendre le train]
         John managed to take the train.

Following previous analysis, we treat the passé composé as a past. As repeatedly observed, in fact, the actuality entailment arises with the simple past as well [28]. We use ACC for the accessibility relation function, which we further elaborate later in the paper. (From now on, we call the proposition p ('prendre le train' in (25)), the prejacent.)

We assume the decomposition that follows:

(26)   PAST(MOD(VP))

As standard, we are going to assume that s is the type for worlds, i the type of times and t the type for truth values.

The meaning of the operators is in (27).

(27)   a. $MOD = \lambda p^{s \to \langle i \to t \rangle} \lambda w^s \lambda t^i \exists w'[w' \in ACC(w)(t) \land p(w')(t)]$
         b. $PAST = \lambda p^{s \to \langle i \to t \rangle} \lambda w^s \lambda t^i \exists t'[t' \prec t \land p(w)(t')]$
         c. $VP = \lambda w^s \lambda t^i p(w)(t)$

(28)   Composition.
         a. $MOD(VP) = \lambda w^s \lambda t^s . \exists w'[w' \in ACC(w)(t) \land [\lambda w^s \lambda t^i . p(w)(t)](w')(t)] =$
                $\lambda w^s \lambda t^i . \exists w'[w' \in ACC(w)(t) \land p(w')(t)]$

b. $PAST(MOD(VP)) = \lambda w^s \lambda t^i . \exists t'[t' \prec t \wedge [\lambda w^s \lambda t^s . \exists w'[w' \in ACC$
     $(w)(t) \wedge p(w')(t)](w)(t')]] =$
     $\lambda w^s \lambda t^i . \exists t'[t' \prec t \wedge \exists w'[w' \in ACC(w)(t') \wedge p(w')(t')]]$

c. t is fixed as t_u and w is the world of evaluation
     Truth conditions: $\exists t'[t' \prec t_u \wedge \exists w'[w' \in ACC(w)(t') \wedge p(w')(t')]]$
     Paraphrase: there is a past time at which there is a world accessible
     from the world of evaluation, at which p is true (e.g. John takes the
     train).

Past fixes both the time of evaluation of the modal and of the prejacent (in
absence of a tense that fixes the time of evaluation of the prejacent independently
of the time of evaluation of the modal, cf. *infra*). [18] refers to this phenomenon
by stating that the tense of the embedded proposition is anaphoric to the higher
tense. In other terms, the time of evaluation of the prejacent (the time at which
e.g. John takes the train) and the time of the evaluation of the modal (i.e. the
time at which the possibility of taking the train occurs) are the same. Note that
this is parallel to what happens with implicative verbs. For a sentence 'John
managed to take the train' the time at which John takes the train and the time
at which John *manages* to take the train, are the same.

Interpreting the sentence in a branching time framework allows us to explain
why the actuality entailment arises when the time t' at which the quantificational
domain of the modal coincides with the time at which the prejacent is evaluated.
In such a past time there is just one world, the actual one. In this configuration
this is the only world of evaluation (Fig. 3).

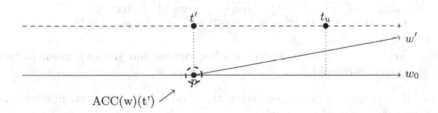

Fig. 3. The domain of quantification of the goal-oriented modality in the past.

Our model theoretic assumptions also allow us to predict that the actuality
entailment disappears with prospectivity (PROSP or FUT in the literature, see
e.g., [1,31]), that this to say, in the configuration where the time of evaluation
of the prejacent follows the time of evaluation of the modality. The actuality
entailment, under our hypothesis, does not arise when there is prospectivity
because the prejacent lies in possibilities that are not actual, as seen from the
time at which the alternatives (the branches) are determined (as we show in
Sect. 3.3 this prediction is borne out).

There is no overt mark of prospective aspect in the French language. However,
we can demonstrate *ex absurdo* that if prospective aspect were present when the

modal is in the passé composé, we would predict the licit use of forward-shifting temporal adverbs like 'tomorrow'. This type of temporal adverb is incompatible with the modal in the past (29) [21].

(29) #Hier   il a pu   rendre son devoir   demain.
        Yesterday he has can.pp return his homework tomorrow.
        '#Yesterday, Pierre managed to return his homework tomorrow.'

We thus assume that prospectivity is absent when the modal is in the passé composé, and because we do not have a past either (past on the infinitive of the embedded predicate is overt in French), we hold that the time at which the modal and the prejacent are evaluated are the same and are fixed by the higher past operator. In this configuration, the entailment arises.

3.3 Prospectivity and the Absence of Entailment of Actuality

The major predictions that we are able to make in adopting the branching time framework, is that the entailment does not arise with prospectivity. This prediction is borne out by cross-linguistic evidence.

Gitksan [31] offers an overt prospective aspect marker *dim*, which suppresses the actuality entailments.

In Gitksan, modals are lexically restricted with respect to the modal bases they allow: *da'akxw* is the circumstantial modal.

With *dim* (which in fact is obligatory with non-epistemic modals) the actuality entailment does not arise.

(30)   da'akxw[-i]-'y   dim   ayee=hl   bax-'y
        POSSIBILITY[-tra]-1sg.II PROSP go.fast=CN run-1sg.II
        'I can run fast'.
        Rejected in context: 'You were a fast runner, but you've become permanently paralyzed.'

In other terms, as predicted, since the modal has future orientation, the actuality entailment does not arise.

The French 'imparfait' features a variety of modal uses that include the counterfactual, as well as the progressive (e.g. [24]). Some (if not all) of these uses have been argued to involve a modal component. [4,20] propose that when the modal is in the imperfective, the modal GEN levels the entailment of actuality.

With [1], for French, we assume that in the counterfactual use of the imperfective, PAST is combined with FUT (here PROSP) - for previous discussion about the counterfactual interpretation of the imperfective, see [3]. (31) is analyzed as in (32) along the lines of ([31]).

(31)   Jean pouvait prendre le train (mais il ne l'a pas pris).
        John can.3sg.impf take the train (but he not that-has taken).
        'John could take the train (but he did not take it).'

(32)   PAST(MOD(PROSP(VP))

On the assumption that PROSP has the semantics in (33), we obtain the truth-conditions in (34-d) for (31). Let t_u be the time of utterance.

(33) $PROSP = \lambda p^{s \to \langle i \to t \rangle} \lambda w^s \lambda t^i \exists t''[t'' \in [t, \infty) \wedge p(w)(t'')]$

(34)   Composition.

    a. $PROSP(VP) = \lambda w^s \lambda t^i . \exists t''[t'' \in [t, \infty) \wedge [\lambda w^s \lambda t^i . p(w)(t)](w)(t'')]$
       $= \lambda w^s \lambda t^i . \exists t''[t'' \in [t, \infty) \wedge p(w)(t'')]$

    b. $MOD(PROSP(VP)) = \lambda w^s \lambda t^i . \exists w'[w' \in ACC(w)(t) \wedge [\lambda w^s \lambda t^i .$
       $\exists t''[t'' \in [t, \infty) \wedge p(w)(t'')]](w')(t)] =$
       $\lambda w^s \lambda t^i . \exists w'[w' \in ACC(w)(t) \wedge \exists t''[t'' \in [t, \infty) \wedge p(w')(t'')]]$

    c. $PAST(MOD(PROSP(VP))) = \lambda w^s \lambda t^i . \exists t'[t' \prec t \wedge [\lambda w^s \lambda t^i . \exists w'[w' \in$
       $ACC(w)(t) \wedge \exists t''[t'' \in [t, \infty) \wedge p(w')(t'')]]](w)(t')] =$
       $\lambda w^s \lambda t^i . \exists t'[t' \prec t \wedge \exists w'[w' \in ACC(w)(t') \wedge \exists t''[t'' \in [t', \infty) \wedge$
       $p(w')(t'')]]]$

    d. t is fixed as t_u and w is the world of evaluation.
       Truth conditions: $\exists t'[t' \prec t_u \wedge \exists w'[w' \in ACC(w)(t') \wedge \exists t''[t'' \in$
       $[t', \infty) \wedge p(w')(t'')]]]$
       Paraphrase: There is a past time t' such that there is a world w' accessible from the actual world at t' such that there is a time t'' **future** with respect to t' such that p is true at t'' in w'.

Since the truth of the prejacent is calculated at a time that follows the time at which the possibilities are projected, given a branching time framework, the actuality entailment does not arises. The prejacent lies in possibilities that are not actual from the perspective of the branching point at which they are projected.

Again, we do not have an overt marking of prospective aspect in French. However, prospectivity is detectable in (35), where forward-shifting temporal adverbs locate the time of the truthiness of the prejacent with the resulting future temporal orientation of the modal.

(35)   Hier il pouvait rendre son devoir demain.
       Yesterday he can.3sg.impf return his homework tomorrow.
       'Yesterday, Pierre could return his homework tomorrow.'

3.4 Accounting for the Contribution of the Modal

In our account so far, the modal turns out to be trivialized in the assertion, as its domain of quantification contains only one world, the actual one.

We must now implement the contribution of the modal, in order to be able to distinguish between the semantics of bare assertions and modalized assertion in the passé composé. We also elaborate on the constraints on the branches.

Let t'' be a contextually determined time. We define what follows:[9]

---

[9] We use here the aristotelian notion of *telos*, which includes both goals and tendecies of natural entities, although here we do not discuss the case of this type of entities.

(36) $MB(w_0)(t'') =$
           $\{w \in \mathcal{I}(w_0)(t'') : $ a relevant entity has a *telos* in w at t'' $\}$

Note from the outset that having the *telos* does not imply actualization of the *telos*.[10] We posit a condition on the modal space, namely that it is not homogeneous and contains both p and $\neg p$ continuations: p worlds are worlds in which the *telos* is achieved and $\neg p$ worlds are worlds in which the *telos* is not achieved.

(37)   Non-homogeneity of the historical modal base (*to be revised*):
           $\exists t' \succ t'' \Big(\exists w' \in MB(w_0)(t'')\big(p(t')(w')\big) \Big) \wedge \Big(\exists w'' \in MB(w_0)(t'')\big(\neg p(t')(w'')\big) \Big)$

Here, the time at which the truthiness of the prejacent is evaluated (t') follows the time at which the alternatives - including p and $\neg p$ worlds - are projected (t'').

This is not sufficient. As we have shown in the data section, it is not only the case that $\neg p$ was metaphysically possible at a time prior to the realization of p (this is always trivially the case, given a metaphysical space). $\neg p$, instead, was the possibility expected to get realized.

We add ordering sources to restrict the metaphysical space, which, recall, contains worlds in which the subject entity has a certain telos. Recall also that the abilitative, deontic and teleological modal are instances of goal-oriented modality. In our account these flavors of goal-oriented modality are implemented as ordering sources.

Second, in order to implement the notion of *expectation*, we use a secondary ordering source, which is epistemic [23,33]. We conceive an expectation as an epistemic object: the speaker selects those worlds among the metaphysical accessible ones that better conform to his/her own beliefs [14].

Following [33], we define ordering of worlds and Best worlds as follows.

(38)   *Ordering of worlds* - [33]
           For any set of propositions X and any worlds $w, v : w \leqslant_X v$ iff for all $p \in X$, if $v \in p$, then $w \in p$.

(39)   For any set of propositions X, Best worlds as per X.
           $\text{Best}_X: \{w' : \forall q \in X(w' \in q)\}$

$\mathcal{B}, \mathcal{D}, \mathcal{A}$ are, respectively the doxastic, deontic and abilitative ordering sources. These are set of propositions that better conforms to the belief of the speaker (including stereotypicality conditions) $(\mathcal{B})$, the orders and the permissions received $(\mathcal{D})$, and the abilities $\mathcal{A}$ (we will not use $\mathcal{A}$ here).

Let us consider the following example, where permissions (hence a deontic ordering source is considered).

---

[10] The only exceptions to this are **natural entities**, whose *telos* (final cause) - e.g., the final cause of the wind is to blow - is necessarily *in acto*. For space reasons, we do not consider natural entities here.

(40)   Jean a pu rentrer à la piscine.
        John could enter in the swimming pool (and he did enter, in virtue of a
        permission).

Ordering sources restrict the set of worlds to be taken into consideration. From the entire metaphysical modal base $\mathcal{M}$ (in our case, this is $MB(w_0)(t'')$), first the deontic ordering source applies.

(41) $\text{Best}_\mathcal{D}$: $\{w' \in \mathcal{M} : \forall q \in \mathcal{D}(w' \in q)\}$

The doxastic ordering source, if any, then further restricts $\text{Best}_\mathcal{D}$.

(42) $\text{Best}_\mathcal{B}$: $\{w' \in \text{Best}_\mathcal{D} : \forall q \in \mathcal{B}(w' \in q)\}$

Let us consider the case of deontically flavored goal-oriented modality (40).[11] Again, recall that the metaphysical modal space is already restricted to the worlds in which a relevant entity has a telos (namely we are dealing with goal oriented modality), $\mathcal{D}$ restricts the initial domain in which a goal is being pursued. Our final analysis is as follows.

We can now modify the lexical entry for MOD in (27-a), as in (43), where $\mathcal{X}$ is an ordering source.

(43)   MOD $= \lambda p^{s \to \langle i \to t \rangle} \lambda w^s \lambda t^t . \exists w'[w' \in \text{Best}_\mathcal{X} \wedge p(w')(t)]$

(44)   a. $[\![\text{PAST(MOD(VP))}]\!]$ is defined if and only if
              there is a contextually determined past time t'' s.t.
              (i) $MB(w_0)(t'') =$
              $\{w \in \mathcal{I}(w_0)(t'') :$ a relevant entity has a *telos* p in w at t'' $\}$
              (ii) $\exists t' \succ t'' \left(\exists w' \in \text{Best}_\mathcal{D} \left(p(t')(w') \right) \right) \wedge$
              $\left(\exists w'' \in \text{Best}_\mathcal{D} \left(\neg p(t')(w'') \right) \right)$
              (iii) $\forall w'' \in \text{Best}_\mathcal{B} \left(p(t')(w') \right)$
        b.   If defined, $[\![\text{PAST(MOD(VP))}]\!] = 1$ iff
              t' defined in (a.-ii.) is such that: $t' \prec t_u$ such that $\exists w'[w' \in \text{Best}_\mathcal{D} \wedge$
              $p(w')(t')]$

The presupposition (44-a) can be paraphrased as follows: there is a contextually determined time at which a certain entity has a telos (John intends to go at the swimming pool). There is a world compatible with the laws such that p is true and a world compatible with the laws such that p is not true, these are the $\text{Best}_\mathcal{D}$ worlds (note that the time of evaluation of p and $\neg p$ follows the time at which alternatives are projected). In all worlds compatible with the expectations (i.e. in the $\text{Best}_\mathcal{B}$), p is not true.

The sentence asserts (44-b) that at a time t' that precedes t_u and follows t'', there is a world compatible with the laws such that p is realized there.

We thus obtain the configuration depicted in Fig. 4. The actual world is the domain of quantification of the modal DQ, determined at the time t'.

---

[11] For abilitative modality the ordering source $\mathcal{A}$ would have been used, instead.

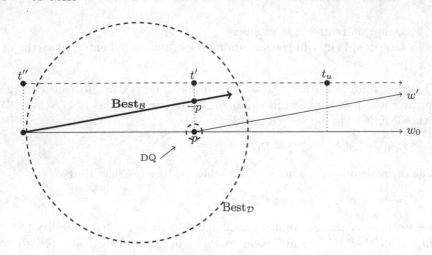

Fig. 4. Domain of quantification DQ, deontic modal base $\mathcal{D}$ and expectations $\mathcal{B}$

Let us add two comments. First we can now provide the final non-homogeneity condition on the deontic flavored goal-oriented modality. This is condition (44-a)-(ii). Generalizing for a set of propositions $\mathcal{X} \in \{\mathcal{D}, \mathcal{A}\}$, we obtain:

(45)   Non-homogeneity of the modal base for past modals (final).
         Let t'' be a contextually determined past time.
$$\exists t'_{t' \succ t''} \Big(\exists w' \in \text{Best}_{\mathcal{X}} \wedge \big(p(t')(w') \big) \Big) \wedge$$
$$\Big(\exists w'' \in \text{Best}_{\mathcal{X}} \wedge \big(\neg p(t')(w'') \big) \Big)$$

As shown in Fig. 4, we can now clearly distinguish between the domain of quantification of the modal from the modal base (this is not a peculiarity of modals giving rise to the entailment, but it is also a well-studied feature of number of modals across languages, see e.g. [15,38]). The domain of quantification contains just one world, the actual one. In the assertion, the modalized and the non-modalized statement are equivalent. However, the modal statement contributes meaning in the presupposition. By adding this layer of meaning, the informativity of the modal sentence becomes higher than the one of the non-modalized sentence, and the Gricean Quality maxim is fulfilled.

To conclude the discussion, let's consider what happens with negation, considering the case in (46).

(46)   Jean n'a pas pu rentrer à la piscine.
         John could not enter in the swimming pool (and he did not enter).

(46) states both that (i) the permission is denied and (ii) John did not enter. By negating (44-b), our analysis is as follows.

(47)   a.   Presupposition. As above.

  b.   (47) is true iff $\neg t'(t' \prec t_u)$ such that $\exists w' \in \text{Best}_{\mathcal{D}} \wedge (p(t')(w'))$

The condition in (47-b) states that there is no time at which a world compatible with the laws is accessed and p is true. This amounts to stating that the permission is not given. Moreover, since there is just one world, the actual one, we conclude that p is not true there.

Note that the presuppositonal content remains unchanged. There are worlds compatible with the permissions and p is true, and worlds compatible with the permission and in which p is not true. Moreover, the expectation that p would not be true is also maintained. As for the non-homogeneity conditions, it will hold in the metaphysical modal base only (see [8]).

This concludes our discussion of modality giving rise to the actuality entailments. Universal modals will use universal quantification on the entire deontic space, projected at a contextually determined time, preceding the time at which p becomes true.

4   Further Discussion

4.1   Anchoring to Times and Opportunity Reading: The Role of the Adverbs

We now consider another interpretation of past modals in French that [29] have labeled as the 'opportunity' reading. This reading typically arises when temporal boundaries at which the possibility holds are overtly specified via a temporal adverb. In these cases, the actuality entailment does not arise.

(48)   Jean a   pu   entrer entre   3 heures et   5 heures, mais il   n'est
  John has can.pp enter   between 3 hours and 5 hours, but   he did
  pas entré.
  not enter.
  'John had the opportunity to enter between 3 and 5, but he did not enter.

We are aware of no formal analysis that addresses the opportunity reading. We propose that the opportunity reading of modals is obtained by *anchoring the modal to the time* introduced by the adverb. The opportunity reading is not parametric to *teloi*; thus, it does not appeal to the historical accessibility relation.

An opportunity can be thought of as a state of affairs that holds over a certain period of time in a certain location. Such states of affairs are indeed what we usually call 'circumstances'. The modal base of the opportunity reading uses circumstantial similarity: it contains those worlds in which the circumstances that obtain in the actual world at the time denoted by the adverb, also obtain, and are such that p is true there.

Leveling the assumption about the identity of worlds and about a settled past and present that the historical accessibility relation introduces allows one to capture the opportunity. In (49), MB_{circ} returns the set of worlds circumstantially accessible from w_0 at t' (having leveled the constraint on identity of worlds, these are not identical to w_0 up to t'). (49) states that there is at least one accessible world in which p is true at the time provided by the adverb.

(49) $[\![(48)]\!] = 1$ iff
$\exists w' \in MB_{circ}(w_0)(\text{between 3 and 5 pm})(p(\text{between 3 and 5 pm})(w'))]$

Without further constraints, the actuality entailment does not obtain as expected.

4.2   Generic Deontic Modality

We now consider more closely the difference between addressee-oriented deontic modality and generic deontic modality in relation to actuality entailments.

In French, deontic modality can be both present and past-oriented (*pace* [32]).

(50)   Pour entrer tu   dois   avoir acheté les billets.
         To   get-in you must.2SG have bought the tickets.
         'You must have bought the tickets to get in.'

(51)   Tu   dois   être un homme pour pouvoir utiliser ces toilettes.
         You must.2SG be   a   male   to   can   use   this restroom.
         'You must be a male to use this restroom.'

In our account, the present is settled and represented as a branching point, the time of the utterance. It is predicted under our account that the actuality entailment is obtained in (50) and (51) as well. This conclusion would prove our account wrong, as in (50), it is not entailed that my addressee has bought the tickets, nor that my addressee is a man in (51).

This criticism rests on the unwarranted premise that all instances of deontic modality are instances of goal-oriented modality.

Deontic modality in the present can be interpreted in at least two different ways (see discussion in [33]). First, it can be *addressee-oriented*. In this case, (50) is felicitous if the addressee still has time to buy the tickets (see [22]) and the speaker is urging him to do so. In other terms, the addressee must be able to make p true. In this context, the buying of the tickets lies at a past time t_l of a future time, such that t_l is in the future of the time of the utterance. The actuality entailment is not obtained because p lies in the future of t_u, and there is not yet an actual future after t_u.

Deontic modality can also have a *generic* interpretation (see [34]). Consider now (50) in the context in which the hearer does not have time to buy the tickets, but the speaker is uttering a general rule, independent of the possible exceptions or correct implementations of the rule. There is no action that the addressee can take to fulfill the rule. We can replicate the observation with (51). Consider a context in which the addressee is a woman. There is no way for the female addressee to change sex instantaneously and become a man. The speaker is thus uttering a rule without expecting the hearer to fulfill it or to have it fulfilled. The same sentence can also be uttered at a male addressee. The addressee can then choose whether or not he wants to use the bathroom. The rule for using the bathroom is provided, but the addressee is not urged to use it.

(52)   a.   Male addressee: Est-ce que je peux utiliser ces toilettes?
              'Can I use this restroom?'
        b.   Speaker: Oui, tu dois être un homme pour pouvoir les utiliser.
              'Yes, you must be a male to use it.'

Here, deontic modality is being used in a generic sentence, where the present tense introduces GEN (see discussion and proposal in [34], *ibid*). In these cases, the accessibility relation is not historical, and there is not an actual *telos* being pursued. We would rather use a circumstantial accessibility relation, without any constraints on the identity of worlds (unlike with goal-oriented modality), and a deontic ordering source ($\mathcal{D}$) that ranks as best those worlds in which the rules are obeyed. Given the presence of GEN, one might also want to add normality ordering sources ($\mathcal{N}$). A bouletic ordering source ($\mathcal{B}$) might also be used to take into account the role of personal choices in connection with deontic modality. We do not provide here a full analysis of generic deontic modality. (53) reveal the spirit of it (see [34] for an extended discussion). ($\mathcal{D}$, $\mathcal{N}$ and $\mathcal{B}$ are each a set of propositions).

(53) $[\![\text{GEN}(\text{MOD}_{deontic})(p)]\!] = 1$ iff
        $\forall w' \in \text{Best}_{\mathcal{D}} \cap \text{Best}_{\mathcal{N}} \cap \text{Best}_{\mathcal{B}} \cap (MB_{circ}(w_0)(t_u))(p(t_u)(w'))$

Since we are not assuming historical accessibility here, the actuality entailment does not arise.[12]

4.3   Past-Oriented Abilitative Modality? A Final Note

To conclude the discussion about past orientation, we would like to raise a potential final concern about whether there are instances of past-oriented abilitative modality and add a brief note. We have argued that goal-oriented modals (including abilitative modality) are inherently future oriented (the time of evaluation of the prejacent is evaluated at a time that follows the time at which the modal base is determined. Note that this is the case in Fig. 4). *Pouvoir* cannot have an abilitative interpretation when past oriented.

(54)   Jean peut avoir déplacé la   table. (epistemic only)
        John can   have moved   the table.
        'John might have move the table.'

Our theory seems thus to deliver a correct prediction.

However, extending it beyond *pouvoir* and *devoir*, one can observe that *être capable de* (*be able to*) can have past orientation (the time of evaluation of the

---

[12] For an overview about the interpretations of GEN, see [30].

prejacent is evaluated in the past with respect to the time at which the modal base is determined). Scenario: Mary has been found dead in her bed.

(55)   Jean est capable de l'avoir   tuée.
        John is   able   of her-have killed.
        'John might have killed her.'

One might want to propose that *être capable de* is the dedicated expression of abilitative modality. This attempt, however, to confine the coverage of *être capable de* to abilitative modality is deemed to fail. Several differences exist between the English *be able to* and its Romance equivalents. In a very recent study of Spanish, [6,7] show that one of the peculiarities of *ser capaz* is its ambiguity between an abilitative and an epistemic interpretation. The new data presented can be straightforwardly duplicated in French - we do not replicate them here for space reasons (see [6] for discussion), and (55) qualify as an epistemic reading of *être capable*. Note that, for (55), we have set up a scenario in which the speaker must infer who the murderer is. The use of this type of contexts is the hallmark of epistemic modality (see [11,15,33]).

In French, *être capable* thus shows the same versatility as *pouvoir*, which, when past oriented, features an epistemic interpretation. As a result, we can conclude that even *être capable*, just like *pouvoir*, can have an abilitative interpretation only when future oriented.

5   Conclusion

In this paper we have shown that modals in the past give rise to actuality entailments and that this is an unexpected phenomenon given the non-implicative behavior of modals in natural language. We have also show that actuality entailments arise with goal-oriented modality only. We have endorsed the view of [2] that goal-oriented modals use historical accessibility with a fixed past and an open future. These model theoretic assumptions have allowed us to spell out the precise modal-temporal configuration in which the actuality entailment arises and our predictions are borne out by the data, cross-linguistically. We have also shown that, when such an assumption about the identity of worlds at branching point is leveled - which appears to be the case with generic deontic and opportunity modals, the actuality entailments disappear.

Finally, we have also shown that modal sentences giving rise to actuality entailments are informative, insofar as the contribution of the modality survives as a presupposition that the modal base is non-homogeneous.

References

1. Anand, P., Hacquard, V.: The role of the imperfect in Romance counterfactuals. In: Prinzhorn, M., Schmitt, V., Zobel, S. (eds.) Proceedings of Sinn und Bedeutung, vol. 14, pp. 37–50 (2010)

2. Belnap, N.D.: Backwards and forwards in the modal logic of agency. Philos. Phenomenol. Res. **51**, 777–807 (1991)
3. Berthonneau, A.-M., Kleiber, G.: Sur l'imparfait contrefactuel. Travaux de linguistique **53**, 7–65 (2006)
4. Bhatt, R.: Covert modality in non-finite contexts. Ph.D. thesis, Philadelphia, University of Pennsylvania (1999)
5. Borgonovo, C., Cummins, S.: Tensed modals. In: Eguren, L., Fernandez, S.O. (eds.) Coreference, Modality, and Focus, pp. 1–18. John Benjamins Publishing Company, Amsterdam/Philadelphia (2007)
6. Castroviejo, E., Oltra-Massuet, I.: On capacities and their epistemic extensions. In: Selected Papers from LSRL 42, John Benjamins, Dordrecht (to appear)
7. Castroviejo, E.: What does be capable tell us about capacities? An answer from Romance. Dispositions Workshop, Stuttgart (2015)
8. Condoravdi, C.: Temporal interpretation of modals: modals for the present and for the past. In: Beaver, D., Kaufmann, S., Clark, B., Casillas, L. (eds.) The Construction of Meaning, pp. 59–88. CSLI, Stanford (2002)
9. Davis, H., Louie, M., Matthewson, L., Paul, I., Reis Silva, A., Peterson, T.: Perfective aspect and actuality entailments: a cross-linguistic approach. In: Proceedings of SULA 5: The Semantics of Under-Represented Languages in the Americas. GLSA, Amherst (2010)
10. von Fintel, K.: Would you believe it? The king of France is back! Presuppositions and truth-value intuitions. In: Reimer, M., Bezuidenhout, A. (eds.) Descriptions and Beyond. Oxford University Press, Oxford (2004)
11. von Fintel, K., Gillies, A.: Must.. stay.. strong!. Nat. Lang. Semant. **18**, 351–383 (2010)
12. Giannakidou, A.: Affective dependencies. Linguist. Philos. **22**, 367–421 (1999)
13. Giannakidou, A., Mari, A.: The future of Greek and Italian: an evidential analysis. In: Proceedings of Sinn und Bedeutung, vol. 17, pp. 255–270 (2013)
14. Giannakidou, A., Mari, A.: A unified analysis of the future as epistemic modality: the view from Greek and Italian. Ms. University of Chicago and Institut Jean Nicod (2015)
15. Giannakidou, A., Mari, A., Epistemic future, epistemic MUST: nonveridicality, evidence, and partial knowledge. In: Blaszack, J., et al. (ed.) Tense, Mood, and Modality: New Perspectives on Old Questions. University of Chicago Press, Chicago (in press)
16. Giannakidou, A., Mari, A.: Emotive-factive and the puzzle of the subjunctive. In: 2015 Proceeding of CLS, vol. 51, pp. 181–195 (2016)
17. Giannakidou, A., Staraki, E.: Ability, action and causation: from pure ability to force. In: Mari, A., Beyssade, C., Del Prete, F. (eds.) Genericity, pp. 250–275. OUP, Oxford (2012)
18. Grano, T.: Control and restructuring at the syntax-semantic interface. Ph.D. University of Chicago (2012)
19. Hacquard, V.: Aspects of modality. Ph.D. thesis, MIT, Cambridge, MA (2006)
20. Hacquard, V.: On the interaction of aspect and modal auxiliaries. Linguist. Philos. **32**, 279–315 (2009)
21. Homer, V.: French modals and perfective: a case of aspectual coercion. In: Proceedings of WCCFL, vol. 28, pp. 106–114 (2010)
22. Kaufmann, M.: Interpreting Imperatives. Springer, Dordrecht (2012)
23. Kratzer, A.: Modality. In: von Stechow, A., Wunderlich, D. (eds.) Semantics: An International Handbook of Contemporary Research, pp. 639–650. de Gruyter, Berlin (1991)

24. Ippolito, M.: Imperfect modality. In: Guéron, J., Lecarme, J. (eds.) The Syntax of Time, pp. 359–387. MIT Press, Cambridge (2004)
25. Mari, A.: Temporal reasoning and modality. Invited talk at temporality: typology and acquisition, University of Paris VIII (2010)
26. Mari, A.: Pouvoir au passé composé: effet épistémique et lecture habilitative. In: de Saussure, L., Rhis, A. (eds.) Etudes de sémantique et de pragmatique Françaises, pp. 67–99. Peter Lang, Geneva (2012)
27. Mari, A.: Each other, asymmetry and reasonable futures. J. Semant. **31**(2), 209–261 (2014)
28. Mari, A.: Modalités et Temps. Des modèles aux données. Peter Lang AG, Bern (2015)
29. Mari, A., Martin, F.: Tense, abilities and actuality entailment. In: Aloni, M., Dekker, P., Roelofsen, F. (eds.) Proceedings of the XVI Amsterdam Colloquium, pp. 151–156. ITLI, University of Amsterdam, Amsterdam (2007)
30. Mari, A., Beyssade, C., Del Prete, F.: Introduction. In: Mari, A., Beyssade, C., Del Prete, F. (eds.) Genericity, pp. 1–92. Oxford University Press, Oxford (2012)
31. Matthewson, L.: On the (non)-future orientation of modals. In: Proceedings of Sinn und Bedeutung, vol. 16, pp. 431–446 (2012)
32. Niñán, D.: Two puzzles about deontic necessity. In: Gajewski, J., Hacquard, V., Nickel, B., Yalcin, S. (eds.) New Work on Modality, MIT Working Papers in Linguistics, vol. 51, pp. 149–178. MIT, Cambridge (2005)
33. Portner, P.: Modality. Oxford University Press, Oxford (2009)
34. Saint Croix, C., Thomason, R.H.: Chisholm's paradox and conditional oughts. In: Cariani, F., Grossi, D., Meheus, J., Parent, X. (eds.) DEON 2014. LNCS (LNAI), vol. 8554, pp. 192–207. Springer, Heidelberg (2014). doi:10.1007/978-3-319-08615-6_15
35. Schlenker, P.: Local contexts. Semant. Pragmat. **2**, 1–78 (2009)
36. Thomason, R.: Combinations of tense and modality. In: Gabbay, D.M., Guenthner, F. (eds.) Handbook of Philosophical Logic: Extensions of Classical Logic, vol. II, pp. 136–165. Reidel, Dordrecht (1984)
37. Tonhauser, J., Beaver, D., Roberts, C., Simons, M.: Toward a taxonomy of projective content. Language **89**(1), 66–109 (2013)
38. Werner, T.: Future and non-future modal sentences. Nat. Lang. Semant. **14**, 235–255 (2006)

Non-crossing Tree Realizations of Ordered Degree Sequences

Laurent Méhats[1] and Lutz Straßburger[2(✉)]

[1] Collège de Guinette, Étampes, France
[2] Inria Saclay, Palaiseau, France
lutz.strassburger@inria.fr

Abstract. We investigate the enumeration of non-crossing tree realizations of integer sequences, and we consider a special case in four parameters, that can be seen as a four-dimensional tetrahedron that generalizes Pascal's triangle and the Catalan numbers. This work is motivated by the study of ambiguities in categorial grammars.

Keywords: Proof nets · Non-crossing trees · Integer sequences · Catalan's triangle · Pascal-Catalan-tetrahedron

1 Introduction

A *non-crossing* tree t is a labeled tree on a *sequence* of vertices $\langle v_0, v_1, \ldots, v_n \rangle$ drawn in counterclockwise order on a circle, and whose edges are straight line segments that do not cross. For any index $0 \leq i \leq n$, let d_i stand for the number of edges incident with v_i (that is the degree of v_i). Then as any other tree on $n+1$ vertices, t satisfies $\sum_{i=0}^{n} d_i = 2n$. Thus, the *sequence* $\langle d_0, d_1, \ldots, d_n \rangle$ defines a *composition* of $2n$ into $n+1$ positive summands (two sequences of integers that differ only in the order of their elements define *distinct* compositions of the same integer). Stated otherwise, t is a non-crossing tree *realization* of the composition $\langle d_0, d_1, \ldots, d_n \rangle$.

For any composition $c = \langle d_0, d_1, \ldots, d_n \rangle$ of $2n$ into $n+1$ positive summands, let $nct(c)$ stand for the number of non-crossing tree realizations of c, that is the number of non-crossing trees on $n+1$ vertices $\langle v_0, v_1, \ldots, v_n \rangle$ such that vertex v_i has degree d_i for any index $0 \leq i \leq n$ (there always exists at least one, see Proposition 2.3). We aim at computing nct. Note that here the input is more specific than the degree partition, as for example in [5].

From Proof Nets to Non-crossing Trees. Our interest for these non-crossing tree realizations comes from linguistics and proof theory. The starting point for this work was the following linguistic problem: How many different readings can an ambiguous sentence at most have? Particularly, which sentence of a given length has the most different readings? When using categorial grammars based on the Lambek calculus [9] or related systems, a parse tree is a formal proof in a

M. Amblard et al. (Eds.): LACL 2016, LNCS 10054, pp. 211–227, 2016.
DOI: 10.1007/978-3-662-53826-5_13

deductive system. Thus, our questions become: How many different formal proofs can a formula have? Particularly, which formula of a given length has the most different formal proofs? In category theoretical terms these questions come down to the cardinality of the Hom-sets in a free non-commutative star-autonomous category [2]. The corresponding logic is a variant of non-commutative intuitionistic linear logic [8,17] for which formal proofs can be represented as planar *proof nets*. It would go too far beyond the scope of this paper to go into the details of this correspondence. However, to give the reader an idea, we have shown in Fig. 1 the transformation of a parse tree into a proof net. The first step transforms the parse tree into a formal proof according to Lambek's work [9]. In the second step, this proof is embedded into a one-sided multiple conclusion system using the binary connectives ⅋ and ⊗ [17]. In order not to lose the information on positive and negative positions in the formulas we use polarities (see, e.g., [8] for details). The final three steps show how this one-sided sequent proof is translated into a proof net by simply drawing the flow graph on the atoms appearing in the proof (for more details, see [3,8,15]). It is a well-known fact of linear logic that such a graph G does indeed correspond to a sequent proof if and only if every switching (that is, every graph obtained from G by removing for each ⅋-node one of the two edges that it to its children) is a connected and

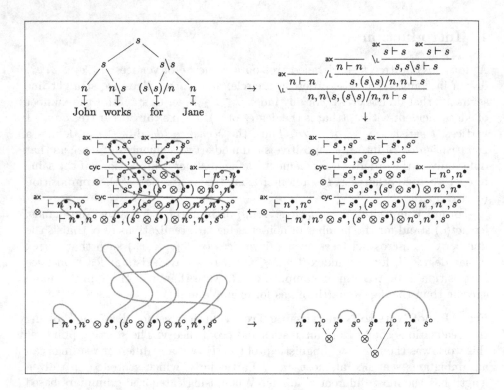

Fig. 1. From parse trees via proof trees to proof nets

acyclic graph [4]. If G does not contain any $\wp$-nodes, as in our example, then G itself has to be connected and acyclic. Furthermore, we have that G is planar if and only if the sequent proof does not contain the exchange rule, as it is the case for the Lambek calculus [17].

To summarize, a sentence corresponds to a sequent, and a parsing of the sentence to a planar proof net for the sequent. Thus, our question of how many readings does a sentence have becomes:

(i) How many different planar proof nets can at most be defined over a given sequent?

Particularly, if we are interested in sentences that are as ambiguous as possible we have to ask:

(ii) Over which sequent of a given length can the most different planar proof nets be defined?

In that respect, we can ignore the names of the atoms, and only $\wp$-*free* sequents are of interest: on the one hand, occurrences of $\wp$ lying above an occurrence of $\otimes$ can moved down by the transformations

$$\begin{array}{ccc} \underset{\substack{\wp\\ \otimes}}{\overset{A\ B}{}C} \to \underset{\substack{\wp\\ \otimes}}{A\overset{B\ C}{}} & \text{and} & A\underset{\substack{\wp\\ \otimes}}{\overset{B\ C}{}} \to \underset{\substack{\wp\\ \otimes}}{\overset{A\ B}{}}C \end{array}$$

which preserve correctness without affecting linkings (see [6,7]); on the other hand, root occurrences of $\wp$ are irrelevant and can be removed. Hence, for every sequent Γ there is a $\wp$-free sequent Γ', such that for Γ' exist *at least* as many different planar proof nets as for Γ.

Finally, up to the associativity of $\otimes$, planar $\wp$-free proof nets are in bijection with non-crossing trees as shown in Fig. 2.

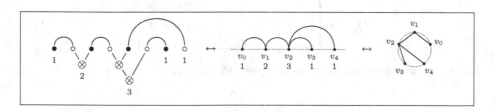

Fig. 2. Planar $\wp$-free proof nets as non-crossing trees

Unfortunately, we were not able to find a closed formula for $nct\langle d_0, d_1, \ldots, d_n \rangle$ depending only on the input composition $\langle d_0, d_1, \ldots, d_n \rangle$, which would be needed to give a clear answer to Question (i) above. When investigating Question (ii), we found through experiments that at least up to $n = 25$, nct is maximized for compositions of the shape

$$1, \quad \underbrace{2,2,\ldots,2}_{p \text{ summands}}, \underbrace{1,3,1,3,\ldots,1,3,1}_{2q \text{ summands}}, \quad \underbrace{2,2,\ldots,2}_{r \text{ summands}}, \underbrace{1,3,1,3,\ldots,1,3}_{2s \text{ summands}} \qquad (1)$$

which we write as $12^p(13)^q12^r(13)^s$ and are the first step of our study (compositions maximizing nct for higher values of n may be of a different shape or involve summands higher then 3). The input is now reduced to four parameters p, q, r and s such that $n+1 = 1+p+2q+1+r+2s$. We write $nct\langle 12^p(13)^q12^r(13)^s\rangle$ as $N_{p,q,r,s}$, and we are interested in computing $N_{p,q,r,s}$.

A Four-Dimensional Generalization of Pascal's and Catalan's Triangles.
Recall that Pascal's triangle $P_{p,r} = \binom{p+r}{p,r}$ can be generated recursively by:

$$P_{p,r} = \begin{cases} 1 & \text{if } p = 0 \text{ or } r = 0 \\ P_{p,r-1} + P_{p-1,r} & \text{if } p > 0 \text{ and } r > 0 \end{cases} \qquad (2)$$

The first few values are shown below (see [14, A007318]):

$$
\begin{array}{ccccccccccccc}
 & & & & & & {}^{p}\swarrow & 1 & \searrow{}^{r} & & & & \\
 & & & & & 1 & & 1 & & & & & \\
 & & & & 1 & & 2 & & 1 & & & & \\
 & & & 1 & & 3 & & 3 & & 1 & & & \\
 & & 1 & & 4 & & 6 & & 4 & & 1 & & \\
 & 1 & & 5 & & 10 & & 10 & & 5 & & 1 & \\
1 & & 6 & & 15 & & 20 & & 15 & & 6 & & 1
\end{array}
\qquad (3)
$$

The Catalan numbers $C_q = \frac{1}{q+1}\binom{2q}{q,q}$ are generated recursively by $C_0 = 1$ and $C_{q+1} = \sum_{j=0}^{q} C_j \cdot C_{q-j}$. A combination of Pascal's triangle and the Catalan numbers is known as Catalan's triangle $Q_{p,q} = \frac{p+1}{p+q+1}\binom{p+2q}{p+q,q}$ which can be generated recursively by:

$$Q_{p,q} = \begin{cases} 1 & \text{if } q = 0 \\ C_q & \text{if } p = 0 \\ Q_{p+1,q-1} + Q_{p-1,q} & \text{if } p,q > 0 \end{cases} \qquad (4)$$

The first few values are shown below (see [14, A009766]):

$$
\begin{array}{ccccccccccccc}
 & & & & & {}^{p}\swarrow & 1 & \searrow{}^{q} & & & & & \\
 & & & & 1 & & 1 & & & & & & \\
 & & & 1 & & 2 & & 2 & & & & & \\
 & & 1 & & 3 & & 5 & & 5 & & & & \\
 & 1 & & 4 & & 9 & & 14 & & 14 & & & \\
1 & & 5 & & 14 & & 28 & & 42 & & 42 & & \\
1 & & 6 & & 20 & & 48 & & 90 & & 132 & & 132
\end{array}
\qquad (5)
$$

It is also possible to generalize the recursive formula of the Catalan numbers into a triangle $R_{q,s}$ generated by (assuming that $C_{-1} = 0$):

$$R_{q,s} = \begin{cases} 1 & \text{if } q = s = 0 \\ \sum_{j=0}^{q} \sum_{l=0}^{s} C_{j+l-1} \cdot R_{q-j,s-l} & \text{if } q + s > 0 \end{cases} \tag{6}$$

The first few values are shown below:

$$\begin{array}{ccccccc} & & & q \diagup & \diagdown s & & \\ & & & 1 & & & \\ & & 1 & & 1 & & \\ & 2 & & 3 & & 2 & \\ 5 & & 9 & & 9 & & 5 \\ 14 & 28 & & 34 & & 28 & 14 \\ 42 & 90 & & 123 & & 123 & 90 & 42 \\ 132 & 297 & 440 & & 497 & 440 & 297 & 132 \end{array} \tag{7}$$

As an example,

$R_{3,2} = 123$

$\quad = C_0 \cdot (28 + 34) + C_1 \cdot (5 + 9 + 9) + C_2 \cdot (2 + 3 + 2) + C_3 \cdot (1 + 1) + C_4 \cdot 1$

We shall establish that $N_{p,q,r,s}$ is a four-dimensional "tetrahedron" that generalizes the three triangles P, Q and R above, insofar as:

$$N_{p,0,r,0} = N_{r,0,p,0} = P_{p,r}, \tag{8}$$

$$N_{p,q,0,0} = N_{0,0,p,q} = Q_{p,q}, \tag{9}$$

$$N_{0,q,0,s} = N_{0,s,0,q} = R_{q,s}. \tag{10}$$

Outline. The organization of this paper is as follows: First, in Sect. 2, we study the general case of enumerating non-crossing tree realizations of integer compositions. Then, in Sects. 3, 4, 5, 6 and 7, we concentrate on the four-parameter case. In particular, we will prove identities (8)–(10) in Sects. 4 and 5. Finally, we will provide the generating function for $N_{p,q,r,s}$ in Sect. 7.

Missing proofs can be found in the technical report [10].

2   General Case

Any labeled tree on a *sequence* of vertices can be drawn in such a way that its vertices lie in counterclockwise order on a circle and its edges are straight line segments lying inside that circle. In that case, of course, some of its edges may cross each other. Let us call such a labeled tree a *crossing* tree. The order of summands in a composition does not matter regarding the number of its labeled tree realizations (there are six for any composition of $2 \cdot 4$ into $4 + 1$ summands in the multiset $\{1, 1, 2, 2, 2\}$). But it does as soon as we distinguish between non-crossing and crossing realizations. As an example, there are one non-crossing

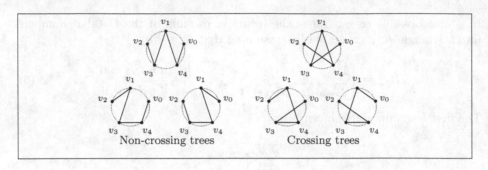

Fig. 3. The six labeled tree realizations of $\langle 1, 2, 1, 2, 2 \rangle$

and five crossing realizations of $\langle 1, 1, 2, 2, 2 \rangle$, while there are three non-crossing and three crossing realizations of $\langle 1, 2, 1, 2, 2 \rangle$ (these are shown on Fig. 3).

Remark 2.1. A proof of Cayley's formula (see e.g., [1]), which asserts that the number of labeled trees on $n + 1$ vertices is $(n+1)^{n-1}$ (see [14, A000272]) relies on:

$$\sum_{\langle d_0, d_1, \ldots, d_n \rangle} \binom{n-1}{d_0 - 1, d_1 - 1, \ldots, d_n - 1} = (n+1)^{n-1} \tag{11}$$

where the sum ranges over the $\binom{2n-1}{n, n-1}$ compositions of $2n$ into $n + 1$ positive summands. Noy established in [11, Corollary 1.2] that the number of non-crossing trees on $n + 1$ vertices is $\frac{1}{2n+1}\binom{3n}{2n, n}$ (see [14, A001764]). Recall that $nct\langle d_0, d_1, \ldots, d_n \rangle$ stands for the number of non-crossing tree realizations of the composition $nct\langle d_0, d_1, \ldots, d_n \rangle$. Then

$$\sum_{\langle d_0, d_1, \ldots, d_n \rangle} nct\langle d_0, d_1, \ldots, d_n \rangle = \frac{1}{2n+1} \binom{3n}{2n, n} \tag{12}$$

where the sum ranges over the $\binom{2n-1}{n, n-1}$ compositions of $2n$ into $n + 1$ positive summands.

The image under rotation of a non-crossing tree t on vertices $\langle v_0, v_1, \ldots, v_n \rangle$ is a non-crossing tree on vertices $\langle v_{k+1}, \ldots, v_n, v_0, \ldots, v_k \rangle$ for some $k \leq n$. Moreover, t realizes a composition $\langle d_0, d_1, \ldots, d_n \rangle$ iff its image under rotation realizes the composition $\langle d_{k+1}, \ldots, d_n, d_0, \ldots, d_k \rangle$. Thus, for any composition $\langle d_0, d_1, \ldots, d_n \rangle$ and any $k \leq n$,

$$nct\langle d_0, d_1, \ldots, d_n \rangle = nct\langle d_{k+1}, \ldots, d_n, d_0, \ldots, d_k \rangle. \tag{13}$$

We shall refer to this property as *stability under rotation*.[1]

---

[1] In that respect, we may focus on *necklace*-compositions, i.e., compositions that are lexicographically minimal under rotation [13].

In the same way, the mirror image of a non-crossing tree t on vertices $\langle v_0, v_1, \ldots, v_n \rangle$ is a non-crossing tree on vertices $\langle v_n, v_{n-1}, \ldots, v_0 \rangle$, and t realizes a composition $\langle d_0, d_1, \ldots, d_n \rangle$ iff its mirror image realizes the composition $\langle d_n, d_{n-1}, \ldots, d_0 \rangle$. Thus, for any composition $\langle d_0, d_1, \ldots, d_n \rangle$,

$$nct\langle d_0, d_1, \ldots, d_n \rangle = nct\langle d_n, d_{n-1}, \ldots, d_0 \rangle. \tag{14}$$

We shall refer to this property as *stability under mirror image*.[2]

We will now establish that for any positive integer n and any composition c of $2n$ into $n + 1$ positive summands, there exists a non-crossing tree realization of c (Proposition 2.3).

Lemma 2.2. *For any positive integer n and any sequence $\langle 1, d_1, \ldots, d_n, d_{n+1} \rangle$ of $n + 2$ positive integers such that $1 + \sum_{i=1}^{n+1} d_i < 2(n+1)$, there is an index $1 \leq k < n + 1$ such that $1 + \sum_{i=1}^{k} d_i = 2k$.*

Proof. For any index $1 \leq l \leq n + 1$, let S_l stand for $1 + \sum_{i=1}^{l} d_i$. We prove the following implication by induction on l: if there is no index $1 \leq k < l$ such that $S_k = 2k$, then $S_l \geq 2l$. Since by hypothesis $S_{n+1} < 2(n+1)$, there must exist an index $1 \leq k < n+1$ such that $S_k = 2k$.
Base. Since $d_1 \geq 1$, $S_1 = 1 + d_1 \geq 2 \cdot 1$ and the stated implication holds trivially.
Induction. Assume that the stated implication holds for l (IH), and that there exists no index $1 \leq k < l + 1$ such that $S_k = 2k$. We reformulate the latter hypothesis as: (i) there exists no index $1 \leq k < l$ such that $S_k = 2k$, and (ii) $S_l \neq 2l$. By (IH) we get from (i), that $S_l \geq 2l$, and from (ii), that $S_l > 2l$, i.e., $S_l \geq 2l + 1$. Since $d_{l+1} \geq 1$, $S_{l+1} = S_l + d_{l+1} \geq 2(l+1)$. $\square$

Proposition 2.3. *For any positive integer n and any composition c of $2n$ into $n + 1$ positive summands, there exists a non-crossing tree realization of c.*

Proof. We proceed by induction on n.
Base. The unique composition $\langle 1, 1 \rangle$ of $2 \cdot 1$ into $1 + 1$ positive summands is realized by the unique (trivially non-crossing) tree on $1 + 1$ vertices.
Induction. Assume that the stated property holds for any positive integer up to n (IH), and let $\langle d_0, d_1, \ldots, d_{n+1} \rangle$ be a composition of $2(n+1)$ into $n+2$ positive summands. Since n is a positive integer, $2(n + 1) > n + 2$ and there must exist at least one summand $d_k > 1$. By stability under rotation, we can assume without loss of generality that d_0 is such a summand, i.e. that $d_0 > 1$. Then $1 + \sum_{i=1}^{n+1} d_i < 2(n+1)$ and by Lemma 2.2, there exists an index $1 \leq k < n+1$ such that $1 + \sum_{i=1}^{k} d_i = 2k$. By difference, $(d_0 - 1) + \sum_{i=k+1}^{n+1} d_i = 2(n - k + 1)$. Then by (IH):

---

[2] In that respect, we may focus on *bracelet*-compositions, i.e., necklace-compositions that are lexicographically minimal under mirror image [12].

– there exists a non-crossing tree on vertices $\langle t_0, t_1, \ldots, t_k \rangle$ realizing the composition $\langle 1, d_1, \ldots, d_k \rangle$ of $2k$ into $k+1$ positive summands,
– there exists a non-crossing tree on vertices $\langle u_0, u_1, \ldots, u_{n-k+1} \rangle$ realizing the composition $\langle d_0 - 1, d_{k+1}, \ldots, d_{n+1} \rangle$ of $2(n-k+1)$ into $n-k+2$ positive summands.

Let T and U stand for the respective edge sets of these non-crossing trees (where edges are defined as couples of vertices). We "merge" t_0 and u_0 into a single vertex v_0 to get a tree on vertices $\langle v_0, v_1, \ldots, v_{n+1} \rangle$ which edge set is defined as

$$\begin{aligned} &\big\{ \{v_i, v_j\} \mid \{t_i, t_j\} \in T \big\} \\ &\cup \big\{ \{v_0, v_{j+k}\} \mid \{u_0, u_j\} \in U \big\} \\ &\cup \big\{ \{v_{i+k}, v_{j+k}\} \mid \{u_i, u_j\} \in U, i > 0, j > 0 \big\} \end{aligned} \tag{15}$$

This tree is non-crossing and it realizes the composition $\langle d_0, d_1, \ldots, d_n, d_{n+1} \rangle$ of $2(n+1)$ into $n+2$ positive summands (see Fig. 4 for an example). $\square$

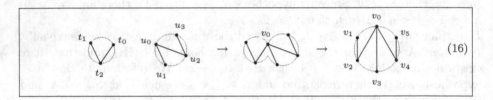

$$\tag{16}$$

Fig. 4. Merging two non-crossing trees into a single one

The previous proof suggests a recursive definition of nct:

– The unique composition $\langle 1, 1 \rangle$ of $2 \cdot 1$ into $1+1$ positive summands is realized by the unique non-crossing tree on $1+1$ vertices. Thus,

$$nct\langle 1, 1 \rangle = 1. \tag{17}$$

– Let n be strictly greater than 1 and $\langle d_0, d_1, \ldots, d_n \rangle$ be a composition of $2n$ into $n+1$ positive summands. Let k be the smallest index such that $d_k > 1$ (there exists at least one). By stability under rotation,

$$nct\langle 1, \ldots, 1, d_k, \ldots, d_n \rangle = nct\langle d_k, \ldots, d_n, 1, \ldots, 1 \rangle. \tag{18}$$

Thus we can assume that $d_0 > 1$. In that case,

$$nct\langle d_0, d_1, \ldots, d_n \rangle = \sum_k \big(nct\langle 1, d_1, \ldots, d_k \rangle \cdot nct\langle d_0 - 1, d_{k+1}, \ldots, d_n \rangle \big) \tag{19}$$

where the sum ranges over the set of indices $1 \le k < n$ such that $1 + \sum_{i=1}^{k} d_i = 2k$ (there exists at least one).

Remark 2.4. Applying the recursive formula in a row to $d_0, d_0 - 1, \ldots, 1$, we get

$$nct\langle d_0, d_1, \ldots, d_n \rangle = \sum_{\langle k_0, \ldots, k_{d_0} \rangle} \prod_{j=1}^{d_0} nct\langle 1, d_{k_{j-1}+1}, \ldots, d_{k_j} \rangle \qquad (20)$$

where the sum ranges over the set of sequences $\langle k_0, \ldots, k_{d_0} \rangle$ of $d_0 + 1$ indices such that $0 = k_0 < \cdots < k_{d_0} = n$ and such that for all $0 < j \le d_0$,

$$1 + \sum_{i=k_{j-1}+1}^{k_j} d_i = 2(k_j - k_{j-1}). \qquad (21)$$

Remark 2.5. The construction in the proof of Proposion 2.3 is similar to the proof of the sequentialization theorem of linear logic using the *splitting tensor*. It is then easy to see that every non-crossing tree can be obtained this way as a merging of smaller ones, as indicated in Fig. 4. However, it is not clear how this insight can be used for counting non-crossing trees since there is no unique decomposion for a given tree.

3   The Four Parameters Case

We focus now on the special case where compositions $\langle d_0, d_1, \ldots, d_n \rangle$ are of the shape

$$1, \underbrace{2, 2, \ldots, 2}_{p}, \underbrace{1, 3, 1, 3, \ldots, 1, 3}_{2q}, 1, \underbrace{2, 2, \ldots, 2}_{r}, \underbrace{1, 3, 1, 3, \ldots, 1, 3}_{2s} \qquad (22)$$

which we write as $12^p(13)^q 12^r (13)^s$. For the following, recall that $N_{p,q,r,s}$ stands for $nct\langle 12^p (13)^q 12^r (13)^s \rangle$.

Lemma 3.1. *For any* p, q, r *and* s, $N_{p,q,r,s} = N_{r,q,p,s} = N_{p,s,r,q}$.

Proof. This follows from stability under rotation and mirror image. We give the formal calculations here in full, because we use similar arguments later on without showing them explicit.

$$\begin{aligned}
N_{p,q,r,s} &= nct\langle 12^p (13)^q 12^r (13)^s \rangle && \text{by definition of } N \\
&= nct\langle 2^p (13)^q 12^r (13)^s 1 \rangle && \text{by stability under rotation} \\
&= nct\langle 2^p 1 (31)^q 2^r 1 (31)^s \rangle && \text{by reparenthesizing} \\
&= nct\langle (13)^s 12^r (13)^q 12^p \rangle && \text{by stability under mirror image} \\
&= nct\langle 12^r (13)^q 12^p (13)^s \rangle && \text{by stability under rotation} \\
&= N_{r,q,p,s} && \text{by definition of } N.
\end{aligned}$$

The same way, $N_{p,q,r,s} = N_{p,s,r,q}$. $\square$

According to Lemma 3.1, so as to get a recursive definition of $N_{p,q,r,s}$, we need to consider only $N_{0,0,0,0}$ on the one hand, $N_{p+1,q,r,s}$ and $N_{p,q,r,s+1}$ on the other hand.

Proposition 3.2. *For any p, q, r and s,*

$$N_{0,0,0,0} = 1 \tag{23}$$

$$N_{p+1,q,r,s} = N_{0,0,0,0} \cdot N_{p,q,r,s} + \sum_{j=1}^{q} \left(N_{0,j,0,0} \cdot N_{p,q-j,r,s} \right)$$
$$+ \sum_{k=1}^{r} \left(N_{0,q,k,0} \cdot N_{p,0,r-k,s} \right) + \sum_{l=1}^{s} \left(N_{0,q,r,l} \cdot N_{p,0,0,s-l} \right) \tag{24}$$

$$N_{p,q,r,s+1} = N_{0,0,0,0} \cdot N_{1+p,q,r,s} + \sum_{i=1}^{p} \left(N_{i,0,0,0} \cdot N_{1+p-i,q,r,s} \right)$$
$$+ \sum_{j=1}^{q} \left(N_{p,j,0,0} \cdot N_{1,q-j,r,s} \right) \tag{25}$$

Proof. We have $N_{0,0,0,0} = nct\langle 12^0 (13)^0 12^0 (13)^0 \rangle = 1$ by definition of N and (17). Next we have $N_{p+1,q,r,s} = nct\langle 12^p 2(13)^q 12^r (13)^s \rangle = nct\langle 21(31)^q 2^r (13)^s 12^p \rangle$ by definition of N, reparenthesizing and stability under rotation. Applying (19) we get

$$N_{p+1,q,r,s} = nct\langle 11 \rangle \cdot nct\langle 1(31)^q 2^r (13)^s 12^p \rangle$$
$$+ \sum_{j=1}^{q} \left(nct\langle 11(31)^j \rangle \cdot nct\langle 1(31)^{q-j} 2^r (13)^s 12^p \rangle \right)$$
$$+ \sum_{k=1}^{r} \left(nct\langle 11(31)^q 2^k \rangle \cdot nct\langle 12^{r-k} (13)^s 12^p \rangle \right)$$
$$+ \sum_{l=1}^{s} \left(nct\langle 11(31)^q 2^r (13)^l \rangle \cdot nct\langle 1(13)^{s-l} 12^p \rangle \right).$$

Notice that there is no other way to "split" $\langle 21(31)^q 2^r (13)^s 12^p \rangle$ into two compositions such that the first one is of the form $\langle 1, d_1, \ldots, d_k \rangle$ and satisfies $1 + \sum_{i=1}^{k} d_i = 2k$. Then we get (24) by reparenthesizing and stability under rotation. A similar argument applies to the proof of (25). □

We have the following immediate consequences.

Corollary 3.3. *For any p and s,*

$$N_{p,0,0,0} = 1 \tag{26}$$
$$N_{0,0,0,s} = C_s \tag{27}$$
$$N_{0,0,r,s+1} = N_{1,0,r,s} \tag{28}$$

where C_s stands for the s-th Catalan number.

Proof. Both (26) and (27) are easily proved by induction, and (28) follows from (25) and (26). □

4 Pascal's and Catalan's Triangles

In this section we are going to establish identities (8) and (9) mentioned in the introduction. First, recall Pascal's second identity

$$\sum_{k=0}^{b} \binom{a+k}{a,k} = \binom{a+1+b}{a+1,b} \tag{29}$$

and as a consequence for all a, b, and c

$$\sum_{k=c+1}^{b} \binom{a+k}{a,k} = \binom{a+1+b}{a+1,b} - \binom{a+1+c}{a+1,c}. \tag{30}$$

Proposition 4.1. *For all p, r and s,*

$$N_{p,0,r,0} = \binom{p+r}{p,r} = P_{p,r} \tag{31}$$

$$N_{p,0,0,s} = \binom{p+2s}{p+s,s} - \frac{s}{p+s+1}\binom{p+2s}{p+s,s} \tag{32}$$

$$= \frac{p+1}{p+s+1}\binom{p+2s}{p+s,s} = Q_{p,s} \tag{33}$$

$$N_{p,0,r,s} = \binom{p+r+2s}{p+s,r+s} - \frac{s}{p+r+s+1}\binom{p+r+2s}{p+r+s,s} \tag{34}$$

A proof can be found in [10]. The triangles defined by $N_{p,0,r,0}$ and $N_{p,0,0,s}$ are known as *Pascal's* triangle and *Catalan's* triangle respectively (A007318, A009766).

Remark 4.2. We may have deduced (27) from (34). Indeed, by (34)

$$N_{0,0,0,s} = \binom{2s}{s,s} - \frac{s}{s+1}\binom{2s}{s,s}$$

$$= \frac{1}{s+1}\binom{2s}{s,s}$$

$$= C_s$$

5 A Triangular Catalan Recurrence

In this section we establish the identity (10) from the introduction. We need the following lemma:

Lemma 5.1. *For any t and u,*

$$\sum_{i=0}^{t}\sum_{j=0}^{u} C_i \cdot C_j \cdot C_{(t-i)+(u-j)} = C_{t+u+1}. \tag{35}$$

Proof. By induction on u (the base case is the usual recurrence for Catalan numbers).   □

Proposition 5.2. (A recurrence for $N_{0,q,0,s}$). *For any q and s,*

$$N_{1,q,0,s} = \sum_{j=0}^{q} \sum_{l=0}^{s} C_{j+l} \cdot N_{0,q-j,0,s-l} \tag{36}$$

$$N_{2,q,0,s} = \sum_{j=0}^{q} \sum_{l=0}^{s} C_j \cdot C_l \cdot N_{1,q-j,0,s-l} \tag{37}$$

$$= \sum_{j=0}^{q} \sum_{l=0}^{s} C_{j+l+1} \cdot N_{0,q-j,0,s-l} \tag{38}$$

$$N_{0,q+1,0,s+1} = N_{1,q+1,0,s} + N_{1,q,0,s+1} - N_{2,q,0,s} \tag{39}$$

$$N_{0,q,0,s} = \begin{cases} 1 & \text{if } q = s = 0 \\ \sum_{j=0}^{q} \sum_{l=0}^{s} C_{j+l-1} \cdot N_{0,q-j,0,s-l} & \text{if } q + s > 0 \end{cases} \tag{40}$$

Hence $N_{0,q,0,s} = R_{q,s}$.

6   Triangles and Tetrahedra

The value of $N_{p,q,r,s}$ depends on four parameters p, q, r, s. If we fix two of them, we can obtain triangles. For example, for $q = s = 0$ we get *Pascal's triangle* (3), and for $r = s = 0$ we get the *Catalan triangle* (5) (which should more precisely be called *Pascal-Catalan triangle*). If we let $p = r = 0$, then we get the triangle (7), which could also be called *Catalan triangle*.

If we fix only one parameter, we obtain a tetrahedron. For example, let $s = 0$, and let us define $T_{p,q,r} = N_{p,q,r,0}$. Then we get from (34) and Lemma 3.1:

$$T_{p,q,r} = \binom{p+r+2q}{p+q,r+q} - \frac{q}{p+r+q+1} \binom{p+r+2q}{p+r+q,q} \tag{41}$$

This defines a tetrahedron where one side is Pascal's triangle and the other two sides are the Catalan triangle. Thus we can call it the *Pascal-Catalan tetrahedron*. We have the following recursive identities:

Proposition 6.1. *For all p, q, and r, we have*

$$T_{p+1,q+1,r+1} = T_{p+1,q+1,r} + T_{p,q+1,r+1} + T_{(p+1)+(r+1),q,0} \tag{42}$$

$$T_{p+1,q,r+1} = T_{p,q+1,r} + T_{(p+1)+(r+1),q,0} \tag{43}$$

$$T_{p+1,q+1,r} = T_{p+2,q,r} + T_{p,q+1,r} \tag{44}$$

Proof. Easy calculation using (41).   □

For $r = 0$, we get the tetrahedron $V_{p,q,s} = N_{p,q,0,s}$, which we can call the *Catalan tetrahedron*, because two of its sides are the Catalan triangle (5) and the third side is the new Catalan triangle (7). Unfortunately, we could not find a closed formula for $V_{p,q,s}$. However, in Sect. 7 we will give the generating function. We also have the following:

Proposition 6.2. *For all p, q, and s, we have:*

$$\sum_{i=2}^{p+2} V_{i,q,s} = \sum_{j=0}^{q} C_j \cdot V_{p,q+1-j,s} \tag{45}$$

Observe that identities (42)–(44) establish close relationships among the triangles with $s = 0$ (or $q = 0$), i.e., the triangles that live inside the tetrahedron $T_{p,q,r}$. For example, from (43) we can get $N_{p+1,q,7,0} = N_{p,q+1,6,0} + N_{p+8,q,0,0}$.

Below, we exhibit some identities between triangles where $q \neq 0$ and $s \neq 0$.

$$N_{1,q,1,s} = N_{0,q,0,s+1} + N_{0,q+1,0,s} \tag{46}$$

$$N_{1,q,2,s} = N_{0,q+1,0,s+1} \tag{47}$$

$$N_{1,q+1,2,s+1} = N_{2,q,2,s} + N_{1,q,1,s+2} + N_{1,q+2,1,s} \tag{48}$$

$$N_{1,q+1,0,s+1} = N_{0,q+1,0,s+1} + N_{3,q,1,s} \tag{49}$$

$$N_{1,q+1,1,s+1} = N_{2,q+1,0,s+1} + N_{2,q,0,s} \tag{50}$$

$$N_{p,q,0,1} = \sum_{i=2}^{p+2} N_{i,q,0,0} \tag{51}$$

They can all be proved by using Lemma 3.1, (24) and (25) by easy but tedious calculations.

Now we can derive another recurrence for the triangle in (7), i.e., different from the one given in (6):

Proposition 6.3. *For all q and s, we have*

$$R_{q,s+2} = \sum_{j=0}^{q} C_{j+1} \cdot \left(R_{q-j,s+1} + R_{q+1-j,s} \right) \tag{52}$$

To see an example for (52) consider again the triangle (7):

$$
\begin{array}{c}
{}^{q}\diagup\quad\diagdown\,{}^{s} \\
1 \\
1\quad 1 \\
2\diagup\ 3\ \ 2 \\
5\quad 9\quad 9\quad 5 \\
14\diagup 28\quad 34\diagup 28\quad 14 \\
42\quad 90\quad 123\ \ 123\quad 90\quad 42 \\
132\ \ 297\ \ 440\ \ 497\ \ 440\ \ 297\ \ 132
\end{array} \tag{53}
$$

We have $123 = C_1 \cdot (28 + 34) + C_2 \cdot (9 + 9) + C_3 \cdot (3 + 2)$.

Remark 6.4. In the next section we will make use of the following identity:

$$N_{2,q,0,s} = \sum_{j=0}^{q} N_{0,j,0,0} \cdot N_{0,q+1-j,0,s} = \sum_{l=0}^{s} N_{0,0,0,l} \cdot N_{0,q,0,s+1-l} \qquad (54)$$

It is a special case of (45), but it can also be shown directly: By (37), we have

$$N_{2,q,0,s} = \sum_{j=0}^{q} \sum_{l=0}^{s} C_j \cdot C_l \cdot N_{1,q-j,0,s-l}$$

Then (54) follows immediately by (25) and (27).

In the remainder of this section we derive a closed formula for the tetrahedron $s = 1$. We need the following observation:

Lemma 6.5. *For any p, r, s and t,*

$$N_{p,0,r,s+t} + N_{p+r+t,0,t,s} = N_{p+t,0,r+t,s} + N_{p+r,0,0,s+t} \qquad (55)$$

Proof. Easy calculation, using (34). □

For any t and n, let $U_t(n)$ be defined as

$$U_t(n) = \frac{1}{t+2n}\binom{t+2n}{t+n,n}. \qquad (56)$$

Notice that for any t,

$$tU_t(0) = \frac{t}{t}\binom{t}{t,0} = 1. \qquad (57)$$

The following two identities are called *Rothe's identities* (see [16, identities 14–15, p. 329]).

$$\sum_{k=0}^{n} U_t(k) \cdot U_u(n-k) = \frac{t+u}{tu}U_{t+u}(n) \qquad (58)$$

and

$$\sum_{k=0}^{n} kU_t(k) \cdot U_u(n-k) = \frac{n}{u}U_{t+u}(n). \qquad (59)$$

As a consequence, we get

$$\sum_{k=0}^{n}(t+k)U_t(k) \cdot uU_u(n-k) = (t+u+n)U_{t+u}(n). \qquad (60)$$

Lemma 6.6. *For any p, r and s,*

$$N_{p,0,r,s} = (p + s + 1)U_{p-r+1}(r + s) - sU_{p+r+1}(s) \tag{61}$$

$$N_{p,0,0,s} = (p + 1)U_{p+1}(s) \tag{62}$$

Proof. The identity (61) follows immediately from (34) and (56). The identity (62) is a special case of (61). □

Lemma 6.7. *For any p, r, s and t,*

$$\sum_{l=0}^{s} N_{p,0,r,l} \cdot N_{t,0,0,s-l} = N_{p+t+1,0,r,s}$$

$$\tag{63}$$

$$- \sum_{l=0}^{r-1} (p - r + 1 + l)U_{p-r+1}(l) \cdot (t + 1)U_{t+1}(r + s - l)$$

Now we can give a closed formula for the $N_{p,q,r,1}$ tetrahedron.

Proposition 6.8. *For all p, q and r,*

$$N_{p,q,r,1} = T_{p,q+1,r} + T_{p+r+1,q+1,0} - T_{0,q+1,r} - T_{p,q+1,0} \tag{64}$$

$$= T_{p+1,q,r+1} + T_{p+r,q,1} - T_{1,q,r} - T_{p,q,1} \tag{65}$$

7 Generating Functions

We can use the identities (24) and (54) for calculating the generating function for $N_{p,q,r,s}$. Recall that we use the following abbreviations:

$$\begin{aligned}
P_{p,r} &= N_{p,0,r,0} & C_q &= N_{0,q,0,0} \\
Q_{p,q} &= N_{p,q,0,0} & T_{p,q,r} &= N_{p,q,r,0} \\
R_{q,s} &= N_{0,q,0,s} & V_{p,q,s} &= N_{p,q,0,s}
\end{aligned} \tag{66}$$

Theorem 7.1. *We have*

(i) $C(y) = \sum_q C_q y^q = \dfrac{1 - \sqrt{1 - 4y}}{2y}$

(ii) $P(x, z) = \sum_{p,r} P_{p,r} x^p z^r = \dfrac{1}{1 - x - z}$

(iii) $Q(x, y) = \sum_{p,q} Q_{p,q} x^p y^q = \dfrac{C(y)}{1 - x \cdot C(y)}$

(iv) $R(y, w) = \sum_{q,s} R_{q,s} y^q w^s = \dfrac{C(y) \cdot C(w) \cdot (w - y)}{w \cdot C(y) - y \cdot C(w)}$

(v) $T(x, y, z) = \sum_{p,q,r} T_{p,q,r} x^p y^q z^r = \dfrac{(1 - x - z - x \cdot z \cdot C(y)) \cdot C(y)}{(1 - x \cdot C(y)) \cdot (1 - z \cdot C(y)) \cdot (1 - x - z)}$

$$= P(x, z) \cdot \left(1 + y \cdot Q(x, y) \cdot Q(z, y)\right)$$

(vi) $V(x, z, w) = \sum_{p,q,s} V_{p,q,s} x^p y^q w^s$

$$= \frac{C(y) \cdot C(w) \cdot (w - y - x \cdot (w \cdot C(y) - y \cdot C(w)))}{(w \cdot C(y) - y \cdot C(w)) \cdot (1 - x \cdot C(y)) \cdot (1 - x \cdot C(w))}$$

$$= Q(x, y) \cdot Q(x, w) \cdot \left(\frac{R(y, w)}{C(y) \cdot C(w)} - x \right)$$

(vii) $N(x, y, z, w) = \sum_{p,q,r,s} N_{p,q,r,s} x^p y^q z^r w^s$

$$= \frac{C(y)C(w)((1-x-z)(w-y) + (wC(y) - yC(w))((1-x-z)(xz(C(y) + C(w)) - x - z) + x^2 z^2 C(y)C(w)))}{(1 - xC(y))(1 - xC(w))(1 - zC(y))(1 - zC(w))(1 - x - z)(wC(y) - yC(w))}$$

$$= Q(x, y)Q(x, y)Q(x, y)Q(x, y) \left(\frac{R(y, w)}{C(y)^2 \cdot C(w)^2} + x^2 z^2 P(x, z) - \frac{x}{C(w)Q(z, y)} - \frac{z}{C(y)Q(x, w)} \right)$$

Proof. The formulas in (i) and (ii) are well-known. For the others, the calculation can be found in [10]. □

Acknowledgments. We thank Mireille Bousquet-Mélou, Christian Retoré, and Gilles Schaeffer for fruitful and instructive discussions, and the anonymous referees for helpful comments for improving the paper.

References

1. Aigner, M., Ziegler, G.M.: Proofs from the Book, 3rd edn. Springer, Heidelberg (2003)
2. Barr, M.: Non-symmetric *-automomous categories. Theoret. Comput. Sci. **139**, 115–130 (1995)
3. Blute, R.: Linear logic, coherence and dinaturality. Theoret. Comput. Sci. **115**, 3–41 (1993)
4. Danos, V., Regnier, L.: The structure of multiplicatives. Ann. Math. Logic **28**, 181–203 (1989)
5. Flajolet, P., Noy, M.: Analytic combinatorics of non-crossing configurations. Discrete Math. **204**(1–3), 203–229 (1999). doi:10.1016/S0012-365X(98)00372-0
6. Guglielmi, A.: A system of interaction and structure. ACM Trans. Comput. Logic **8**(1), 1 (2007)
7. Guglielmi, A., Straßburger, L.: Non-commutativity and MELL in the calculus of structures. In: Fribourg, L. (ed.) CSL 2001. LNCS, vol. 2142, pp. 54–68. Springer, Heidelberg (2001). doi:10.1007/3-540-44802-0_5
8. Lamarche, F., Retoré, C.: Proof nets for the Lambek-calculus – an overview. In: Michele Abrusci, V., Casadio, C., (eds.) Proceedings of the Third Roma Workshop "Proofs and Linguistic Categories", pp. 241–262. CLUEB, Bologna (1996)
9. Lambek, J.: The mathematics of sentence structure. Am. Math. Mon. **65**, 154–169 (1958)
10. Méhats, L., Straßburger, L.: Non-crossing tree realizations of ordered degree sequences. Technical report, Inria (2009). https://hal.inria.fr/hal-00649591
11. Noy, M.: Enumeration of noncrossing trees on a circle. Discrete Math. **180**(1–3), 301–313 (1998). doi:10.1016/S0012-365X(97)00121-0

12. Sawada, J.: Generating bracelets in constant amortized time. SIAM J. Comput. **31**(1), 259–268 (2001)
13. Sawada, J.: A fast algorithm to generate necklaces with fixed content. Theoret. Comput. Sci. **301**(1–3), 477–489 (2003)
14. Sloane, N.J.A.: The On-Line Encyclopedia of Integer Sequences (2010). https://oeis.org/
15. Straßburger, L.: Proof nets and the identity of proofs. Research report 6013, INRIA, 10 2006. Lecture notes for ESSLLI 2006
16. Strehl, V.: Identities of Rothe-Abel-Schläfli-Hurwitz-type. Discrete Math. **99**(1–3), 321–340 (1992). doi:10.1016/0012-365X(92)90379-T
17. Yetter, D.N.: Quantales and (noncommutative) linear logic. J. Symbolic Logic **55**(1), 41–64 (1990)

On the Logic of Expansion in Natural Language

Glyn Morrill[(⊠)] and Oriol Valentín[(⊠)]

Departament of Computer Science,
Universitat Politècnica de Catalunya, Barcelona, Spain
{morrill,ovalentin}@cs.upc.edu

Abstract. We consider, for intuitionistic categorial grammar, an iteration modality with a rule of Mingle and an infinitary left rule, similar to infinitary action logic. Newly, we give Curry-Howard labelling for the iteration modality, in terms of lists, and we prove soundness and completeness of displacement calculus with additives and this modality, for phase semantics. This result has as a corollary semantic Cut-elimination. We review linguistic application of the iteration modality to unbounded addicity iterated coordination, and we present an application of a calibrated version of the iteration modality to an unbounded addicity respectively construction, this being to our knowledge the first account of respectively taking care of cases $n > 2$.

Keywords: Expansion · Exponentials · Iterated coordination · Mingle · Phase semantics · Respectively construction · Semantic Cut-elimination

1 Introduction

In standard logic information does not have multiplicity. Thus where $+$ is the notion of addition of information and $\leq$ is the notion of inclusion of information we have $x+x \leq x$ and $x \leq x+x$; together these two properties amount to idempotency: $x+x = x$. These properties are expressed by the rules of inference of Contraction and Expansion:

$$(1) \quad \frac{\Delta(A, A) \Rightarrow B}{\Delta(A) \Rightarrow B} \text{ Contraction} \qquad \frac{\Delta(A) \Rightarrow B}{\Delta(A, A) \Rightarrow B} \text{ Expansion}$$

In general linguistic resources do not have these properties: grammaticality is not often preserved under addition or removal of copies of expressions. However, there are some constructions manifesting something similar. In this paper we investigate categorial logic and expansion.

Iterated coordination has a kind of expansion, of unbounded addicity:

(2) John likes, Mary dislikes, ... and Bill loves London.

Likewise an unbounded addicity *respectively* construction:

(3) Tom, Dick, ... and Harry walk, talk, ... and sing respectively.

© Springer-Verlag GmbH Germany 2016
M. Amblard et al. (Eds.): LACL 2016, LNCS 10054, pp. 228–246, 2016.
DOI: 10.1007/978-3-662-53826-5_14

That is, in logical grammar a *controlled* use of expansion is motivated. Girard [4] introduced exponentials for control of structural rules. For the use of nonlinearity for iterated coordination in categorial grammar see Morrill [13] and Morrill and Valentín [11].

The iteration modality is closely related to the Kleene star modality of the infinitary action logic of Buszkowski and Palka [2].[1] Our new results include the Curry-Howard annotation of the iteration modality, with (non-empty) lists, combination with the full displacement calculus, and a strong completeness result à la Okada [14], namely soundness and completeness with respect to phase semantics (Girard [4]), and as a by product of this there is a semantic proof of Cut-elimination, which differs from the syntactic Cut-elimination of Palka [15]. Linguistic applications include for the first time in categorial grammar syntactic and semantic analysis of an unbounded addicity *respectively* construction.[2]

In Sect. 2 we define a displacement calculus **DA?** with additives, and an existential exponential with a Mingle structural rule (Kamide [5]) and an infinitary left rule, which entail expansion. In Sect. 3 we give a sound and complete phase semantics for **DA?**. The completeness has as a corollary semantic Cut-elimination. In Sect. 4 we present a calibrated version of the Mingle modality and present a linguistic fragment including iterated coordination and the *respectively* construction with analyses generated by a version of the categorial parser/theorem-prover CatLog2.[3]

2   The Categorial Logic

The multiplicative basis is the displacement calculus of Morrill et al. [12]; in addition there are additives, and the existential exponential. The syntactic types of the categorial logic are sorted according to the number of points of discontinuity their expressions contain. Each type predicate letter has a sort and an arity which are naturals, and a corresponding semantic type. Assuming ordinary terms to be already given, where P is a type predicate letter of sort i and arity n and $t_1, \ldots, t_n$ are terms, $Pt_1 \ldots t_n$ is an (atomic) type of sort i of the corresponding semantic type. Compound types are formed by connectives as in Fig. 1.[4]

For a type A, its sort $s(A)$ is the i such that $A \in \mathcal{F}_i$. Tree-based sequent calculus is as follows. Configurations are defined by:[5]

---

[1] We can define the Kleene star modality $*$ in terms of our modality? by: $A^* = I \oplus ?A$.

[2] In the type logical literature iteration has been considered in Bechet et al. [1] who propose syntactic pregroup analyses but without enjoying intuitionistic Curry-Howard labelling, nor algebraic models.

[3] https://www.cs.upc.edu/~morrill/CatLog/CatLog2/index.php.

[4] Observe that the iteration modality ? only applies to types of sort 0 because otherwise expansion would not preserve the equality of antecedent and succedent sorts.

[5] Note that the colons in the fourth clause of the definition punctuate the list of configurations intercalating the points of discontinuity of $\mathcal{F}_{i>0}$ of sort i; this is entirely distinct from (the standard) use of colons in type assignments made later.

(4) $\mathcal{O} ::= \Lambda$
$\mathcal{O} ::= 1, \mathcal{O}$
$\mathcal{O} ::= \mathcal{F}_0, \mathcal{O}$
$\mathcal{O} ::= \mathcal{F}_{i>0}\{\underbrace{\mathcal{O} : \ldots : \mathcal{O}}_{i\ \mathcal{O}'s}\}, \mathcal{O}$

For a configuration Δ we define the *type-equivalent* $\Delta^\bullet$, which is a type which has the same algebraic meaning as Δ. Via the BNF formulation of $\mathcal{O}$ in (4) one defines recursively $\Delta^\bullet$ as follows:

(5) $\Lambda^\bullet \overset{def}{=} I$
$(1, \Gamma)^\bullet \overset{def}{=} J \bullet \Gamma^\bullet$
$(A, \Gamma)^\bullet \overset{def}{=} A \bullet \Gamma^\bullet$, if $s(A) = 0$
$(A\{\Delta_1 : \ldots : \Delta_{s(A)}\}, \Gamma)^\bullet \overset{def}{=}$
$\quad ((\cdots(A \odot_1 \Delta_1^\bullet) \cdots) \odot_{1+s(\Delta_1)+\cdots+s(\Delta_{s(A)})} \Delta_{s(A)}^\bullet) \bullet \Gamma^\bullet$, if $s(A) > 0$

For a configuration Γ, its sort $s(\Gamma)$ is $|\Gamma|_1$, i.e. the number of metalinguistic separators 1 which it contains. A sequent $\Gamma \Rightarrow A$ comprises an antecedent configuration Γ and a succedent type A such that $s(\Gamma) = s(A)$. The figure $\overrightarrow{A}$ of a type A is defined by:

(6) $\overrightarrow{A} = \begin{cases} A & \text{if } sA = 0 \\ A\{\underbrace{1 : \ldots : 1}_{sA\ 1's}\} & \text{if } sA > 0 \end{cases}$

Where Γ is a configuration of sort i and $\Delta_1, \ldots, \Delta_i$ are configurations, the fold $\Gamma \otimes \langle \Delta_1 : \ldots : \Delta_i \rangle$ is the result of replacing the successive 1's in Γ by $\Delta_1, \ldots, \Delta_i$ respectively. Where Δ is a configuration of sort $i > 0$ and Γ is a configuration, the kth metalinguistic wrap $\Delta |_k \Gamma$, $1 \leq k \leq i$, is given by

(7) $\Delta |_k \Gamma =_{df} \Delta \otimes \langle \underbrace{1 : \ldots : 1}_{k-1\ 1's} : \Gamma : \underbrace{1 : \ldots : 1}_{i-k\ 1's} \rangle$

i.e. the kth metalinguistic wrap $\Delta |_k \Gamma$ is the configuration resulting from replacing by Γ the kth separator in Δ.

1.	$\mathcal{F}_i ::= \mathcal{F}_{i+j}/\mathcal{F}_j$	$T(C/B) = T(B){\rightarrow}T(C)$	over [9]
2.	$\mathcal{F}_j ::= \mathcal{F}_i\backslash\mathcal{F}_{i+j}$	$T(A\backslash C) = T(A){\rightarrow}T(C)$	under [9]
3.	$\mathcal{F}_{i+j} ::= \mathcal{F}_i \bullet \mathcal{F}_j$	$T(A \bullet B) = T(A)\&T(B)$	continuous product [9]
4.	$\mathcal{F}_0 ::= I$	$T(I) = \top$	continuous unit [8]
5, k.	$\mathcal{F}_{i+1} ::= \mathcal{F}_{i+j}{\uparrow}_k\mathcal{F}_j, 1 \leq k \leq i+j$	$T(C{\uparrow}_k B) = T(B){\rightarrow}T(C)$	extract [12]
6, k.	$\mathcal{F}_j ::= \mathcal{F}_{i+1}{\downarrow}_k\mathcal{F}_{i+j}, 1 \leq k \leq i+1$	$T(A{\downarrow}_k C) = T(A){\rightarrow}T(C)$	infix [12]
7, k.	$\mathcal{F}_{i+j} ::= \mathcal{F}_{i+1}\odot_k\mathcal{F}_j, 1 \leq k \leq i+1$	$T(A\odot_k B) = T(A)\&T(B)$	discontinuous product [12]
8.	$\mathcal{F}_1 ::= J$	$T(J) = \top$	discontinuous unit [12]
9.	$\mathcal{F}_i ::= \mathcal{F}_i\&\mathcal{F}_i$	$T(A\&B) = T(A)\&T(B)$	additive conjunction [7, 10]
10.	$\mathcal{F}_i ::= \mathcal{F}_i\oplus\mathcal{F}_i$	$T(A\oplus B) = T(A)+T(B)$	additive disjunction [7, 10]
18.	$\mathcal{F}_0 ::= ?\mathcal{F}_0$	$T(?A) = T(A)^+$	existential exponential [13]

Fig. 1. Categorial logic types of **DA?**

$$\frac{\Gamma \Rightarrow B:\psi \qquad \Delta\langle \overrightarrow{C:z}\rangle \Rightarrow D:\omega}{\Delta\langle \overrightarrow{C/B}:x,\Gamma\rangle \Rightarrow D:\omega\{(x\ \phi)/z\}}\ /L \qquad \frac{\Gamma,\overrightarrow{B}:y \Rightarrow C:\chi}{\Gamma \Rightarrow C/B:\lambda y\chi}\ /R$$

$$\frac{\Gamma \Rightarrow A:\phi \qquad \Delta\langle \overrightarrow{C:z}\rangle \Rightarrow D:\omega}{\Delta\langle \Gamma,\overrightarrow{A\backslash C}:y\rangle \Rightarrow D:\{(y\ \phi)/z\}}\ \backslash L \qquad \frac{\overrightarrow{A}:x,\Gamma \Rightarrow C:\chi}{\Gamma \Rightarrow A\backslash C:\lambda x\chi}\ \backslash R$$

$$\frac{\Delta\langle \overrightarrow{A}:x,\overrightarrow{B}:y\rangle \Rightarrow D:\omega}{\Delta\langle \overrightarrow{A\bullet B}:z\rangle \Rightarrow D:\omega\{\pi_1 z/x,\pi_2 z/y\}}\ \bullet L \qquad \frac{\Gamma_1 \Rightarrow A:\phi \qquad \Gamma_2 \Rightarrow B:\psi}{\Gamma_1,\Gamma_2 \Rightarrow A\bullet B:(\phi,\psi)}\ \bullet R$$

$$\frac{\Delta\langle\Lambda\rangle \Rightarrow A:\phi}{\Delta\langle \overrightarrow{I}:x\rangle \Rightarrow A:\phi}\ IL \qquad \frac{}{\Lambda \Rightarrow I:0}\ IR$$

$$\frac{\Gamma \Rightarrow B:\psi \qquad \Delta\langle \overrightarrow{C:z}\rangle \Rightarrow D:\omega}{\Delta\langle \overrightarrow{C\uparrow_k B}:x\,|_k\,\Gamma\rangle \Rightarrow D:\omega\{(x\ \psi)/z\}}\ \uparrow_k L \qquad \frac{\Gamma\,|_k\,\overrightarrow{B}:y \Rightarrow C:\chi}{\Gamma \Rightarrow C\uparrow_k B:\lambda y\chi}\ \uparrow_k R$$

$$\frac{\Gamma \Rightarrow A:\phi \qquad \Delta\langle \overrightarrow{C:z}\rangle \Rightarrow D:\omega}{\Delta\langle \Gamma\,|_k\,\overrightarrow{A\downarrow_k C}:y\rangle \Rightarrow D:\omega\{(y\ \phi)/z\}}\ \downarrow_k L \qquad \frac{\overrightarrow{A}:x\,|_k\,\Gamma \Rightarrow C:\chi}{\Gamma \Rightarrow A\downarrow_k C:\lambda x\chi}\ \downarrow_k R$$

$$\frac{\Delta\langle \overrightarrow{A}:x\,|_k\,\overrightarrow{B}:y\rangle \Rightarrow D:\omega}{\Delta\langle \overrightarrow{A\odot_k B}:z\rangle \Rightarrow D:\omega\{\pi_1 z/x,\pi_2 z/y\}}\ \odot_k L \qquad \frac{\Gamma_1 \Rightarrow A:\phi \qquad \Gamma_2 \Rightarrow B:\psi}{\Gamma_1\,|_k\,\Gamma_2 \Rightarrow A\odot_k B:(\phi,\psi)}\ \odot_k R$$

$$\frac{\Delta\langle 1\rangle \Rightarrow A:\phi}{\Delta\langle \overrightarrow{J}:x\rangle \Rightarrow A:\phi}\ JL \qquad \frac{}{1 \Rightarrow J:0}\ JR$$

Fig. 2. Multiplicative rules of **DA?**

Where the notation $\Xi(\Omega)$ signifies a configuration Ξ with a distinguished subconfiguration Ω, the notation $\Delta\langle\Gamma\rangle$ abbreviates $\Delta_0(\Gamma \otimes \langle\Delta_1 : \ldots : \Delta_n\rangle)$, i.e. a configuration with a potentially discontinuous distinguished subconfiguration Γ with external context Δ_0 and internal context $\Delta_1,\ldots,\Delta_n$.

The semantically annotated identity axiom *id* and *Cut* rule are:

$$(8)\quad \frac{}{P:x \Rightarrow P:x}\ id,\ P\ \text{atomic} \qquad \frac{\Gamma \Rightarrow A:\phi \qquad \Delta\langle \overrightarrow{A}:x\rangle \Rightarrow B:\beta}{\Delta\langle\Gamma\rangle \Rightarrow B:\beta\{\phi/x\}}\ Cut$$

The semantically annotated multiplicative rules of **DA?** are given in Fig. 2. The semantically annotated additive and exponential rules are given in Fig. 3.[6]

---

[6] Notice that although the sequent calculus is infinitary and has possibly infinite proofs, the proveable sequents are always finite. The system is undecidable by a result of Buszkowski and Palka [2] but a linguistically sufficient fragment, without antedent iteration modalities, is decidable.

The expansion rule with iteration modalities is derivable by the following reasoning. Given an arbitrary type A of sort 0, for every $i > 0$ and a fixed index $j_0 > 0$,

$$\frac{\Gamma\langle \vec{A}:x\rangle \Rightarrow C:\chi}{\Gamma\langle \overrightarrow{A\&B}:z\rangle \Rightarrow C:\chi\{\pi_1 z/x\}}\&L_1 \qquad \frac{\Gamma\langle \vec{B}:y\rangle \Rightarrow C:\chi}{\Gamma\langle \overrightarrow{A\&B}:z\rangle \Rightarrow C\chi\{\pi_2 z/y\}}\&L_2$$

$$\frac{\Gamma \Rightarrow A:\phi \qquad \Gamma \Rightarrow B:\psi}{\Gamma \Rightarrow A\&B:(\phi,\psi)}\&R$$

$$\frac{\Gamma\langle \vec{A}:x\rangle \Rightarrow C:\chi_1 \qquad \Gamma\langle \vec{B}:y\rangle \Rightarrow C:\chi_2}{\Gamma\langle \overrightarrow{A\oplus B}:z\rangle \Rightarrow C:z \to x.\chi_1; y.\chi_2}\oplus L$$

$$\frac{\Gamma \Rightarrow A:\phi}{\Gamma \Rightarrow A\oplus B:\iota_1\phi}\oplus R_1 \qquad \frac{\Gamma \Rightarrow B:\psi}{\Gamma \Rightarrow A\oplus B:\iota_2\phi}\oplus R_2$$

$$\frac{\Delta(A:x) \Rightarrow B:\psi([x]) \qquad \Delta(A:x,A:y) \Rightarrow B:\psi([x,y]) \qquad \cdots}{\Delta(?A:z) \Rightarrow B:\psi(z)}?L$$

$$\frac{\Gamma \Rightarrow A:\phi}{\Gamma \Rightarrow ?A:[\phi]}?R \qquad \frac{\Gamma \Rightarrow A:\phi \qquad \Delta \Rightarrow ?A:\phi'}{\Gamma,\Delta \Rightarrow [\phi|\phi']:?A}?M$$

Fig. 3. Additive and exponential rules of **DA?**

3 Phase Semantics

DA? incorporates the useful language-theoretic concept of *iteration*. This is done by means of an (existential) exponential modality, notated ? which licenses the structural rule of Mingle, which entails expansion.

Let i, j and k range over the set of natural numbers ω. Where A is a type of sort 0, and $i > 0$, A^i denotes $\underbrace{A,\ldots,A}_{i\,\text{times}}$. A^0 is the empty string Λ.

3.1 Semantic Interpretation

In the following, we describe the phase space machinery in order to give a result of strong completeness in the style of Okada [14]. Phase spaces from linear logic (Girard [4]) are based on (commutative) monoids. Likewise, the proper algebras for the displacement calculus **D** are the so-called *displacement algebras* (DA

---

by one application of $?R$ and a finite number of applications of the Mingle rule we get the infinite provable sequents indexed by i $(i > 0)$ $A^i, A^{j_0} \Rightarrow ?A$. We can then apply the $?L$ rule, obtaining $?A, A^{j_0} \Rightarrow ?A$. Since j_0 is a positive natural, we have that for every $j > 0$, $?A, A^j \Rightarrow ?A$. We can apply again then the $?R$ rule, whence $?A, ?A \Rightarrow ?A$. This proves the expansion rule.

for short) (see Valentín [17]) which can be seen as a generalisation of (non-commutative) monoids where the operations of k-th intercalation in a punctuated string are incorporated. In Valentín [17] it is proved that DAs can be axiomatised; see Fig. 4). We can define the class of residuated DAs (Valentín [18]), and therefore models.

Given a mapping $v : \mathrm{Pr} \to \mathbf{A}$ where $\mathbf{A}$ is a residuated DA, there exists a unique ω-sorted homomorphism $\widehat{v}$ which extends v as follows: $\widehat{v} : \mathbf{Tp} \to \mathbf{A}$ and $\widehat{v}(p) = v(p)$ for any primitive type. Needless to say, since we are working in an ω-sorted setting, equations, inequations and mapping and so on, are to be understood modulo sorting; in order to give a smoother reading of formulas we always avoid if possible the explicit reference to sorts.

Continuous associativity

$x + (y + z) \approx (x + y) + z$

Discontinuous associativity

$x \times_i (y \times_j z) \approx (x \times_i y) \times_{i+j-1} z$

$(x \times_i y) \times_j z \approx x \times_i (y \times_{j-i+1} z)$ if $i \leq j \leq 1 + s(y) - 1$

Mixed permutation

$(x \times_i y) \times_j z \approx (x \times_{j-s(y)+1} z) \times_i y$ if $j > i + s(y) - 1$

$(x \times_i z) \times_j y \approx (x \times_j y) \times_{i+s(y)-1} z$ if $j < i$

Mixed associativity

$(x + y) \times_i z \approx (x \times_i z) + y$ if $1 \leq i \leq s(x)$

$(x + y) \times_i z \approx x + (y \times_{i-s(x)} z)$ if $x + 1 \leq i \leq s(x) + s(y)$

Continuous unit and discontinuous unit

$0 + x \approx x \approx x + 0$ and $1 \times_1 x \approx x \approx x \times_i 1$

Fig. 4. Axiomatisation of a DA

A subset B of the carrier set A of a DA is called a *same-sort* subset iff there exists an $i \in \omega$ such that for every $a \in B$, $s(a) = i$. Notice that $\emptyset$ vacuously satisfies the *same-sort* condition. $\mathcal{P}(\mathrm{A})$ is in fact an ω-sorted subset $(\mathcal{P}(\mathrm{A})_i)_{i \in \omega}$ where for every i, $\mathcal{P}(\mathrm{A})_i = \{X : X \text{ is a same-sort subset of sort } i\}$.

Definition 1. *A displacement phase space* $\mathbf{P} = (\mathbf{A}, \mathbf{Closed})$ *is a structure partially ordered by the relation of subset inclusion such that:*

1. $\mathbf{A}$ is a DA.

2. $\mathbf{Closed} = (\mathbf{Closed}_i)_i$ is a set of subsets such that $\mathbf{Closed}_i \subseteq \mathcal{P}(\mathrm{A})_i$, $\mathbf{Closed}_i \cap \mathbf{Closed}_j = \{\emptyset\}$ iff $i \neq j$, and:

   (a) For every $F \in \mathbf{Closed}_i$, F is called a closed *subset.*

   (b) $\mathbf{Closed}$ is closed by intersections of arbitrary families of same-sort subsets. In particular, the intersection of the empty family of closed subsets of sort i is A_i which belongs to $\mathbf{Closed}_i$.

(d) *For all $F \in \mathbf{Closed}_i$, and for all $x \in A_j$:*

$$x \backslash F \ \in \mathbf{Closed}_{i-j} \quad F/x \ \in \mathbf{Closed}_{i-j}$$
$$F \uparrow_k x \in \mathbf{Closed}_{i-j+1} \quad x \downarrow_k F \in \mathbf{Closed}_{i-j+1}$$

Closed is also called (an ω-sorted) *closure system*.

Where F, G denote subsets of A of sort i, we define the ω-sorted closure operator cl_i:

(9) $cl_i(G) \overset{def}{=} \bigcap \{F \in \mathbf{Closed}_i : G \subseteq F\}$

We write $\overline{G}^i$ for $cl_i(G)$. If the context is clear we omit the subscript.

Where F and G are same-sort subsets, it is readily seen that:

(i) $\overline{F}$ is the least closed set of sort $s(F)$ such that $F \subseteq \overline{F}$.
(ii) $cl(\cdot)$ is extensive, i.e.: $G \subseteq \overline{G}$.
(iii) $cl(\cdot)$ is monotone, i.e.: if $G_1 \subseteq G_2$ then $\overline{G_1} \subseteq \overline{G_2}$.
(iv) $cl(\cdot)$ is idempotent, i.e.: $cl^2(G) = cl(G)$.

We define the following operators at the level of same-sort subsets:

- $F \circ G \overset{def}{=} \{f + g : f \in F \text{ and } g \in G\}$
- $F \circ_i G \overset{def}{=} \{f \times_i g : f \in F \text{ and } g \in G\}$
- $f \circ G \overset{def}{=} \{f\} \circ G$ and $F \circ g \overset{def}{=} F \circ \{g\}$
- $f \circ_i G \overset{def}{=} \{f\} \circ_i G$ and $F \circ_i g \overset{def}{=} F \circ_i \{g\}$
- $G//F \overset{def}{=} \{h : \forall f \in F, h + f \in G\}$ and similarly for $F \backslash\backslash G$
- $G \uparrow\uparrow_i F \overset{def}{=} \{h : \forall f \in F, h \times_i f \in G\}$ and similarly for $F \downarrow\downarrow_i G$
- $G//f \overset{def}{=} G//\{f\}$ and similarly for $f \backslash\backslash G$
- $G \uparrow\uparrow_i f \overset{def}{=} G \uparrow\uparrow_i \{f\}$ and similarly for $f \downarrow\downarrow_i G$

The following basic properties for ω-sorted closure operators are evident:

Lemma 1.

- $F \circ G \subseteq H$ *iff* $F \subseteq H//G$ *iff* $G \subseteq F \backslash\backslash H$.
- $F \circ_i G \subseteq H$ *iff* $F \subseteq H \uparrow\uparrow_i G$ *iff* $G \subseteq F \downarrow\downarrow_i H$.
- *By construction,* $\overline{F}$ *is the least closed subset such that* $F \subseteq \overline{F}$. *Hence:*
- *If* $A \subseteq F$ *and* $\overline{F} = F$ *then* $\overline{A} \subseteq \overline{F}$.

Lemma 2. *If A is closed, then:*

- $A//F, F\backslash\backslash A, A\uparrow\uparrow_i F$, *and* $F \downarrow\downarrow_i A$ *are closed.*
  Proof: $A\uparrow\uparrow_i F = \bigcap_{x \in F} A\uparrow\uparrow_i x$, *whence* $A\uparrow\uparrow_i F$ *is closed.* $\square$
- *Similarly for the other implicative operations.*
- $cl(F) \circ cl(G) \subseteq cl(F \circ G)$. *Similarly,* $cl(F) \circ_i cl(G) \subseteq cl(F \circ_i G)$
- *Hence,* $\overline{F} \circ \overline{G} \subseteq \overline{F \circ G}$, *and* $\overline{F} \circ_i \overline{G} \subseteq \overline{F \circ_i G}$

- *It follows that $cl(cl(F) \circ cl(G)) = cl(F \circ G)$ and $cl(cl(F) \circ_i cl(G)) = cl(F \circ_i G)$*
  Proof: Let us see the case of $\circ_i$. $F \circ_i G \subseteq \overline{F \circ_i G}$. By residuation, $F \subseteq \overline{F \circ_i G} \uparrow\uparrow_i G$.
  $\overline{F \circ_i G} \uparrow\uparrow_i G$ is a closed subset (see previous proof). Hence, $\overline{F} \subseteq \overline{F \circ_i G} \uparrow\uparrow_i G$.
  Applying again residuation, we have $\overline{F} \circ_i G \subseteq \overline{F \circ G}$
  We repeat the process with G, obtaining $\overline{G} \subseteq \overline{F} \downarrow\downarrow_i \overline{F \circ_i G}$. It follows that:
  $\overline{F} \circ_i \overline{G} \subseteq \overline{F \circ_i G}$. Hence, $\overline{\overline{F} \circ_i \overline{G}} \subseteq \overline{F \circ_i G}$

We see now operations on closed subsets which return values into the set of closed subsets. This paves the way to the definition of valuations from the set of types into phase spaces, concretely into the set of closed sets. Given F, G closed sets:

(10) $F \overline{\circ} G \overset{def}{=} \overline{F \circ G}$

$F \overline{\circ_i} G \overset{def}{=} \overline{F \circ_i G}$

$F \& G \overset{def}{=} F \cap G$. In general we write $F \cap G$.

$F \overline{\cup} G \overset{def}{=} \overline{F \cup G}$.

$G \overline{\uparrow\uparrow}_i F \overset{def}{=} \overline{G \uparrow\uparrow_i F}$. In general we write $\uparrow\uparrow_i$ avoiding the use of $\overline{\uparrow\uparrow}_i$.
Similarly for the other implications.

$\overline{\mathbb{I}} \overset{def}{=} \overline{\{0\}}$.

$\overline{\mathbb{J}} \overset{def}{=} \overline{\{1\}}$.

Valuations in phase spaces are mappings between the set of types into the set of closed sets. More concretely, given a valuation $v : \mathrm{Pr} \longrightarrow \mathbf{Closed}$, where $\mathbf{P} = (\mathbf{A}, \mathbf{Closed})$ is a phase space, we see the interpretation of v and its recursive extension $\widehat{v}$ w.r.t. any type in the set of primitive types by using the closed operation on the set of closed subsets defined in (10):[7]

- $v(p)$ is a closed subset of A_i where p is primitive of sort i.
  We extend recursively v to $\widehat{v}$:
- $\widehat{v}(B \uparrow_i A) \overset{def}{=} \widehat{v}(B) \uparrow\uparrow_i \widehat{v}(A)$. Similarly for the other implications.
- $\widehat{v}(A \bullet B) \overset{def}{=} \widehat{v}(A) \overline{\circ} \widehat{v}(B)$. $\qquad \widehat{v}(A \odot_i B) \overset{def}{=} \widehat{v}(A) \overline{\circ_i} \widehat{v}(B)$.
- $v(A \oplus B) \overset{def}{=} v(A) \overline{\cup} v(B)$. $\qquad v(A \& B) \overset{def}{=} v(A) \cap v(B)$.
- $\widehat{v}(I) \overset{def}{=} \overline{\mathbb{I}}$. $\qquad \widehat{v}(J) \overset{def}{=} \overline{\mathbb{J}}$.

Notice that for any type A, $v(A)$ is a closed subset.

3.2 The Semantics of the Iteration Connective

Given a phase space model $(\mathbf{P}, v)$, we define $\widehat{v}(?A)$ as:

(11) $\widehat{v}(?A) \overset{def}{=} \overline{\bigcup_{i>0} \widehat{v}(A)^i}$

---

[7] The semantic interpretation of a configuration Δ (for a given valuation v) is $\widehat{v}(\Delta) \overset{def}{=} \widehat{v}(\Delta^\bullet)$.

Lemma 3. *Where $(F_i)_{i\in\omega} \subseteq P$, $F, G \subseteq P$, and A is a type of sort 0
We have:*
$$\overline{\bigcup_{i\in\omega} F_i} = \bigcup_{i\in\omega} \overline{F_i}$$

Proof. $\boxed{\subseteq}$ is obvious.

$\boxed{\supseteq}$ For every $k \in \omega$, $F_k \subset \overline{\bigcup_{i\in\omega} F_i}$. Hence, $\overline{F_k} \subset \overline{\bigcup_{i\in\omega} F_i}$ for every k. Therefore, $\bigcup_{i\in\omega} \overline{F_i} \subseteq \overline{\bigcup_{i\in\omega} F_i}$. Taking closure, we obtain $\overline{\bigcup_{i\in\omega} \overline{F_i}} \subseteq \overline{\bigcup_{i\in\omega} F_i}$. $\square$

Let $(\mathbf{P}, v)$ be a phase space model. We know that $\Delta\langle\Gamma\rangle$ abbreviates $\Delta_0|_k(\Gamma\otimes\langle\Delta_1;\ldots;\Delta_{s(\Gamma)}\rangle)$ for a certain Δ_0, Δ_i, and $k > 0$. We recall that
$$\widehat{v}\,(\Gamma\otimes\langle\Delta_1;\ldots;\Delta_{s(\Gamma)}\rangle) \stackrel{def}{=} \widehat{v}\,(\Gamma) \times_1 \widehat{v}\,(\Delta_1)\ldots \times_{1+s(\Delta_1)+\ldots+s(\Delta_{s(\Gamma)})} \widehat{v}\,(\Delta_{s(\Gamma)}).$$

(12) $\widehat{v}\,(\Gamma\otimes\langle\Delta_1;\ldots;\Delta_{s(\Gamma)}\rangle) \stackrel{def}{=} (\ldots(\widehat{v}\,(\Gamma) \times_1 \widehat{v}\,(\Delta_1))\ldots) \times_{1+s(\Delta_1)+\ldots+s(\Delta_{s(\Gamma)})} \widehat{v}\,(\Delta_{s(\Gamma)})$

The rhs of (12) is abbreviated overloading the symbol $\otimes$, i.e.:
$$\widehat{v}\,(\Gamma\otimes\langle\Delta_1;\ldots;\Delta_{s(\Gamma)}\rangle) \stackrel{def}{=} \widehat{v}\,(\Gamma) \otimes \langle\widehat{v}\,(\Delta_1);\ldots;\widehat{v}\,(\Delta)\rangle.$$

In order to prove soundness for phase semantics it is useful to directly compute configurations w.r.t. valuations without the use of type-equivalence. We have:

(13) $\widehat{v}\,(\Lambda) \stackrel{def}{=} \widehat{v}\,(I)$

$\widehat{v}\,(1,\Gamma) \stackrel{def}{=} \widehat{v}\,(J)\overline{\circ}\widehat{v}\,(\Gamma)$

$\widehat{v}\,(A,\Gamma) \stackrel{def}{=} \widehat{v}\,(A)\overline{\circ}\widehat{v}\,(\Gamma)$, if $s(A) = 0$

$\widehat{v}\,(A\{\Delta_1:\ldots:\Delta_{s(A)}\},\Gamma) \stackrel{def}{=}$
$((\cdots(\widehat{v}\,(A)\overline{\circ_1}\Delta_1)\cdots)\overline{\circ_{1+s(\Delta_1)+\cdots+s(\Delta_{s(A)})}\Delta_{s(A)}}\widehat{v}\,(\Delta_{s(A)})\overline{\circ}\widehat{v}\,(\Gamma)$, if $s(A) > 0$

But how do we interpret $\Delta\langle\Gamma\rangle$? As said before, $\Delta\langle\Gamma\rangle$ abbreviates $\Delta_0\langle\Gamma \otimes \langle\Delta_1;\ldots;\Delta_{s(\Gamma)}\rangle\rangle$. $\Gamma \otimes \langle\Delta_1;\ldots;\Delta_{s(\Gamma)}\rangle$ is a configuration. We have:

(14) $\widehat{v}\,(\Gamma \otimes \langle\Delta_1;\ldots;\Delta_{s(\Gamma)}\rangle) \stackrel{def}{=} (\cdots(\widehat{v}\,(A)\overline{\circ_1}\Delta_1)\cdots)\overline{\circ_{1+s(\Delta_1)+\cdots+s(\Delta_{s(A)})}\Delta_{s(A)}}\widehat{v}\,(\Delta_{s(\Gamma)})$
$\stackrel{\text{by Lemma 2}}{=} \overline{(\cdots(\widehat{v}\,(A)\circ_1\Delta_1)\cdots)\circ_{1+s(\Delta_1)+\cdots+s(\Delta_{s(A)})}\Delta_{s(A)}}\widehat{v}\,(\Delta_{s(A)})$

We abbreviate (14) as $\widehat{v}\,(\Gamma) \overline{\otimes} \langle\widehat{v}\,(\Delta_1);\ldots;\widehat{v}\,(\Delta_{s(\Gamma)})\rangle$ and by Lemma 2 as $\overline{\widehat{v}\,(\Gamma) \otimes \langle\widehat{v}\,(\Delta_1);\ldots;\widehat{v}\,(\Delta_{s(\Gamma)})\rangle}$. So $\widehat{v}\,(\Delta\langle\Gamma\rangle) = \widehat{v}\,(\Delta_0)\circ_k \overline{(\widehat{v}\,(\Gamma) \otimes \langle\widehat{v}\,(\Delta_1);\ldots;\widehat{v}\,(\Delta_{s(\Gamma)})\rangle)} = \widehat{v}\,(\Delta_0)\circ_k(\widehat{v}\,(\Gamma) \otimes \langle\widehat{v}\,(\Delta_1);\ldots;\widehat{v}\,(\Delta_{s(\Gamma)})\rangle)$, for a certain $k > 0$, and where the last equality is due to Lemma 2. We abbreviate $\widehat{v}\,(\Delta\langle\Gamma\rangle)$ as $\widehat{v}\,(\Delta)(\widehat{v}\,(\Gamma))$. By simple tonicity properties we have that if $\widehat{v}\,(\Gamma_1) \subseteq \widehat{v}\,(\Gamma_2)$ then $\widehat{v}\,(\Delta)(\widehat{v}\,(\Gamma)_1) \subseteq \widehat{v}\,(\Delta)(\widehat{v}\,(\Gamma)_2)$.

Theorem 1. $\mathbf{DA}?$ *is sound w.r.t. phase semantics.*

Proof. By induction on the derivation of $\mathbf{DA}?$ sequents. For reasons of space we omit the proof cases of the remaining multiplicative and additive connectives, and units, and we only prove a representative case of the discontinuous implicative extract connective, and the case of the iteration connective.

Case of $\uparrow_k L$ $k > 0$ (similar for the $\downarrow_k$ connective) we have:

$$(15) \quad \frac{\Gamma \Rightarrow A \qquad \Delta\langle \overrightarrow{B} \rangle \Rightarrow C}{\Delta\langle \overline{C \uparrow_k \overrightarrow{B}}|_k \Gamma \rangle \Rightarrow C} \uparrow_k L$$

By induction hypothesis (i.h.), $\hat{v}(\Gamma) \subseteq \hat{v}(A)$. We have $\hat{v}(\overrightarrow{B \uparrow_k \overrightarrow{A}}|_k \Gamma) = \hat{v}(\overrightarrow{B \uparrow_k \overrightarrow{A}}) \circ_k \hat{v}(\Gamma) \subseteq \hat{v}(B)$. Hence $\hat{v}(\Delta)(\hat{v}(\overrightarrow{B \uparrow_k \overrightarrow{A}}|_k \Gamma)) \subseteq \hat{v}(\Delta)(\hat{v}(\overrightarrow{B}) \subseteq \hat{v}(C)$, where the last equality follows from the i.h.

Let us see rule $?L$. By i.h. for every $i > 0$ $\hat{v}(\Delta\langle A^i \rangle) \subseteq \hat{v}(B)$. $\hat{v}(\Delta\langle A^i \rangle) = \overline{\hat{v}(\Delta) \circ_k \hat{v}(A)^i}$, for a certain $k > 0$. $\hat{v}(\Delta) \circ_k \hat{v}(A)^i \subseteq \overline{\hat{v}(\Delta) \circ_k \hat{v}(A)^i}$. Hence $\bigcup_{i>0} \hat{v}(\Delta) \circ_k \hat{v}(A)^i \subseteq \hat{v}(B)$. But $\bigcup_{i>0} \hat{v}(\Delta) \circ_k \hat{v}(A)^i = \hat{v}(\Delta) \circ_k \bigcup_{i>0} \hat{v}(A)^i$. Taking closure $\overline{\hat{v}(\Delta) \circ_k \bigcup_{i>0} \hat{v}(A)^i} \; =^{\text{lemma } 3} = \; \overline{\hat{v}(\Delta) \circ_k \overline{\bigcup_{i>0} \hat{v}(A)^i}} = \overline{\hat{v}(\Delta) \circ_k \hat{v}(?A)} = \hat{v}(\Delta(?A)) \subseteq \hat{v}(B)$.

Rule $?R$ soundness is due to the fact that by i.h. $\hat{v}(\Delta) = \hat{v}(A) \subseteq \bigcup_{i>0} \hat{v}(A)^i = \hat{v}(?A)$.

Finally, let us see the Mingle rule $?M$:

$$(16) \quad \frac{\Gamma_1 \Rightarrow A \qquad \Gamma_2 \Rightarrow ?A}{\Gamma_1, \Gamma_2 \Rightarrow ?A} \, ?M$$

By i.h $\hat{v}(\Gamma_1) \subseteq A$ and $\hat{v}(\Gamma_2) \subseteq \hat{v}(?A)$. $\hat{v}(\Gamma_1) \circ \hat{v}(\Gamma_2) \subseteq \hat{v}(A) \circ \bigcup_{i>0} v(A)^i \subseteq \bigcup_{i>0} v(A)^i$. Taking closure we obtain $\overline{\hat{v}(\Gamma_1) \circ \hat{v}(\Gamma_2)} \subseteq \overline{\bigcup_{i>0} v(A)^i} = \overline{\bigcup_{i>0} v(A^i)} = \hat{v}(?A)$. $\qquad \square$

Let us use the following notation:

(17) For any type A, $[A] \overset{def}{=} \{\Delta \in \mathcal{O} : \Delta \Rightarrow {}^- A\}$

where $\Rightarrow {}^-$ means provability without Cut

The strategy of the proof of strong completeness is to construct a canonical model which we call the syntactic phase space. Its underlying DA is the DA of configurations $\mathcal{O}$ with its operations of concatenation and intercalation, so that we define the phase space $(\mathbf{M}, \mathbf{cl})$ where $\mathbf{M} = (\mathcal{O}, conc, (interc_i)_{i>0}, \Lambda, 1)$. $\mathbf{cl}$ is the least ω-sorted closure system such that it is generated by the family $([D])_{D \in \mathbf{Tp}}$. The condition $(2.d)$ from Definition 1 is satisfied (by way of example we prove it only for one discontinuous implication): Let F be a closed set and Γ be a configuration. Let us see that $F \uparrow \uparrow_i \Gamma$ is a closed set. By definition there exists a same-sort family of types $\mathcal{G}$ such that $F = \bigcap_{D \in \mathcal{G}} [D]$. We have $\Delta \in F \uparrow \uparrow_i \Gamma$ iff $\Delta|_i \Gamma \in F$ iff for any $D \in \mathcal{G}$ $\Delta|_i \Gamma \in [D]$ iff $D \in \mathcal{G}$ $\Delta|_i \Gamma^\bullet \in [D]$ iff for any $D \in \mathcal{G}$ $\Delta \in [D \uparrow_i \Gamma^\bullet]$. Therefore since $F \uparrow \uparrow_i \Gamma$ is the intersection of a same-sort family of sets, it is a closed set.

Lemma 4. *Let v be the valuation $v : \mathbf{Pr} \to \mathbf{cl}$ such that $v(p) \overset{def}{=} [p]$ for any primitive type p. There holds:*

(18) $\overrightarrow{A} \in \hat{v}(A) \subseteq [A]$ *for any type A*

Proof. By induction on the structure of type A:

- If $A = p$ where p is a primitive type, we have by definition $v(A) = [A]$. Hence, $\vec{A} \in v(A) \subseteq [A]$.
- Case $A = J$ (the discontinuous unit). By the JR rule, $1 \in [J]$, i.e. $\{1\} \subseteq [J]$. Applying closure $\hat{v}(J) = \overline{\{1\}} \subseteq [J]$.
  On the other hand $\hat{v}(J) = \bigcap_{D \in \mathcal{G}}$ for a certain family $\mathcal{G}$. $1 \in \hat{v}(J)$, i.e., for every $D \in \mathcal{G}$, $1 \in [D]$. By JL rule, $\vec{J} \in [D]$. Therefore $\vec{J} \in \hat{v}(J)$.
- Suppose $A = B \odot_i C$. $v(B) \circ_i v(C) = \{\Gamma_B|_i \Gamma_C : \Gamma_B \in \hat{v}(B)$, and $\Gamma_C \in \hat{v}(C)\}$. By i.h. $v(B) \subseteq [B]$ and $\hat{v}(C) \subseteq [C]$. Hence, by application of $\odot_i L$ $\hat{v}(B) \circ_i \hat{v}(C) \subseteq [B \odot_i C]$. Hence, $\overline{\hat{v}(B) \circ_i \hat{v}(C)} \subseteq [B \odot_i C]$. This proves $\hat{v}(B \odot_i C) \subseteq [B \odot_i C]$. On the other hand, $\hat{v}(B \odot_i C) = \bigcap_{D \in \mathcal{G}}[D]$ for a certain $\mathcal{G}$. By i.h. $\vec{B} \in \hat{v}(B)$ and $\vec{C} \in \hat{v}(C)$. Hence $\vec{B}|_i\vec{C} \in \hat{v}(B) \circ_i \hat{v}(C) \subseteq \hat{v}(B \odot_i C)$. Then, for every $D \in \mathcal{G}$ $\vec{B}|_i\vec{C} \in [D]$. By application of $\odot_i L$, $\overrightarrow{B \odot_i C} \in [D]$. Hence, $\overrightarrow{B \odot_i C} \in \hat{v}(B \odot_i C)$.
- Suppose $A = C\uparrow_i B$. The case for the other implicative connectives is completely similar. Let $\Gamma \in v(C)\uparrow\uparrow_i v(B)$. By i.h., $\vec{B} \in v(B)$. We have $\Gamma|_i \vec{B} \Rightarrow v(C)$ and $v(C) \subseteq [C]$ by i.h. Hence, $\Gamma|_i \vec{B} \subseteq [C]$, and by application of $\uparrow_i R$, $\Gamma \in [C\uparrow_i B]$.
- $v(C) = \bigcap_{D \in \mathcal{G}}[D]$ for some $\mathcal{G}$. By i.h., $\vec{C} \in v(C)$. Applying $\uparrow_i L$, we get $\overrightarrow{C\uparrow_i B}|_i \Gamma_B \in [D]$ for all $\Gamma_B \in \hat{v}(B)$ (by i.h. $\hat{v}(B)[B]$). We have then that $\overrightarrow{C\uparrow_i B} \circ_i \hat{v}(B) \subseteq [D]$ for all $D \in \mathcal{G}$, whence $\overrightarrow{C\uparrow_i B} \circ_i \hat{v}(B) \subseteq \hat{v}(C)$. By applying residuation, $\overrightarrow{C\uparrow_i B} \in \hat{v}(C)\uparrow\uparrow_i \hat{v}(B) = \hat{v}(C\uparrow_i B)$.
- Case $A = B \oplus C$. By i.h. $v(B) \subseteq [B]$ and $v(C) \subseteq [C]$. Hence, $v(B) \cup v(C) \subseteq cl([B] \cup [C]) \subseteq [B \oplus C]$. The first inclusion is due to the monotony property and properties of cl. In fact, we have $[B] \cup [C] \subseteq [B \oplus C]$. For, $[B] \subseteq [B \oplus C]$ and $[C] \subseteq [B \oplus C]$ by $\oplus iR$ ($i = 1, 2$). It follows that $cl(v(B) \cup v(C)) \subseteq [B \oplus C]$.
- On the other hand, $v(B \oplus C) = \bigcap_{D \in \mathcal{G}}[D]$ for a certain $\mathcal{G}$. By i.h $\vec{B} \in v(B)$. Hence, $\vec{B} \subseteq cl(v(B) \cup v(C))$. Similarly, $\vec{C} \subseteq cl(v(B) \cup v(C))$. Therefore, for any $D \in \mathcal{G}$, $\vec{B} \in [D]$ and $\vec{C} \in [D]$. By $\oplus L$ we get $\overrightarrow{B \oplus C} \in [D]$. It follows that $\overrightarrow{B \oplus C} \subseteq v(B \oplus C)$.
- Case $C = ?A$

$$(19) \quad \cfrac{\Gamma_{i-1} \Rightarrow A \quad \cfrac{\cfrac{\Gamma_i \Rightarrow A}{\Gamma_i \Rightarrow ?A}\,?R}{\vdots}\,?M}{\cfrac{\Gamma_1 \Rightarrow A \quad \Gamma_2, \ldots, \Gamma_i \Rightarrow ?A}{\Gamma_1, \ldots, \Gamma_i \Rightarrow ?A}\,?M}\,?M$$

The proof above shows that for every $i > 0$ $\hat{v}(A)^i \subseteq [?A]$. We have then $\bigcup_{i>0} \hat{v}(A)^i \subseteq [?A]$. Applying the closure map we get $\overline{\bigcup_{i>0} \hat{v}(A)^i} \subseteq [?A]$, whence $\hat{v}(?A) \subseteq [?A]$.

We prove now $?A \in \hat{v}(?A)$. We know that $\hat{v}(?A) = \bigcap_{D \in \mathcal{G}}[D]$, for a certain family of closed sets $\mathcal{G}$. By i.h. $A \in \hat{v}(A)$. It follows that for every $i > 0$ $A^i \in \hat{v}(A^i)$, whence $A^i \in \bigcup_{k>0} \hat{v}(A^i) \subseteq \hat{v}(?A)$. We have therefore:

For every $i > 0$ $A^i \in \widehat{v}\,(?A)$ iff For every $i > 0$, and for every $D \in \mathcal{G}, A^i \in [D]$
iff For every $D \in \mathcal{G}, ?A \in [D]$, by application of $?R$
iff $?A \in \widehat{v}\,(?A)$

$\square$

Theorem 2 (Strong Completeness à la Okada). *Let $\Delta \Rightarrow A$ be such that for every $(\mathbf{P}, v)$, $(\mathbf{P}, v) \models \Delta \Rightarrow B$. It follows that $\Delta \Rightarrow {}^-B$.*

Proof. In particular, this sequent holds in the syntactic phase displacement model. By the previous lemma, for any A, $\overrightarrow{A} \in \widehat{v}\,(A)$. Hence $\Delta \in \widehat{v}\,(\Delta)$. By soundness, for every $(\mathbf{P}, w)$ $\widehat{w}\,(\Delta) \subseteq \widehat{w}\,(B)$. Therefore we have that $\widehat{v}\,(\Delta) \subseteq \widehat{v}\,(B)$. Since $\Delta \in \widehat{v}\,(\Delta)$, $\Delta \in \widehat{v}\,(A)$, which entails (by the truth lemma) that $\Delta \in [A]$, i.e. $\Delta \Rightarrow {}^-A$. $\square$

By the previous theorem $\Delta \Rightarrow A$ is provable without Cut, whence:

Corollary 1 (Cut admissibility). *The Cut rule is admissible.* $\square$

4 CatLog2 Analyses

In Fig. 5 we give a mini-lexicon for a fragment. The heart of the analysis of iterated coordination is the assignment to a coordinator of types of the form $(?A\backslash A)/A$. For a *respectively* construction we employ in conjunction with displacement connectives a calibrated version $?_n$ of the Mingle exponential as follows, with list Curry-Howard labelling:

$$\frac{\Delta(A_1\!:\!x_1, \ldots, A_n\!:\!x_n) \Rightarrow B\!:\!\psi([x_1, \ldots, x_n])}{\Delta(?_n A\!:\!z) \Rightarrow B\!:\!\psi(z)} \ ?_n L$$

$$\frac{\Gamma \Rightarrow A\!:\!\phi}{\Gamma \Rightarrow ?_1 A\!:\![\phi]} \ ?_n R \qquad \frac{\Gamma \Rightarrow A\!:\!\phi \quad \Delta \Rightarrow ?_n A\!:\!\phi'}{\Gamma, \Delta \Rightarrow [\phi|\phi']\!:\!?_{n+1} A} \ ?_n M$$

A crucial aspect of what makes the respectively construction work here is the information sharing between two $?_A$ connectives in the type assignment to *respectively*—an implicit quantification over the natural A in the type: i.e. a kind of dependent type.

The output of a version of CatLog2 for some examples is as follows:

4.1 Iterated Coordination

To express the lexical semantics of (iterated) coordination, including iterated coordination and various arities (zeroary e.g. sentence, unary e.g. verb phrase, binary e.g. transtive verb, . . .), we use combinators: a non-empty list map apply α^+, a non-empty list list apply β^+, and a non-empty list map $\mathbf{\Phi^n}$ combinator $\mathbf{\Phi^{n+}}$.[8]

---

[8] For the list map apply cf. Schiehlen [16]. The combinator $\mathbf{\Phi}$ is such that $\mathbf{\Phi}\,x\,y\,z\,w = x\,(y\,w)\,(z\,w)$ (Curry and Feys [3]).

and : $(?Sf\backslash Sf)/Sf : (\Phi^{n+} \, 0 \, and)$
and : $(?(Sf/NA)\backslash(Sf/NA))/(Sf/NA) : (\Phi^{n+} \, (s \, 0) \, and)$
and : $(?(Sf/!NA)\backslash(Sf/!NA))/(Sf/!NA) : (\Phi^{n+} \, (s \, 0) \, and)$
and : $(?(NA\backslash Sf)\backslash(NA\backslash Sf))/(NA\backslash Sf) : (\Phi^{n+} \, (s \, 0) \, and)$
and : $(?((NA\backslash Sf)/NB)\backslash((NA\backslash Sf)/NB))/((NA\backslash Sf)/NB) : (\Phi^{n+} \, (s \, (s \, 0)) \, and)$
and : $(?((NA\backslash Sf)/!NB)\backslash((NA\backslash Sf)/!NB))/((NA\backslash Sf)/!NB) : (\Phi^{n+} \, (s \, (s \, 0)) \, and)$
and+1+and+1+respectively $:?_A NB\backslash((SC\uparrow(ND\backslash SC))\uparrow(NE\bullet?_A(NF\backslash SC))) :$
$\lambda G\lambda H\lambda I(((\Phi^{n+} \, 0 \, and) \, (I \, \pi_1 H)) \, (\beta^+ \, \pi_2 H \, G))$
Bill : $Nt(s(m)) : b$
danced : $NA\backslash Sf : \lambda B(Past \, (dance \, B))$
John : $Nt(s(m)) : j$
Mary : $Nt(s(f)) : m$
laughed : $NA\backslash Sf : \lambda B(Past \, (laugh \, B))$
likes : $(Nt(s(A))\backslash Sf)/NB : like$
London : $\blacksquare Nt(s(n)) : l$
love : $(NA\backslash Sb)/NB : love$
praised : $(NA\backslash Sf)/NB : \lambda C\lambda D(Past \, ((praise \, C) \, D))$
sang : $NA\backslash Sf : \lambda B(Past \, (sing \, B))$
sings : $Nt(s(A))\backslash Sf : sing$
talks : $Nt(s(A))\backslash Sf : talk$
walks : $Nt(s(A))\backslash Sf : walk$
will : $(NA\backslash Sf)/(NA\backslash Sb) : \lambda B\lambda C(Fut \, (B \, C))$

Fig. 5. Lexicon

The non-empty list map apply combinator α^+ is as follows:

(20) $(\alpha^+ \, [x] \, y) = [(x \, y)]$
    $(\alpha^+ \, [x,y|z] \, w) = [(x \, w)|(\alpha^+ \, [y|z] \, w)]$

The non-empty list list apply combinator α^+ is as follows:

(21) $(\alpha^+ \, [x] \, [y]) = [(x \, y)]$
    $(\alpha^+ \, [x|y] \, [z|w]) = [(x \, z)|(\alpha^+ \, y \, w)]$

The non-empty list map Φ^n combinator Φ^{n+} is thus:

(22) $(((\Phi^{n+} \, 0 \, and) \, x) \, [y]) = [y \wedge x]$
    $(((\Phi^{n+} \, 0 \, and) \, x) \, [y,z|w]) = [y \wedge (((\Phi^{n+} \, 0 \, and) \, x) \, [z|w])]$
    $((((\Phi^{n+} \, (s \, n) \, c) \, x) \, y) \, z) = (((\Phi^{n+} \, n \, c) \, (x \, z)) \, (\alpha^+ \, y \, z))$

These equations mean that in semantic evaluation any subterm of the form on the left is to be replaced by that on the right, successively.

The first example is of iterated sentence coordination:

(23) **John+walks+Mary+talks+and | Bill+sings** : Sf

Lexical lookup yields the following annotated sequent:

$Nt(s(m)) : j, Nt(s(A))\backslash Sf : walk, Nt(s(f)) : m, Nt(s(B))\backslash Sf : talk, (?Sf\backslash Sf)/$
$Sf : (\Phi^{n+}\ 0\ and), Nt(s(m)) : b, Nt(s(C))\backslash Sf : sing \Rightarrow Sf$

The derivation is given in Fig. 6. This delivers semantics:

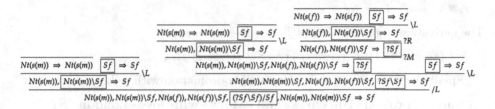

Fig. 6. Derivation of *John walks, Mary talks, and Bill sings*

$[(walk\ j) \wedge [(talk\ m) \wedge (sing\ b)]]$

The second example is of iterated verb phrase coordination:

(24) **John+walks+talks+and+sings** : Sf

Lexical lookup yields:

$Nt(s(m)) : j, Nt(s(A))\backslash Sf : walk, Nt(s(B))\backslash Sf : talk,$
$(?(NC\backslash Sf)\backslash(NC\backslash Sf))/(NC\backslash Sf) : (\Phi^{n+}\ (s\ 0)\ and), Nt(s(D))\backslash Sf : sing \Rightarrow$
Sf

The derivation is given in Fig. 7. This delivers semantics:
$[(walk\ j) \wedge [(talk\ j) \wedge (sing\ j)]]$

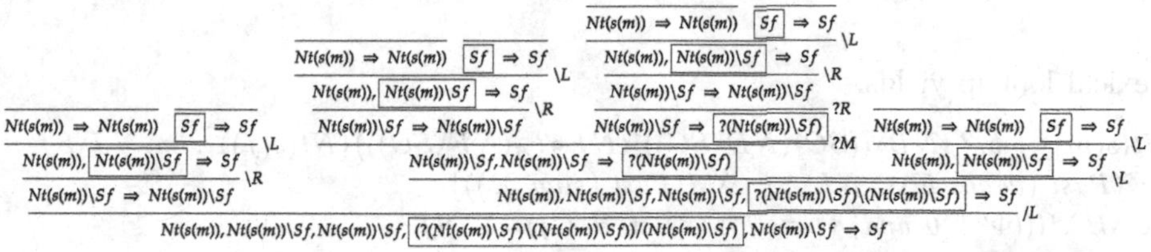

Fig. 7. Derivation of *John walks, talks, and sings*

The next example is of iterated transitive verb coordination, with a non-standard constituent in the right hand conjunct:

(25) **John+praised+likes+and+will+love+London** : Sf

Lexical lookup yields:

$Nt(s(m))$: $j, (NA\backslash Sf)/NB$: $\lambda C\lambda D(Past\ ((praise\ C)D)), (Nt(s(E))\backslash Sf)/NF$: $like, (?((NG\backslash Sf)/NH)\backslash ((NG\backslash Sf)/NH))/((NG\backslash Sf)/NH)$: $(\Phi^{n+}\ (s\ (s\ 0))\ and), (NI\backslash Sf)/(NI\backslash Sb)$: $\lambda J\lambda K(Fut\ (J\ K)), (NL\backslash Sb)/NM$: $love, \blacksquare Nt(s(n)) : l\ \Rightarrow\ Sf$

The derivation is given in Fig. 8. This delivers semantics:

$[(Past\ ((praise\ l)\ j)) \wedge [(((like\ l)\ j) \wedge (Fut\ ((love\ l)\ j)))]]$

Finally we have an example of iterated coordination with right node raising:

(26) **John+praised+Bill+likes+and+Mary+will+love+London** : Sf

Lexical lookup yields:

$Nt(s(m))$: $j, (NA\backslash Sf)/NB$: $\lambda C\lambda D(Past\ ((praise\ C)\ D)), Nt(s(m))$: $b, (Nt\ (s(E))\backslash Sf)/NF : like, (?(Sf/NG)\backslash (Sf/NG))/(Sf/NG)$: $(\Phi^{n+}\ (s\ 0)\ and), Nt(s(f))$: $m, (NH\backslash Sf)/(NH\backslash Sb)$: $\lambda I\lambda J(Fut\ (I\ J)), (NK\backslash Sb)/NL$: $love, \blacksquare Nt(s(n)) : l\ \Rightarrow\ Sf$

There is the derivation in Fig. 9. This delivers semantics:

$[(Past\ ((praise\ l)\ j)) \wedge [(((like\ l)\ b) \wedge (Fut\ ((love\ l)\ m)))]]$

4.2   The *Respectively* Construction

Kubota and Levine [6] provide a type logical account of binary *respectively* constructions using empty operators. By contrast we account here for unbounded addicity *respectively* constructions, without empty operators.

Our first example synchronises parallel pairs of items:

(27) **Bill+and+Mary+danced+and+sang+respectively** : Sf

Lexical lookup yields:

$Nt(s(m))$: $b, ?_A NB\backslash ((SC^\uparrow (ND\backslash SC))^\uparrow (NE\bullet ?_A (NF\backslash SC)))\{Nt(s(f))$: $m, NJ\backslash Sf$: $\lambda K(Past\ (dance\ K)) : NL\backslash Sf : \lambda M(Past\ (sing\ M))\}$: $\lambda G\lambda H\lambda I(((\Phi^{n+}\ 0\ and)\ (I\ \pi_1 H))\ (\beta^+\ \pi_2 H\ G)) \Rightarrow Sf$

There is the derivation given in Fig. 10. This delivers semantics:

$[(Past\ (dance\ b)) \wedge (Past\ (sing\ m))]$

Our other example of the *respectively* construction synchronises parallel triples of items:

Fig. 8. Derivation for *John praised, likes, and will love, London*

244   G. Morrill and O. Valentín

Fig. 9. Derivation for *John praised, Bill likes, and Mary will love, London*

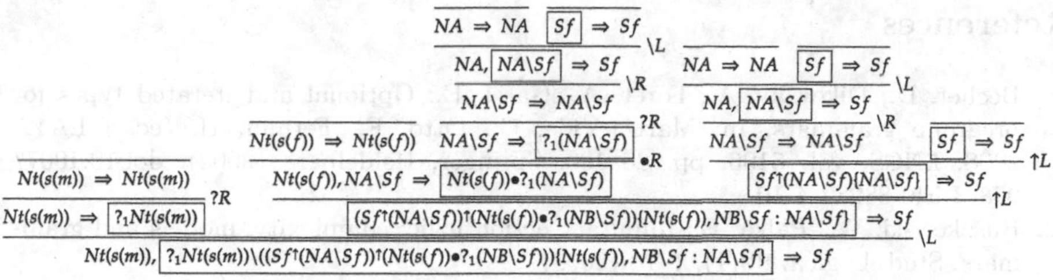

$$\cfrac{\cfrac{\cfrac{NA \Rightarrow NA \quad \boxed{Sf} \Rightarrow Sf}{NA, \boxed{NA\backslash Sf} \Rightarrow Sf} \backslash L}{NA\backslash Sf \Rightarrow NA\backslash Sf} \backslash R}{\dots}$$

Fig. 10. Derivation for *Bill and Mary danced and sang respectively*

(28) **John+Bill+and+Mary+laughed+danced+and+sang+ respectively** $: Sf$

Lexical lookup yields the following:

$Nt(s(m)) : j, Nt(s(m)) : b, ?_A NB\backslash((SC^\uparrow(ND\backslash SC))^\uparrow(NE\bullet?_A(NF\backslash SC)))\{Nt(s(f)) : m, NJ\backslash Sf : \lambda K(Past\ (laugh\ K)), NL\backslash Sf : \lambda M(Past\ (dance\ M)) : NN\backslash Sf : \lambda O(Past\ (sing\ O))\} : \lambda G\lambda H\lambda I(((\Phi^{n+}\ 0\ and)\ (I\ \pi_1 H))\ (\beta^+\ \pi_2 H\ G)) \Rightarrow Sf$

There is the derivation given in Fig. 11. This delivers semantics:

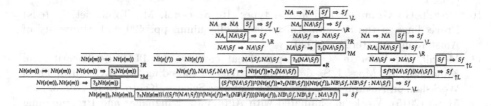

Fig. 11. Derivation for *John, Bill, and Mary laughed, danced, and sang, respectively*

$$[(Past\ (laugh\ j)) \wedge [(Past\ (dance\ b)) \wedge (Past\ (sing\ m))]]$$

Interestingly, our account syntactically blocks examples of the kind *John and Peter walk, talk, and sing, respectively* since the calibrated numbers of occurrences are not equal. A variation of our account with uncalibrated modalities would need to appeal to a semantic anomaly in relation to the combinators.

Acknowlegements. Research partially supported by an ICREA Acadèmia 2012 to the alphabetically first author, and SGR2014-890 (MACDA) of the Generalitat de Catalunya and MINECO project APCOM (TIN2014-57226-P). We thank anonymous LACL referees for valuable comments and suggestions. All errors are our own.

References

1. Béchet, D., Dikovsky, A., Foret, A., Garel, E.: Optional and iterated types for pregroup grammars. In: Martín-Vide, C., Otto, F., Fernau, H. (eds.) LATA 2008. LNCS, vol. 5196, pp. 88–100. Springer, Heidelberg (2008). doi:10.1007/978-3-540-88282-4_10
2. Buszkowski, W., Palka, E.: Infinitary action logic: complexity, models and grammars. Stud. Logica. **89**(1), 1–18 (2008)
3. Curry, H.B., Feys, R.: Combinatory Logic, vol. I. North-Holland, Amsterdam (1958)
4. Girard, J.-Y.: Linear logic. Theor. Comput. Sci. **50**, 1–102 (1987)
5. Kamide, N.: Substructural logics with mingle. J. Logic Lang. Inf. **11**(2), 227–249 (2002)
6. Kubota, Y., Levine, B.: The syntax-semantics interface of respective predication: a unified analysis in hybrid type-logical categorial grammar. Nat. Lang. Linguist. Theor. **34**(3), 911–973 (2016)
7. Lambek, J.: On the calculus of syntactic types. In: Jakobson, R. (ed.) Structure of Language and its Mathematical Aspects. Proceedings of the Symposia in Applied Mathematics XII, pp. 166–178. American Mathematical Society, Providence (1961)
8. Lambek, J.: Categorial and categorical grammars. In: Oehrle, R.T., Bach, E., Wheeler, D. (eds.) Categorial Grammars and Natural Language Structures. Studies in Linguistics and Philosophy, vol. 32, pp. 297–317. D. Reidel, Dordrecht (1988)
9. Lambek, J.: The mathematics of sentence structure. Am. Math. Mon. **65**, 154–170 (1958)
10. Morrill, G.: Grammar and logical types. In: Stockhof, M., Torenvliet, L. (eds) Proceedings of the Seventh Amsterdam Colloquium, pp. 429–450. University of Amsterdam, Amsterdam (1990)
11. Morrill, G., Valentín, O.: Computational coverage of TLG: nonlinearity. In: Kanazawa, M., Moss, L.S., de Paiva, V., (eds.) Proceedings of NLCS 2015, Third Workshop on Natural Language and Computer Science, Kyoto. EPiC, vol. 32, pp. 51–63 (2015). Workshop affiliated with Automata, Languages and Programming (ICALP) and Logic in Computer Science (LICS)
12. Morrill, G., Valentín, O., Fadda, M.: The displacement calculus. J. Logic Lang. Inf. **20**(1), 1–48 (2011)
13. Morrill, G.V.: Type Logical Grammar: Categorial Logic of Signs. Kluwer Academic Publishers, Dordrecht (1994)
14. Okada, M.: Phase semantic Cut-elimination and normalization proofs of first- and higher-order linear logic. Theor. Comput. Sci. **227**(1–2), 333–396 (1999)
15. Palka, E.: An infinitary sequent system for the equational theory of *-continuous action lattices. Fundam. Inf. **78**(2), 295–309 (2007)
16. Schiehlen, M.: The role of lists in a categorial analysis of coordination. In: Dekker, P., Franke, M (eds.) Proceedings of the 15th Amsterdam Colloquium, pp. 221–226 (2005)
17. Valentín, O.: Theory of discontinuous Lambek calculus. Ph.D. thesis, Universitat Autònoma de Barcelona, Barcelona (2012)
18. Valentín, O.: Models for the displacement calculus. In: Foret, A., Morrill, G., Muskens, R., Osswald, R., Pogodalla, S. (eds.) FG 2015-2016. LNCS, vol. 9804, pp. 147–163. Springer, Heidelberg (2016). doi:10.1007/978-3-662-53042-9_9

Context Update for Lambdas and Vectors

Reinhard Muskens[1(✉)] and Mehrnoosh Sadrzadeh[2]

[1] Department of Philosophy, Tilburg University, Tilburg, The Netherlands
r.a.muskens@gmail.com
[2] School of Electronic Engineering and Computer Science,
Queen Mary University of London, London, UK
mehrnoosh.sadrzadeh@qmul.ac.uk

Abstract. Vector models of language are based on the contextual aspects of words and how they co-occur in text. Truth conditional models focus on the logical aspects of language, the denotations of phrases, and their compositional properties. In the latter approach the denotation of a sentence determines its truth conditions and can be taken to be a truth value, a set of possible worlds, a context change potential, or similar. In this short paper, we develop a vector semantics for language based on the simply typed lambda calculus. Our semantics uses techniques familiar from the truth conditional tradition and is based on a form of dynamic interpretation inspired by Heim's context updates.

Keywords: Vector semantics · Simply typed lambda calculus · Context update · Context change potential · Compositionality

1 Introduction

Vector semantic models, otherwise known as distributional models, are based on the contextual aspects of language, the company each word keeps, and patterns of use in corpora of documents. Truth conditional models focus on the logical and denotational aspects of language, sets of objects with certain properties and application and composition of functions. Vector semantics and truth conditional models are based on different philosophies; in recent years there has been much effort to bring them together under one umbrella, see for example [1–3,8,9].

In a recent abstract [14], we sketched an approach to semantics that assigned vector meanings to linguistic phrases using a simply typed lambda calculus in the tradition of [10]. Our previous system was guided by a truth conditional interpretation and provided vector semantics very similar to the approaches of [1–3,8,9]. The difference was that the starting points of these latter approaches are categorial logics such as Pregroup Grammars and Combinatorial Categorial Grammar (CCG). Our reasoning for the use of lambda calculus was that it directly relates our semantics to higher order logic and makes standard ways of

Support by EPSRC for Career Acceleration Fellowship EP/J002607/1 is gratefully acknowledged by M. Sadrzadeh.

© Springer-Verlag GmbH Germany 2016
M. Amblard et al. (Eds.): LACL 2016, LNCS 10054, pp. 247–254, 2016.
DOI: 10.1007/978-3-662-53826-5_15

treating long distance dependencies and coordination accessible to vector-based semantics. In this short account, we follow the same lines as in our previous work. But whereas in previous work we worked with a static interpretation of distributions, here, we focus on a dynamic interpretation.

The lambda calculus approach we use is based on the Lambda Grammars of [11,12], which were independently introduced as Abstract Categorial Grammars (ACGs) in [5]. The theory developed here, however, can be based on any syntax-semantics interface that works with a lambda calculus based semantics. Our approach is agnostic as to the choice of a syntactic theory. Lambda Grammars/ACGs are just a framework for thinking about type and term homomorphisms and we are using them entirely in semantics here. In a longer paper we will show in more detail how lambda logical forms (the abstract terms) can be obtained: (1) from standard linguistic trees with the help of a procedure that is essentially that of Heim and Kratzer [7]; (2) from LFG f-structures by means of a 'glue logic'; (3) from Lambek proofs by means of semantic recipes; (4) and from CCG derivations by means of using the combinators associated with CCG rules.

The dynamic interpretation we work with here is the "context change potential" of [6]. We believe other dynamic approaches, such the update semantics of [16] and the continuation-based semantics of [4], can also be used; we aim to do these in future.

2   Heim's Files and Distributional Contexts

Heim describes her contexts as files that have some kind of information written on (or in) them. Context changes are operations that update these files, e.g. by adding or deleting information from the files. Formally, a context is taken to be a set of sequence-world pairs, in which the sequences come from some domain $\mathcal{D}_I$ of individuals, as follows:

$$ctx \subseteq \{(g, w) \mid g \colon \mathbb{N} \to \mathcal{D}_I, w \text{ a possible world}\}$$

(We follow Heim [6] here in letting the sequences in her sequence-world-pairs be infinite, although they are best thought of as finite.)

Sentence meanings are *context change potentials* (CCPs) in Heim's work, functions from contexts to contexts. A sentence S comes provided with a sequence of instructions that, given any context ctx, updates its information so that a new context denoted as

$$ctx + S$$

results. The sequence of instructions that brings about this update is derived compositionally from the constituents of S.

In distributional semantics, contexts are words somehow related to each other via their patterns of use, e.g. by co-occurring in a neighbourhood word window of a fixed size or via a dependency relation. In practice, one builds a context matrix M over $\mathbb{R}^2$, with rows and columns labeled by words from a vocabulary Σ and with entries taking values from $\mathbb{R}$, for a full description see [15]. M can be seen as the set of its vectors:

$$\{\overrightarrow{v} \mid \overrightarrow{v} : \Sigma \to \mathbb{R}\}$$

where each $\overrightarrow{v}$ is a row or column in M.

If we take Heim's domain of individuals $\mathcal{D}_I$ be the vocabulary of a distributional model of meaning, that is $\mathcal{D}_I := \Sigma$, then a context matrix can be seen as a so-called *quantized* version of a Heim context:

$$\{(\overrightarrow{g}, w) \mid \overrightarrow{g} : \Sigma \to \mathbb{R}, w \text{ a possible world}\}$$

Thus a distributional context matrix is obtainable by endowing Heim's contexts with $\mathbb{R}$. In other words, we are assuming that not only a file has a set of individuals, but also that these individuals take some kind of values, e.g. from reals.

The role of possible worlds in a distributional semantics is arguable, as vectors retrieved from a corpus are not naturally truth conditional. Keeping the possible worlds in the picture provides a machinery to assign a proposition to a distributional vector by other means and can become very useful. But for the rest of this abstract, we shall deprive ourselves from this advantage and only work with the following set as our context:

$$\{\overrightarrow{g} \mid \overrightarrow{g} : \Sigma \to \mathbb{R}, \overrightarrow{g} \in M\}$$

Distributional versions of Heim's CCP's can be defined based on the intuitions and definitions of Heim. In what follows we pan out how these instructions let contexts thread through vectorial semantics in a compositional manner.

3   Vectors, Matrices, Lambdas

Lambda Grammars of [11,12] were independently introduced as Abstract Categorial Grammars (ACGs) in [5]. An ACG generates two languages, an *abstract* language and an *object* language. The abstract language will simply consist of all linear lambda terms (each lambda binder binds exactly one variable occurrence) over a given vocabulary typed with *abstract types*. The object language has its own vocabulary and its own types. It results from (1) specifying a *type homomorphism* from abstract types to object types and (2) specifying a *term homomorphism* from abstract terms to object terms. The term homomorphism must respect the type homomorphism. For more information about the procedure of obtaining an object language from an abstract language, see the papers mentioned or the explanation in [13].

Let the basic abstract types of our setting be D (for determiner phrases), S (for sentences), and N (for nominal phrases). Let the basic object types be I and R. The domain $\mathcal{D}_I$ corresponding to I can be thought of as a vocabulary, $\mathcal{D}_R$ models the set of reals $\mathbb{R}$. The usual operations on $\mathbb{R}$ can be defined using

Tarski's axioms (in full models that satisfy these axioms $\mathcal{D}_R = \mathbb{R}$ will hold; in generalised models we get what boils down to a first-order approximation of $\mathbb{R}$). Objects of type $I \to R$ are abbreviated to IR; these are identified with *vectors* with a fixed basis.

We will associate simple words like names, nouns and verbs with vectors, i.e. with objects of type IR and will denote these with constants like $\overrightarrow{\text{woman}}$, $\overrightarrow{\text{smoke}}$, etc. The typed lambda calculus will be used to build certain functions with the help of these vectors that will then function as the meanings of those words. The meanings of content words will typically be functions that are completely given by some vector, but they will not (necessarily) be identified with vectors (see also Table 1 below).

Sentences will be *context change potentials*. A context for us is a matrix, thus it has type I^2R. A sentence takes the type $(I^2R)(I^2R)$. We abbreviate IR as V, I^2R as M and the sentence type MM as U (for 'update'). Verbs take a vector for each of their arguments, plus an input context, and return a context as their output. For instance, an intransitive verb takes a vector for its subject plus a context and returns a modified context. Thus it takes type $VMM = VU$. A transitive verb takes a vector for its subject, a vector for its object and a context and returns a context. Thus it has type VVU. Nouns are essentially treated as vectors (V), but, since they must be made capable of dynamic behaviour, they are 'lifted' to the higher type $(VU)U$. Our dynamic type homomorphism ρ is defined by letting $\rho(N) = (VU)U$, $\rho(D) = V$ and $\rho(S) = U$. Some consequences of this definition can be found in Table 1.

Table 1. Some abstract constants a typed with abstract types τ and their term homomorphic images $H(a)$ typed by $\rho(\tau)$ (where ρ is a type homomorphism, i.e. $\rho(AB) = \rho(A)\rho(B)$). Here Z is a variable of type VU, Q is of type $(VU)U$, v of type V, c of type M, and p and q are of type U. The functions F, G, I, and J are explained in the main text. In the schematic entry for *and*, we write $\rho(\overline{\alpha})$ for $\rho(\alpha_1) \cdots \rho(\alpha_n)$, if $\overline{\alpha} = \alpha_1 \cdots \alpha_n$.

a	τ	$H(a)$	$\rho(\tau)$
Anna	$(DS)S$	$\lambda Z.Z\overrightarrow{\text{anna}}$	$(VU)U$
woman	N	$\lambda Z.Z\overrightarrow{\text{woman}}$	$(VU)U$
tall	NN	$\lambda QZ.Q(\lambda vc.ZvF(\overrightarrow{\text{tall}}, v, c))$	$((VU)U)(VU)U$
smokes	DS	$\lambda vc.G(\overrightarrow{\text{smoke}}, v, c)$	VU
loves	DDS	$\lambda uvc.I(\overrightarrow{\text{love}}, u, v, c)$	VVU
knows	SDS	$\lambda pvc.pJ(\overrightarrow{\text{know}}, v, c)$	UVU
every	$N(DS)S$	$\lambda Q.Q$	$((VU)U)(VU)U$
who	$(DS)NN$	$\lambda Z'QZ.Q(\lambda vc.Zv(QZ'c))$	$(VU)((VU)U)(VU)U$
and	$(\overline{\alpha}S)(\overline{\alpha}S)(\overline{\alpha}S)$	$\lambda R'\lambda R\lambda \overline{X}\lambda c.R'\overline{X}(R\overline{X}c)$	$(\rho(\overline{\alpha})U)(\rho(\overline{\alpha})U)(\rho(\overline{\alpha})U)$

4 Context Update for Lambda Binders

Object terms corresponding to a content word a may update a context matrix c with the information in $\overrightarrow{a}$ and the information in the vectors of arguments of a. The result is a new context matrix c', with different value entries.

$$
\begin{pmatrix}
m_{11} & \cdots & m_{1k} \\
m_{21} & \cdots & m_{2k} \\
\vdots & & \\
m_{n1} & \cdots & m_{nk}
\end{pmatrix}
+ \overrightarrow{a}, u, v, \cdots =
\begin{pmatrix}
m'_{11} & \cdots & m'_{1k} \\
m'_{21} & \cdots & m'_{2k} \\
\vdots & & \\
m'_{n1} & \cdots & m'_{nk}
\end{pmatrix}
$$

An example of a set of elementary update instructions may be as follows.

- The function denoted by $\lambda vc.G(\overrightarrow{\mathsf{smoke}}, v, c)$ increases the value entry of m_{ij} of c, for i and j indices of smoke and its subject v.
- The function denoted by $\lambda uv.\lambda c.I(\overrightarrow{\mathsf{love}}, u, v, c)$ increases the value entries of m_{ij}, m_{jk}, and m_{ik} of c, for i, j, k indices of loves, its subject u and its object v.
- The function denoted by $\lambda vc.F(\overrightarrow{\mathsf{tall}}, v, c)$ increases the value entry of m_{ij} of c, for i and j indices of tall and its modified noun v. The entry for $tall$ in Table 1 uses this function, but allows for further update of context.
- The function denoted by $\lambda vc.J(\overrightarrow{\mathsf{know}}, v, c)$ increases the value entry of m_{ij} of c, for i and j indices of know and its subject v. The updated matrix is made the input for further update (by the context change potential of the sentence that is known) in Table 1.

Logical words such as $every$ and and are often treated as noise in distributional semantics and not included in the context matrix. We have partly followed this approach here by treating $every$ as the identity function (the noun already has the required 'quantifier' type $(VU)U$). To see this, note that the entry for 'every', $\lambda Q.Q$, is the identity function; it takes a Q and then spits it out again. The alternative would be to have an entry along the lines of that of 'tall', but this would not make a lot of sense. It is the content words that seem to be important in a distributional setting, not the function words.

The word and does have a function here though—it is treated as a generalised form of function composition. The entry for the word in Table 1 is schematic, as and does not only conjoin sentences, but also other phrases of any category. So, the type of the abstract constant connected with the word is $(\overline{\alpha}S)(\overline{\alpha}S)(\overline{\alpha}S)$, in which $\overline{\alpha}$ can be any sequence of abstract types. Ignoring this generalisation for the moment, we obtain SSS as the abstract type for sentence conjunction, with a corresponding object type UUU, and meaning $\lambda pqc.p(qc)$, which is just function composition. This is defined in a way such that the context updated by and's left argument will be further updated by its right argument. So 'Sally smokes and John eats bananas' will, given an initial matrix c, first update c to $G(\mathsf{Sally}, \mathsf{smoke}, c)$, which is a matrix, and then update this further with 'John eats bananas' to $I(\mathsf{eat}, \mathsf{John}, \mathsf{bananas}, G(\mathsf{smoke}, \mathsf{Sally}, c))$.

This treatment is easily extended to coordination in all categories. For example, the reader may check that and admires loves (which corresponds to *loves and admires*) has $\lambda uvc.I(\overrightarrow{\text{admire}}, u, v, I(\overrightarrow{\text{love}}, u, v, c))$ as its homomorphic image.

The update instructions fall through the semantics of phrases and sentences compositionally. The sentence *every tall woman smokes*, for example, will be associated with the following lambda expression:

$$(\text{every tall woman})\lambda\zeta.(\text{smokes } \zeta)$$

This in its turn has a term homomorphic image that is β-equivalent with the following:

$$\lambda c.G\left(\overrightarrow{\text{smoke}}, \overrightarrow{\text{woman}}, F(\overrightarrow{\text{tall}}, \overrightarrow{\text{woman}}, c)\right)$$

which describes a distributional context update for it. This term describes a first update of the context c according to the rule for the constant tall, and then a second update according to the rule for the constant smokes. As a result of these, the value entries at the crossings of ⟨tall, woman⟩ and ⟨woman, smokes⟩ get increased. Much longer chains of context updates can be 'threaded' in this way.

In the following we give some examples. In each case the a. sentence is followed by an abstract term in b. which captures its syntactic structure. The update potential that follows in c. is the homomorphic image of this abstract term.

(1) a. Sue loves and admires a stockbroker
     b. (a stockbroker)$\lambda\xi$.Sue(and admires loves ξ)
     c. $\lambda c.I(\overrightarrow{\text{admire}}, \overrightarrow{\text{stockbroker}}, \overrightarrow{\text{sue}}, I(\overrightarrow{\text{love}}, \overrightarrow{\text{stockbroker}}, \overrightarrow{\text{sue}}, c))$

(2) a. Bill admires but Anna despises every cop
     b. (every cop)$\lambda\xi$.and(Anna(despise ξ))(Bill(admire ξ))
     c. $\lambda c.I(\overrightarrow{\text{despise}}, \overrightarrow{\text{cop}}, \overrightarrow{\text{anna}}, I(\overrightarrow{\text{admire}}, \overrightarrow{\text{cop}}, \overrightarrow{\text{bill}}, c))$

(3) a. The witch who Bill claims Anna saw disappeared
     b. the(who($\lambda\xi$.Bill(claims(Anna(saw ξ)))))witch)disappears
     c. $\lambda c.G(\overrightarrow{\text{disappear}}, \overrightarrow{\text{witch}}, I(\overrightarrow{\text{see}}, \overrightarrow{\text{witch}}, \overrightarrow{\text{anna}}, J(\overrightarrow{\text{claim}}, \overrightarrow{\text{bill}}, c)))$

5   Conclusion and Future Directions

In previous work, we showed how a static interpretation of the lambdas will provide vectors for phrases and sentences of language. There, the object type of the vector of a word depended on its abstract type and could be an atomic vector, a matrix, or a cube, or a tensor of higher rank. Means of combinations thereof then varied based on the tensor rank of the type of each word. For instance one could take the matrix multiplication of the matrix of an intransitive verb with the vector of its subject, whereas for a transitive verb the sequence of operations were a contraction between the cube of the verb and the vector of its object followed by a matrix multiplication between the resulting matrix and the vector of the subject. A toolkit of functions needed to perform these operations was defined in previous work. That toolkit can be restated here for the type I^2R, rather than the previous IR, to provide means of combining matrices and their updates, if needed.

In this work, we show how a dynamic interpretation of the lambdas will also provide vectors for phrases and sentences of language. Truth conditional and vector models of language follow two very different philosophies. The vector models are based on contexts, the truth models on denotations. The dynamic interpretations of language, e.g. the approach of Heim, are also based on context update, hence these seem a more appropriate choice. In this paper, we showed how Heim's files can be turned into vector contexts and how her context change potentials can be used to provide vector interpretations for phrases and sentences. Our context update instructions were defined such that they would let contexts thread through vector semantics in a compositional manner.

Amongst the things that remain to be done in a long paper is to develop a vector semantics for the lambda terms obtained via other syntactic models, e.g. CCG, LFG, and Lambek Grammars, as listed at the end of the introduction section. We also aim to work with other update semantics, such as continuation-based approaches. One could also have a general formalisation wherein both the static approach of previous work and the dynamic one of this work cohabit. This can be done by working out a second pair of type-term homomorphisms that will also work with Heim's possible world part of the contexts. In this setting, the two concepts of meaning: truth theoretic and contextual, each with its own uses and possibilities, can work in tandem.

Acknowledgements. We wish to thank the anonymous referees for excellent feedback.

References

1. Baroni, M., Bernardi, R., Zamparelli, R.: Frege in space: a program for compositional distributional semantics. Linguist. Issues Lang. Technol. **9**, 5–110 (2014)
2. Coecke, B., Sadrzadeh, M., Clark, S.: Mathematical foundations for distributed compositional model of meaning. Lambek Festschrift. Linguist. Anal. **36**, 345–384 (2010)
3. Grefenstette, E., Sadrzadeh, M.: Concrete models and empirical evaluations for the categorical compositional distributional model of meaning. Comput. Linguist. **41**, 71–118 (2015)
4. de Groote, P.: Towards a Montagovian account of dynamics. In: Proceedings of 16th Semantics and Linguistic Theory Conference (SALT 2016), pp. 1–16 (2006)
5. de Groote, P.: Towards abstract categorial grammars. association for computational linguistics. In: Proceedings of the Conference on 39th Annual Meeting and 10th Conference of the European Chapter, pp. 148–155. ACL, Toulouse (2001)
6. Heim, I.: On the projection problem for presuppositions. In: Portner, P., Partee, B.H. (eds.) Formal Semantics - The Essential Readings, pp. 249–260. Blackwell, Hoboken (1983)

7. Heim, I., Kratzer, A.: Semantics in Generative Grammar. Blackwell Textbooks in Linguistics. Blackwell Publishers, Cambridge (1998)
8. Krishnamurthy, J., Mitchell, T.M.: Vector space semantic parsing: a framework for compositional vector space models. In: Proceedings of 2013 ACL Workshop on Continuous Vector Space Models and their Compositionality (2013)
9. Maillard, J., Clark, S., Grefenstette, E.: A type-driven tensor-based semantics for CCG. In: Proceedings of EACL 2014 Type Theory and Natural Language Semantics Workshop (2014)
10. Montague, R.: The proper treatment of quantification in ordinary English. In: Thomason, R. (ed.) Formal Philosophy. Selected Papers of Richard Montague, pp. 247–270. Yale University Press, New Haven (1974)
11. Muskens, R.A.: Categorial grammar and lexical-functional grammar. In: Butt, M., King, T.H. (eds.) Proceedings of LFG 2001 Conference, University of Hong Kong, pp. 259–279. CSLI Publications, Stanford (2001). http://tinyurl.com/jrc3nnw
12. Muskens, R.A.: Language, lambdas, and logic. In: Kruijff, G.J., Oehrle, R. (eds.) Resource Sensitivity in Binding and Anaphora. Kluwer, Studies in Linguistics and Philosophy, vol. 80, pp. 23–54. Springer, Dordrecht (2003)
13. Muskens, R.: New directions in type-theoretic grammars. J. Log. Lang. Inf. **19**(2), 129–136 (2010)
14. Muskens, R., Sadrzadeh, M.: Lambdas and vectors. In: Workshop on Distributional Semantics and Linguistic Theory (DSALT), 28th European Summer School in Logic, Language and Information (ESSLLI). Free University of Bozen, Bolzano, August 2016
15. Rubenstein, H., Goodenough, J.: Contextual correlates of synonymy. Commun. ACM **8**(10), 627–633 (1965)
16. Veltman, F.: Defaults in update semantics. J. Philos. Log. **25**(3), 221–261 (1996)

XMG 2: Describing Description Languages

Simon Petitjean[1], Denys Duchier[2], and Yannick Parmentier[2(✉)]

[1] Heinrich-Heine-Universität Düsseldorf, Universitaetsstr. 1,
D-40225 Düsseldorf, Germany
simon.petitjean@hhu.de
[2] LIFO – Université d'Orléans, 6, Rue Léonard de Vinci, F-45067 Orléans, France
{denys.duchier,yannick.parmentier}@univ-orleans.fr

Abstract. This paper introduces XMG 2, a modular and extensible tool for various linguistic description tasks. Based on the notion of *meta-compilation* (that is, compilation of compilers), XMG 2 reuses the main concepts underlying XMG, namely *logic programming* and *constraint satisfaction*, to generate *on-demand* XMG-like compilers by assembling elementary units called *bricks*. This brick-based definition of compilers permits users to design description languages in a highly flexible way. In particular, it makes it possible to support several levels of linguistic description (e.g. syntax, morphology) within a single description language. XMG 2 aims to offer means for users to easily define description languages that fit as much as possible the linguistic intuition.

Keywords: Formal grammar · Meta-grammar · Compilation · Logic programming

1 Introduction

Various NLP tasks such as Automatic Summarization, Machine Translation or Dialogue Systems benefit from precision linguistic resources (e.g. grammars, lexicons, semantic representations). Alas these can hardly be obtained automatically without any loss of quality (in terms of supported linguistic phenomena or structural correctness), and building hand-crafted precision resources is a very costly task. As an illustration, let us consider syntax. Building resources describing the syntax of natural languages such as French or English (that is, electronic grammars) can take many person-years (see e.g. [1,25]). A common way to reduce this cost consists in using description languages to semi-automatically generate these precision grammars. These description languages provide users (e.g. linguists) with means to define abstractions over linguistic structures in order to capture redundancy. Users thus no longer have to describe actual linguistic structures (e.g. grammar rules), but abstractions over these. Such abstractions are then processed automatically to generate the full set of redundant structures.

Many description languages were successfully used over the past decades to generate linguistic resources ranging from small-size lexicons to real-size grammars (see e.g. [5,10,21,23,24]). Each of these description languages was tailored

© Springer-Verlag GmbH Germany 2016
M. Amblard et al. (Eds.): LACL 2016, LNCS 10054, pp. 255–272, 2016.
DOI: 10.1007/978-3-662-53826-5_16

for handling specific linguistic objects. For instance, LKB [5] was designed for describing typed feature-structures, while LexOrg [24] was designed for representing syntactic trees. Furthermore, description languages were extended with features specific of a target grammar formalism (e.g. HPSG's Head Feature Principle [19] for LKB) making these languages formalism-dependent. Users end up with plethora of specific description languages (and corresponding implementations, i.e. compilers).

These description languages often support a single linguistic framework (e.g. grammar formalism)[1] and a limited number of levels of description (mainly syntax). Depending on the target linguistic objects, the user chooses the most adequate tool (which provides a description language together with a compiler for this language), and uses it to describe and produce an actual precision linguistic resource. The consequences of this are in particular that (i) there is little information sharing between precision linguistic resources (can abstractions defined for a given target formalism be applied to other formalisms?), and (ii) should several levels of description be needed (e.g. syntax and morphology for morphologically rich languages), several tools have to be learned and combined (if possible) or a single tool has to be tinkered with.

The work presented here aims at changing this by offering a common framework which would make it possible for users to define description languages in a modular and extensible way. We built on previous work on modularity made within the eXtensible MetaGrammar (XMG) description language [6]. The paper is organized as follows. In Sect. 2, we introduce XMG and show how it laid down the bases for extensible and customizable description languages. In Sect. 3, we present (i) the concept of assembling description languages underlying XMG 2, and (ii) a compilation architecture based on logic programming and which permits a modular and extensible description (and meta-compilation)[2] of description languages (called hereafter *Domain Specific Languages*, DSL, for the sake of coherency with the terminology used in Compilation theory). In Sect. 4, we show how to use XMG 2 to dynamically assemble the original XMG language while adding a morphological layer so that it can be used to generate not only syntactic trees or flat semantic representations but also morphological representations (i.e. inflected forms). Finally, in Sect. 5, we compare our approach with related work and in Sect. 6 we conclude and present future work.

2   Compiling Extensible Metagrammars

In this section, we present the eXtensible MetaGrammar (XMG) framework [6], on which this work builds. XMG refers to both a description language used to describe tree grammars and a compiler for this language. In the XMG approach, the description of the *linguistic* grammar is seen as a *formal* grammar. XMG

---

[1] A few attempts at multi-formalism within grammar generation have been proposed such as [4] but only focused on toy implementations, see Sect. 5 on related work.

[2] Meta-compilation is meant as the compilation of compilers for description languages.

users thus describe grammar rules by writing a formal grammar (so-called *meta-grammar*). The metagrammar is in our case a Definite Clause Grammar (DCG) [17] (i.e. a logic program), which is compiled and executed by the XMG compiler to produce an actual tree grammar (i.e. a set of trees). Let us briefly define what an XMG metagrammar is, and explain how it is processed (compiled) to generate syntactic trees.

Metagrammars as Logic Programs. Intuitively, an XMG metagrammar consists of (conjunctive and/or disjunctive) combinations of reusable tree fragments. Hence the XMG language provides means to define abstractions over tree fragments along with operators to combine these abstractions conjunctively or disjunctively. These 3 concepts (abstraction, conjunction, disjunction) are already available within DCGs, which are formally defined as follows:

$$Clause \quad ::= \quad Name \rightarrow Goal \tag{1}$$

$$Goal \quad ::= \quad Description \mid Name \mid Goal \vee Goal \mid Goal \wedge Goal \tag{2}$$

Indeed, clauses allow to associate goals (e.g. descriptions) with names, hence providing *abstraction*. Goals can be made of *conjunctions* or *disjunctions* of goals. In DCG, descriptions usually refer to *facts*. In our case, descriptions correspond to formulas of a tree description logic (TDL) defined as follows:

$$Description ::= \quad x \rightarrow y \mid x \rightarrow^* y \mid x \prec y \mid x \prec^+ y \mid x[f{:}E]$$

where x, y range over node variables, $\rightarrow$ represents immediate dominance, $\rightarrow^*$ its reflexive transitive closure, $\prec$ immediate precedence, and $\prec^+$ its transitive closure. $x[f{:}E]$ constrains feature f on node x.

Finally, *axioms* of the DCG indicate clauses which correspond to *complete* tree descriptions:

$$Axiom \quad ::= \quad Name$$

Grammars as Logic Program Executions. The compilation of the input metagrammar is summarized below:

Tokenizer ⟶ Pars ⟶ Unfolder ⟶ Code generator

The metagrammar is first tokenized, then it is parsed to produce an abstract syntax tree (AST). This AST is processed to unfold the statements composing the metagrammar and produce flat structures (instructions for a kernel language). These flat structures are then interpreted by a code generator to produce instructions for a virtual machine (in our case code for a Prolog interpreter). The code is finally executed by the interpreter to produce terms. These terms are accumulations of descriptive constraints. More precisely, code execution outputs sets of conjunctions of input TDL formulas (where logic variables have been unified according to the clause instantiations defined in the metagrammar). There is one such set per derivation of the axiom of the metagrammar. To get actual trees instead of tree description logic formulas, the latter need to be solved using a tree description solver such as [8]. The interpreter is thus enriched with such a solver as a post-processor. The result of this metagrammar compilation and execution is, for each axiom, a set of trees (minimal models of the input tree description).

The architecture above makes it possible to define declarative and concise descriptions of tree grammars. Still, it supports a single level of description, namely syntax. In order to allow users to describe other levels such as semantics, the XMG language has been extended by using DCGs with multiple accumulators (so-called Extended Definite Clause Grammar, EDCG [22]). In (2) above, *Description* is replaced with:

$$<Dimension>\{\ Description\ \}$$

Depending on the dimension being used, there are two distinct description languages: one for describing syntactic trees (dimension syn) and one for describing semantic predicate structures (dimension sem).

XMG thus offers some extensibility in so far as it supports two distinct levels of description.[3] Still XMG does not allow users to describe other linguistic structures than trees or predicates. Nevertheless, XMG's modular architecture and the multiple accumulators provided by EDCG offer an adequate backbone for on-demand composition of description languages by assembling elementary description languages (called hereafter language *bricks*).

In the next section, we will present the concept of assembling language bricks and show how to extend the XMG architecture so that users can define their own description language (or Domain Specific Language, DSL) and compile the corresponding compiler. The output of this meta-compilation is a metagrammar compiler[4] (i) whose architecture follows XMG's architecture introduced above, and (ii) which can be used by linguists to describe actual language resources.

3   Assembling a Domain Specific Language

It is the foundational philosophy of XMG 2 that the tool should be easily customizable to the user's specific requirements in expressivity rather than the user be forced to cast her intuitions in terms of a rigidly predefined framework not necessarily well suited to the task.

Thus XMG 2 aims at facilitating the definition of DSLs for building linguistic resources such as grammars or lexicons. Typically a DSL allows for describing some data structure using a concrete syntax. However, even if the same data structure may be used, under the hood, for different applications, different DSLs may still be desirable: for example, a decorated tree is a very versatile data structure and could potentially be used for tree-based description of syntax and for representing agglutinative morphology; yet the two tasks have quite different requirements and attempt to capture intuitions and generalizations of dissimilar nature.

---

[3] Actually, XMG supports three levels of description: syn, sem and dyn for specifying dynamic interfaces between syn and sem (i.e., shared unification variables).

[4] Also called *meta-executor* since it both compiles a metagrammar and executes it.

3.1   Defining a Modular DSL by Assembling Bricks of Language

Our approach is predicated on assembling DSLs by composing *bricks* from an extensible library. A brick binds together a fragment of context-free syntax with some underlying data structure and some processing instructions to operate on it.

A brick can be viewed as a module: it exports a non-terminal which is the axiom of its language fragment, and it defines sockets which are non-terminals for which rules may be provided by other bricks.

Let us illustrate this brick-based description of linguistic resources by considering feature structures. Feature structures are elements that are used in many grammatical formalisms. The rules describing feature structures would consequently be added to the context free grammar of all DSLs that would be designed to describe these formalisms. This would lead to some redundancy as CFG rules would be repeated several times.

To avoid this redundancy, we propose to divide description languages into reusable and composable fragments called *language bricks*. For example, feature structures (also called Attribute-Value Matrices, AVM for short) use the following concrete syntax:

$$AVM \quad ::= \quad [Feats] \tag{3}$$

$$Feats \quad ::= \quad Feat \mid Feat, Feats \tag{4}$$

$$Feat \quad ::= \quad id = \_Value \tag{5}$$

The axiom of this brick is the non-terminal *AVM*. Note that the brick provides no production for the non-terminal _*Value*: this is what we call an *external* non-terminal and serves as the socket mentioned earlier. A production for _*Value* is obtained by plugging the axiom of another brick into this socket. To this end, let us consider a Value brick:

$$Value \quad ::= \quad id \mid \textbf{bool} \mid \textbf{int} \mid \textbf{string} \mid \_Else \tag{6}$$

The external non-terminal _Else makes it possible to plug in additional kinds of values. Now we can plug the *Value* axiom into the AVM brick's _*Value* socket (to define admissible feature values) and the AVM brick's axiom into the Value brick's _*Else* socket (to allow for recursive AVMs, that is, AVMs whose feature values can be AVMs). Plugging an axiom into a socket is realized by adding a production of the following form:

$$\_Value \quad ::= \quad Value$$

$$\_Else \quad ::= \quad AVM$$

An external non-terminal may have any number of connections, including none. One production is added for each connection: if there are many, then the external non-terminal has alternative expansions; if there are none, then it has no expansion and does not contribute to the generated language. This process can be illustrated graphically as follows (only bricks' axioms and sockets are displayed):

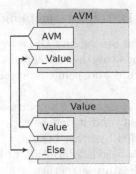

There is a cycle in this graph because we have assembled the concrete syntax for an inductively defined type. It is possible to create several instances of the same brick and to connect each instance differently. The method that we propose to instantiate and connect bricks is concretely based on a configuration file using the YAML[5] syntax. For our last example, the configuration file would contain the following code:

```
avm :
  _Value : value
value :
  _Else : avm
```

where avm and value are instances of the language bricks presented earlier (multiple instances can be distinguished using a suffix). For each one, we give the list of connections, as defined above. The context free grammar generated by this construction is the following:

$$
\begin{aligned}
AVM &::= [Feats] \\
Feats &::= Feat \mid Feat, Feats \\
Feat &::= id = \_Value \\
\_Value &::= id \mid \mathbf{bool} \mid \mathbf{int} \mid \mathbf{string} \mid \_Else \\
\_Else &::= AVM
\end{aligned}
$$

3.2 Meta-Compiling a Modular DSL

Now that we have defined a way to assemble a DSL from a single configuration file, let us see how to assemble the whole processing chain for this DSL from this file.

Let us first have a look at XMG 2's architecture, which is given in Fig. 1.

XMG 2 can be used by three different *types* of users (hereafter called *profiles*). The first profile, called *User* on Fig. 1, corresponds to a linguist, whose aim is to write a description of a linguistic resource (that is, a metagrammar)

---

[5] YAML Ain't Markup Language. See e.g. http://yaml.org.

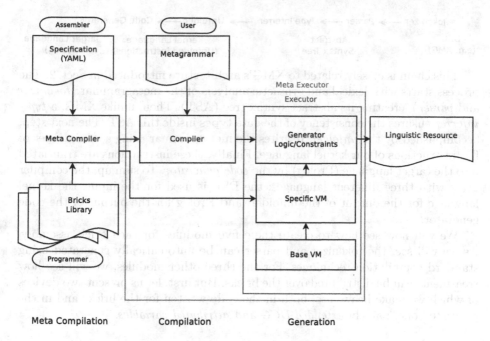

Fig. 1. Architecture of XMG 2.

and feed it to a metagrammar compiler (so-called `Meta Executor`). This tool compiles and executes the input metagrammar to generate the corresponding linguistic resource. Concretely, XMG 1 is an instance of `Meta Executor`.

A new step towards modularity is that new instances of `Meta Executors` can be easily assembled, by an *Assembler*. This type of user writes simple specifications using reusable bricks as shown previously, and a tool called `Meta Compiler` automatically produces the whole processing chain (that is, the `Meta Executor`) for the corresponding assembled DSL.

Bricks used for this assembly are picked from a brick library, which can be extended by a *Programmer*. This profile is the only one which requires programming skills. Creating a new brick consists in giving the context free grammar of the DSL and implementing the processing chain for it.

Let us now see how brick assembly and meta-compilation work in practice. As shown on Fig. 1, the processing chain is divided in two main parts, compilation and generation (performed by the executor), for which we will now detail the modular construction.

Assembling the Compiler. The type of compilers we want to assemble aims at transforming a program written in a DSL into a logic program. The processing chain has the following shape:

Tokenizer ⟶ Parser ⟶ Type Inferrer ⟶ Unfolder ⟶ Code Generator

DSL   Abstract   Kernel Language   Target Language
(e.g. XMG)   Syntax Tree   (Unfolded instructions)   (Prolog)

This chain is closely related to XMG's architecture introduced in Sect. 2. The process starts with lexical and syntactic analysis of the metagrammar (*tokenizer* and *parser*), creating its abstract syntax tree (AST). Then, unlike XMG, a *type-inferrer* ensures the consistency of the data types inside this AST. The next step, accomplished by the *unfolder*, rewrites the metagrammar using a minimal set of flat instructions of our kernel language. Finally, these instructions are translated into the target language (Prolog) by the *code generator*. To sum up, the compiler deals with three different languages: the DSL is used for the input, the kernel language for the output of the unfolder, and Prolog for the output of the code generator.

We will now see how to create these five modules for each brick assembly. As we will see, the tokenizer and parser can be automatically generated using standard compilation techniques. For the three other modules, we will see how treatments can be distributed over the bricks. But first, let us present two devices of which we make heavy use, both in the code written for the bricks and in the generated one, namely *extended DCG* and *attributed variables*.

Extended DCG. In a compiler, global and modifiable data structures are necessary (e.g. tables modeling the context). This is problematic here because Prolog does not offer such structures. The classic way to address this need in logic programming is to use additional pairs of arguments inside predicates, one to represent the structure before the application of the predicate, the other one to represent it after the application. Such pairs are usually called an accumulator. DCGs offer syntactic sugar to make handling accumulators easier. In case an arbitrary number of accumulators is needed, one can use Extended DCG [22]. This is what is done in XMG 2, which contains a library based on EDCG where one can declare accumulators, associate them with predicates, and trigger actions on the accumulations inside these predicates.

Accumulators are accessible by their identifier, and are manipulated by the application of the operations defined for them. For example, `acc::add(H)` applies the operation `add` with the argument `H` to the accumulator `acc`: the element `H` is added to the structure `acc`.

Attributed Variables. In Prolog, manipulated terms have a fixed arity, which is a limit in our case: we wish to manipulate structures that can be constrained incrementaly during compilation. It is for instance the case for feature structures, whose size can be augmented by open unification. For this reason, the feature structures brick includes a module containing a dedicated library. This library uses the concept of attributed variables [14], which allow to associate a Prolog variable with a set of attributes, and to design a dedicated unification algorithm over these attributes when two variables of the same type are unified.

In practice, to handle attributed variables, we use the YAP[6] library atts, which provides two predicates to modify the attribute of a variable or to consult it (put_atts and get_atts). Two other predicates are defined by the user: verify_attributes, which is called during unification, and attribute_goal, which converts an attribute into a goal. In our module for feature structures, a variable is associated to an attribute, this attribute being the list of attribute-value pairs composing the structure.

Lexical and Syntactic Analyses. Every brick used to build a DSL includes a file named lang.def, which contains context free rules as those previously shown. Each such rule is associated with a semantic action. This semantic action specifies which Prolog term will be built when this rule is parsed. For instance, the brick for feature structures comes with the following lang.def file:

```
AVM : '[' (Feat // ',')* ']' {$$=avm:avm($2)};
Feat : id '=' _Value {$$=avm:feat($1,$3)};
```

The second line means that when a Feat is parsed, an avm:feat/2 term is created, its two arguments being the result of the parsing of id and the one of _Value. The parser for the DSL can be created from the lang.def file similarly to what is done by the parser generator Yacc, where the rules for a LALR parser are inferred from the language definition. The tokenizer for the language is also created by extending a generic tokenizer with the punctuation and keywords specific to the brick.

Type Inference. The type inference of the DSL program has to cope with the particular context of our tools, that is to say constrained data structures (structures with partial information). A central problem is the typing of feature structures, for which we follow the ideas of [16] for the typing of records, adapting it for variable arity. The modular type inference of XMG 2 is based on two predicates, xmg:type_stmt(Stmt,Type) and xmg:type_expr(Expr,Type). A brick needs to provide new clauses for these two predicates, for all of their local constructors.

As an illustration, the following clause is given by the brick for AVMs:

```
xmg:type_expr(avm:feat(Attr,Value),Type):--
        feat_type(Attr,UAttr,Type,TypeAttr),
        type_def(TypeAttr,TypeDef),
        value_type(Value,TypeDef),
        extend_type(Type,UAttr,TypeDef),
        !.
```

where avm:feat is a constructor of the AVM brick, Attr and Value are variables representing a feature name and value respectively. TypeDef is the type of the feature, which is used to check the corresponding value. The predicate

---

[6] http://www.dcc.fc.up.pt/~vsc/yap/.

`extend_type` updates the feature type (for instance in case it refers to an AVM).

Unfolding. The kernel language is a minimal set of flat instructions (terms of depth one), which will be easily translated into the target language of compilation. The unfolder rewrites abstract syntax trees (terms of arbitrary depth) into instructions of the kernel language. The modularity is given in the same way as for the type checker, thanks to two predicates, `xmg:unfold_stmt(Stmt)` and `xmg:unfold_expr(Expr,`**Var)**, for which every brick has to provide new clauses. The following clause gives the support for the unfolding of the avm constructor, as provided by the avm brick:

```
xmg:unfold_expr(avm:avm(Coord, Feats), Target) :--
    constraints::enq((Target,avm:avm(Coord))),
    unfold_feats(Feats, Target).
```

where `Target` is the variable which will represent the feature structure, `constraints` is the accumulator where the instructions are accumulated, `enq` is the enqueuing operation on this accumulator, and `unfold_feats` a local predicate handling the unfolding of the features.

Code Generation. We follow the same pattern for code generation. A brick must provide clauses of the predicate `xmg:generate_instr(Instr)` for every instruction of its kernel language. This predicate triggers the accumulation of prolog code. The following example shows the implementation of the code generation rules for feature structures:

```
xmg:generate_instr((v(T),avm:avm(Coord))) :--
    decls::tget(T,Var),
    code::enq(xmg_brick_avm_avm:avm(Var,[])),
    !.
```

where `decls` is the table associating every variable identifier in the kernel language with a Prolog variable, `tget` is the operation allowing to access variables in this table, and `code::enq` is the accumulation operation for the generated instructions.

Assembling the Executor. As shown on Fig. 1, the execution phase is handled by three components: the `Generator`, which transforms Prolog code into a linguistic resource and runs thanks to Prolog's `base virtual machine` (`VM`). The `Specific` VM extends the base VM to fit the linguistic description task (e.g. performing additional treatments when solving linguistic descriptions).

Concretely, the non-deterministic program created by the `Compiler` is first executed. Each successful execution of this program produces a set of accumulations of constraints inside the dimensions. For each dimension which requires solving the accumulated descriptions to obtain linguistic structures, the task of the executor is to extract all valid models described in the corresponding accumulation.

Note that for a given execution, each accumulation (that is dimension) is handled by a specific *solver*. The solving may produce zero, one or more solutions for each execution. The solutions, expressed as terms, are given to the *externalizer* (still specific to the dimension) where they are translated into the target language (XML or JSON) for storage or display.

Solving. Solving descriptions relies on the following sub processing chain:

$$\text{Preparer} \longrightarrow \text{Solver} \longrightarrow \text{Extractor}$$

Each accumulation created by the execution is given to the *preparer*, which transforms the accumulation into a constraint satisfaction problem. The set of constraints is then translated into executable code (Prolog code with bindings to the C++ Gecode library[7]) by the solver. The *extractor* computes all the solutions to the problem and translates them into a term.

Note that solvers are also packaged in bricks (but these do not provide a `lang.def` file since the input language they deal with is the language of constraints which can be accumulated in a dimension). Note also that these solvers can be extended by defining plug-ins to apply specific constraints on the models being computed (e.g. natural language-dependent constraints such as clitic order in French).

4   Application: Designing a Language to Describe Syntax, Semantics *and/or* Morphology

As an illustration of the DSL assembly and meta-compilation techniques introduced above, let us see how the XMG language [6] presented in Sect. 2 can be assembled and enriched with another level of description, namely morphology, so that one can use the same framework to describe various dimensions of language.

4.1   Defining Language Bricks for Describing Tree Grammars

As mentioned above, XMG was designed to describe tree grammars such as Feature-Based Tree-Adjoining Grammars (FB-TAG) (see e.g. [11]). Basically, a FB-TAG is made of elementary trees whose nodes are labelled with feature-structures. These feature-structures associate features with either values or unification variables. The latter can be shared between syntactic and semantic representations.

Recall from (1) and (2) defined on page 3, that an XMG description is made of *clauses* containing either (i) *Descriptions* or (ii) conjunctive or disjunctive *combinations* of these *Descriptions*. In our case, *Descriptions* belong to a *Dimension* which is either syntax or semantics. Syntactic descriptions are tree fragments (defined as formulas of a tree description logic), and semantic descriptions formulas of a predicate logic.

---

[7] The bindings to Gecode have been developed for the needs of XMG 2.

Descriptions. Let us first define a language brick syn for defining syntactic descriptions (that is, syntactic statements). Such a brick contains both the definition of the syntax of the language (that is, a CFG), and instructions for processing (compiling) statements belonging to this language.

$$SynStmt \quad ::= \quad \textbf{node} \; id \quad | \quad \textbf{node} \; id \; \_AVM$$
$$| \quad id \; \texttt{->} \; id \quad | \quad id \; \texttt{->*} \; id \quad | \quad id \; \texttt{<} \; id \quad | \quad id \; \texttt{<+} \; id$$

A syntactic statement (*SynStmt*) is either the definition of a node (identified by the value *id*), the definition of a node labelled with some feature-structure (_AVM), or the definition of a relation between node identifiers (`->` for dominance, `<` for precedence)[8]. An AVM is described using the bricks defined by (3), (4), (5) and (6) page 5. Finally, semantic descriptions are defined using the following brick sem:

$$SemStmt \quad ::= \quad \ell\texttt{:}p\texttt{(}id,\ldots,id\texttt{)} \quad | \quad id \; \texttt{<<} \; id$$

where *p* refers to a predicate, *id* to unification variables representing *p*'s arguments, ℓ to a predicate label (these are all identifiers), and `<<` to a scope constraint.[9]

Combinations. Combinations of descriptions are realized by means of a parameterized brick constructor Dim_X defined as follows (*X* is a lexical keyword, here syn or sem):

$$Dim_X \quad ::= \quad \texttt{<}\textbf{x}\texttt{>}\{Stmt\}$$

This brick constructor is used to instantiate bricks of the following form:[10]

$$Stmt \quad ::= \quad \_Stmt \quad | \quad Stmt \; \texttt{;} \; \_Stmt \quad | \quad Stmt \; \texttt{||} \; \_Stmt$$

which allow to describe conjunctive or disjunctive combinations of statements. The external non-terminal _Stmt makes it possible to connect either syntactic or semantic statements (which are thus accumulated separately):

$$Dim_{\mathrm{syn}}.\_Stmt \quad ::= \quad SynStmt$$
$$Dim_{\mathrm{sem}}.\_Stmt \quad ::= \quad SemStmt$$

4.2   Assembling Language Bricks for Describing Tree Grammars

From the language bricks defined above, it is possible to assemble the XMG compiler. Concretely, this amounts to defining the needed assembly (that is, to

---

[8] The `+` refers to the relation's transitive closure and `*` to its transitive and reflexive closure.

[9] See [12] for more details about flat semantic representations.

[10] In our concrete syntax, `;` refers to the logical operator AND, and `||` to XOR.

writting the YAML configuration file where is defined which bricks to load and how these interact).

The YAML file (named `compiler.yaml`) defining how to automatically assemble the XMG compiler is given below:

```
mg:
    _Stmt: combination
combination:
    _Stmt: dim_syn dim_sem
avm:
    _Value: value
value:
    _Else: avm
```

```
dim_syn:
    tag: "syn"
    solver: "tree"
    _Stmt: syn
syn:
    _AVM: avm
dim_sem:
    tag: "sem"
    _Stmt: sem
```

Concretely, what this YAML file says is the following. The target metagrammatical DSL (that is, the XMG language) corresponds to a brick named mg where statements are defined in the brick `combination`. The brick `combination` contains statements of type either `dim_syn` or `dim_sem` (these are parameters of the `combination` brick). The `dim_syn` brick contains statements defined in the brick `syn`, introduced by the keyword (tag) `"syn"`, and solved using a solver named `"tree"`. Statements of the brick `dim_sem` are introduced by the keyword `"sem"` and defined in the brick `sem`. Semantic statements do not need to be solved (hence the absence of any solver feature). The brick `syn` contains AVMs defined in the brick `avm` and expressions defined in `value`. The brick `avm` is parameterized by `value` and vice versa, as mentioned in Sect. 3.

4.3 Adding a Morphological Layer

So far, we showed how to assemble the DSL corresponding to XMG from a library of language bricks using a configuration file in YAML format. From this file, the XMG compiler is automatically built. Let us now see how to add morphological descriptions to this DSL (and recompile the corresponding meta-executor).

The morphological descriptions we will consider here are inspired by work on Ikota, a Bantu language spoken in Gabon [7]. The idea is to describe inflected verbal forms as (i) concatenations of ordered morphological fields (namely subject, tense, root, aspect, active and proximal) and (ii) morphological features associated with these fields (e.g. person, number, tense, verbal class, etc.).

The metagrammar of verbal forms contains for each field alternative possible realizations (that is, a disjunction of elementary descriptions). A verb is then described as the conjunction of all morphological fields. The metagrammar compiler will compute all combinations of values of these fields (that is, all elements of the cartesian product *subject* × *tense* × *root* × *aspect* × *active* × *proximal*) and keep those where there is no unification failure between morphological features. As an illustration, consider the successful combination for the inflected form ò+é+ʤ+ÀK+À+nÁ *(you will eat)* below, where fields are numbered from 1 to 6.[11]

$Verb \rightarrow$ $Subject(1) \wedge Tense(2) \wedge Root(3) \wedge Aspect(4) \wedge Active(5) \wedge Proximal(6)$

$\rightarrow$
$$
\begin{array}{l}
1 \leftarrow \text{ò} \\
[\text{p} = 2, \\
\text{n} = \text{sg}]
\end{array}
\wedge
\begin{array}{l}
2 \leftarrow \text{é} \\
[[\text{tense} = \text{f}]]
\end{array}
\wedge
\begin{array}{l}
3 \leftarrow \text{ʤ} \\
[\text{vclass} = \text{g1}]
\end{array}
\wedge
\begin{array}{l}
4 \leftarrow \text{ÀK} \\
[\text{tense} = \text{f}, \\
\text{prog} = \text{-}]
\end{array}
\wedge
\begin{array}{l}
5 \leftarrow \text{À} \\
[\text{active} = \text{+}, \\
\text{prog} = \text{-}]
\end{array}
\wedge
\begin{array}{l}
6 \leftarrow \text{nÁ} \\
[\text{prox} = \text{day}]
\end{array}
$$

$\rightarrow$
$$
\begin{array}{llllll}
1 \leftarrow \text{ò} & 2 \leftarrow \text{é} & 3 \leftarrow \text{ʤ} & 4 \leftarrow \text{ÀK} & 5 \leftarrow \text{À} & 6 \leftarrow \text{nÁ}
\end{array}
$$
$$
[\text{p} = 2, \quad \text{prog} = \text{-}, \quad \text{tense} = \text{f}, \quad \text{vclass} = \text{g1},
$$
$$
\text{n} = \text{sg}, \quad \text{active} = \text{+}, \quad \text{prox} = \text{day}]
$$

To extend the DSL defined above with such descriptions, we need a language brick allowing to define morphological fields, to associate them with a lexical form (and potentially also features), and to order them:

$MorphStmt$ ::= **field** id id | **field** id id _AVM | id **>>** id

Note that this brick reuses the avm and value (e.g. for identifiers) bricks already defined above. To assemble this DSL, the YAML configuration file from Sect. 4.2 needs to be extended as illustrated below:

```
mg:
  _Stmt: combination
combination:
  _Stmt: dim_syn dim_sem dim_morph
```

```
dim_morph:
  tag: "morph"
  _Stmt: morph
morph:
  _AVM: avm
```

Basically, combinations no longer only contain syntactic or semantic statements, but also statements defined in the brick dim_morph. These statements are introduced by the keyword "morph" and are of type morph. Finally, AVMs contained in morphological statements are of type avm. From this extended compiler.yaml file, a new XMG-like meta-executor can be compiled. This tool can be used to describe (within the same metagrammar or not) not only tree grammars with flat semantic representations, but also inflected forms.

Note that the DSL assembly and meta-compilation techniques introduced here do not have as a main goal to provide users with means to design meta-grammatical DSLs which would fit several dimensions of language at once (even if it may be technically possible). This extension of the XMG DSL is given for illustration purposes. The motivation underlying XMG 2 is that, depending on the target linguistic resource, one should be able to easily define and use appropriate DSLs. XMG 2 should thus provide users with means to easily assemble and build compilers for such DSLs (no matter which and how many of these are needed).

---

[11] Note that this verbal form does not correspond to the final surface form. A post-processing is applied to (i) replace lexical A with a and K with k, and (ii) to delete vowels to finally obtain the expected form óʤàkàná.

5   Related Work

The meta-compilation architecture presented here exhibits two particularly interesting properties in the context of linguistic resource engineering, namely *modularity* and *extensibility*. It is inspired by previous work on compilation, illustrated by systems such as LISA [13], JastAdd [9], or Neverlang [3]. All these systems allow users to relatively easily extend compilers by defining modules.

Still, their methodology differ from ours. They all aim at offering software designers with means to develop their own DSL by defining formal language specifications. These specifications are often complex (for instance, in Neverlang, assembling elementary bricks requires to solve graph dependencies), while XMG 2 allows for easy configuration using the YAML format.

Furthermore, these systems are used to extend or recreate existing substantial compilers. As an illustration, both JastAdd and Neverlang were used to build extensible Java Virtual Machines. We are mainly interested in providing users with easy-to-assemble *dedicated specific* languages. XMG 2's philosophy is to provide adequate DSLs, that fit the linguistic description tasks. In particular, users should be able to reconfigure their DSL according to their needs, without having to support a large machinery.

Also, as mentioned above, XMG 2 provides three user profiles which make it different from other approaches. Indeed, the expected JastAdd, LISA or Neverlang users are skilled programmers. In XMG 2, contributing to an assembly of language bricks is such an easy task that users who do not know programming can define their own DSL.

Finally, unlike previous approaches, XMG 2 is based on logic programming, which makes it particularly appropriate for describing linguistic structures since these often use unification variables.

Apart from these approaches, to our knowledge, there are very few attempts (in particular in the NLP community) at providing users with such an extensible and modular linguistic description framework. One may cite work on cross-formalism language description by [4]. In their approach, the authors use a meta-grammar compiler primarily designed for Tree-Adjoining Grammars (TAG) to derive both TAG and Lexical-Functional Grammar (LFG) rules from a single linguistic description. Their work was made possible by the fact that TAG and LFG both relies on syntactic trees. Should users be interested in describing less related structures within the same framework, their approach would not permit this.

Another interesting approach is that of *Grammatical Framework* (GF) [18]. GF is a system for designing grammars for various languages. It is based on an abstract syntax which can be mapped to several concrete syntaxes (hence languages). A GF grammar is modular, and can be interpreted by the GF system to parse or generate sentences. That is, GF provides a modular and extensible way to design a specific type of linguistic resources (multi-lingual morpho-syntactic grammars), while XMG 2 tries to its best to remain agnostic regarding the linguistic structures it describes.

6   Conclusion

In this paper, we showed how to design Domain Specific Languages for describing linguistic resources, by assembling elementary language bricks. We then presented how to concretely implement a meta-compiler which would take as input a library of language bricks together with a configuration file defining how to assemble a given DSL, and would produce automatically the corresponding compiler.

This meta-compilation from a DSL specification has been used for instance to produce a Prolog version of the XMG compiler. The resulting compiler was successfully used to (re)compile existing large scale tree grammars for French, English and German [6] (the generated resources are identical to the ones produced by the original XMG). The development of new syntactic resources was also initiated using XMG 2's modular architecture (including works on São Tomense [20] and Arabic [2]).

XMG 2's extensibility made it possible to create new language bricks, and thus new compilers. These were used for various description tasks, including the development of morphological resources (lexicon of inflected forms) for Ikota [7], the definition of new syntax-semantics and morpho-semantics interfaces (see e.g. [15]). This work paved the way for new uses of description languages. Future work includes extending the library of language bricks (and corresponding solvers) to support these new uses.

Several paths remains to be explored in the context of this work, both on the theoretical and on the practical side. First, the expressive power of certain types of DSL needs to be further studied. As an example, we saw that DSLs are used to describe tree-based grammars in the TAG formalism. In our case, formulas of a dominance-based tree description logic are used and solved using Constraint Satisfaction techniques. Alternatively formulas of a monadic second order (MSO) logic could be considered and solved using automaton-based techniques. More generally, the link between the input metagrammatical description language (DSL) and the target grammar formalism requires more attention. Interesting questions include the definition of the class of grammar formalisms (resp. of formal languages) which can be captured by XMG-like metagrammatical descriptions.

Second, on the practical side this approach to linguistic resource production made the relation between software engineering and linguistic resource design clearer. Designing precision resource is very close to designing software (one has to deal with relatively complex formal statements and expressions) and should thus benefit from the same kind of integrated development environments. Future work on metagrammar meta-compilation must take this analogy into account by providing user with facilities such as debuggers, regression tests, etc.

References

1. Abeillé, A., Candito, M., Kinyon, A.: FTAG: current status and parsing scheme. In: Proceedings of Vextal-1999, Venice, Italy (1999)
2. Ben Khelil, C., Duchier, D., Parmentier, Y., Zribi, C., Ben Fraj, F.: ArabTAG: from a handcrafted to a semi-automatically generated TAG. In: Proceedings of 12th International Workshop on Tree Adjoining Grammars and Related Formalisms (TAG+12), Düsseldorf, Germany, pp. 18–26. (2016). http://aclweb.org/anthology/W16-3302
3. Cazzola, W.: Domain-specific languages in few steps. In: Gschwind, T., Paoli, F., Gruhn, V., Book, M. (eds.) SC 2012. LNCS, vol. 7306, pp. 162–177. Springer, Heidelberg (2012). doi:10.1007/978-3-642-30564-1_11
4. Clément, L., Kinyon, A.: Generating parallel multilingual LFG-TAG grammars from a MetaGrammar. In: Proceedings of 41st Annual Meeting of the Association for Computational Linguistics, Sapporo, Japan, pp. 184–191 (2003). http://dx.doi.org/10.3115/1075096.1075120
5. Copestake, A., Sanfilippo, A., Briscoe, T., de Paiva, V.: The ACQUILEX LKB: an introduction. In: Briscoe, T., de Paiva, V., Copestake, A. (eds.) Inheritance, Defaults, and the Lexicon, pp. 148–163. Cambridge University Press, Cambridge (1993)
6. Crabbé, B., Duchier, D., Gardent, C., Le Roux, J., Parmentier, Y.: XMG: eXtensible MetaGrammar. Comput. Linguist. **39**(3), 1–66 (2013). http://dx.doi.org/10.1162/COLI_a_00144
7. Duchier, D., Magnana Ekoukou, B., Parmentier, Y., Petitjean, S., Schang, E.: Describing morphologically-rich languages using Metagrammars: a look at verbs in Ikota. In: Workshop on "Language Technology for Normalisation of Less-resourced Languages", pp. 55–60. LREC, Istanbul (2012)
8. Duchier, D., Niehren, J.: Dominance constraints with set operators. In: Lloyd, J., et al. (eds.) CL 2000. LNCS (LNAI), vol. 1861, pp. 326–341. Springer, Heidelberg (2000). doi:10.1007/3-540-44957-4_22
9. Ekman, T., Hedin, G.: The JastAdd system modular extensible compiler construction. Sci. Comput. Program. **69**(13), 14–26 (2007). http://dx.doi.org/10.1016/j.scico.2007.02.003
10. Evans, R., Gazdar, G.: DATR: a language for lexical knowledge representation. Comput. Linguist. **22**(2), 167–216 (1996). http://www.aclweb.org/anthology/J/J96/J96-2002.pdf
11. Gardent, C.: Integrating a unification-based semantics in a large scale lexicalised tree adjoining grammar for French. In: Proceedings of 22nd International Conference on Computational Linguistics (Coling 2008), Manchester, UK, pp. 249–256 (2008). http://www.aclweb.org/anthology/C08-1032
12. Gardent, C., Kallmeyer, L.: Semantic construction in FTAG. In: EACL 2003, 10th Conference of the European Chapter of the Association for Computational Linguistics, pp. 123–130. Budapest, Hungary (2003). http://dx.doi.org/10.3115/1067807.1067825
13. Henriques, P.R., Pereira, M.J.V., Mernik, M., Lenic, M., Gray, J., Wu, H.: Automatic generation of language-based tools using the LISA system. IEE Proc. Softw. **152**(2), 54–69 (2005). http://dx.doi.org/10.1049/ip-sen:20041317
14. Holzbaur, C.: Metastructures vs. attributed variables in the context of extensible unification. In: Bruynooghe, M., Wirsing, M. (eds.) PLILP 1992. LNCS, vol. 631, pp. 260–268. Springer, Heidelberg (1992). doi:10.1007/3-540-55844-6_141

15. Lichte, T., Petitjean, S.: Implementing semantic frames as typed feature structures with XMG. J. Lang. Model. **3**(1), 185–228 (2015). http://dx.doi.org/10.15398/jlm.v3i1.96
16. Ohori, A.: A polymorphic record calculus and its compilation. ACM Trans. Program. Lang. Syst. **17**(6), 844–895 (1995). http://doi.acm.org/10.1145/218570.218572
17. Pereira, F., Warren, D.: Definite clause grammars for language analysis – a survey of the formalism and a comparison to augmented transition networks. Artif. Intell. **13**, 231–278 (1980). http://dx.doi.org/10.1016/0004-3702(80)90003-X
18. Ranta, A.: Modular grammar engineering in GF. Res. Lang. Comput. **5**(2), 133–158 (2007). http://dx.doi.org/10.1007/s11168-007-9030-6
19. Sag, I., Wasow, T.: Syntactic Theory. A Formal Introduction. CSLI Publications, Stanford (1999)
20. Schang, E., Duchier, D., Magnana Ekoukou, B., Parmentier, Y., Petitjean, S.: Describing São tomense using a tree-adjoining meta-grammar. In: 11th International Workshop on Tree Adjoining Grammars and Related Formalisms (TAG+11), Paris, France, pp. 82–89 (2012). http://www.aclweb.org/anthology/W12-4610
21. Shieber, S.M.: The design of a computer language for linguistic information. In: 10th International Conference on Computational Linguistics (COLING) and 22nd Annual Meeting of the Association for Computational Linguistics (ACL), pp. 362–366 (1984). http://aclweb.org/anthology/P84-1075
22. Van Roy, P.: Extended DCG notation: a tool for applicative programming in prolog. Technical report UCB/CSD 90/583, UC Berkeley (1990). http://www2.eecs.berkeley.edu/Pubs/TechRpts/1990/5471.html
23. Villemonte De La Clergerie, É.: Building factorized TAGs with meta-grammars. In: The 10th International Conference on Tree Adjoining Grammars and Related Formalisms - TAG+10, pp. 111–118. New Haven (2010). http://www.aclweb.org/anthology/W10-4414
24. Xia, F.: Automatic grammar generation from two different perspectives. Ph.D. thesis, University of Pennsylvania (2001)
25. XTAG Research Group: A lexicalized tree adjoining grammar for English. Technical report IRCS-01-03, IRCS, University of Pennsylvania (2001)

Minimalist Grammar Transition-Based Parsing

Miloš Stanojević[✉]

Institute for Logic, Language and Computation, University of Amsterdam,
Amsterdam, The Netherlands
m.stanojevic@uva.nl

Abstract. Current chart-based parsers of Minimalist Grammars exhibit prohibitively high polynomial complexity that makes them unusable in practice. This paper presents a transition-based parser for Minimalist Grammars that approximately searches through the space of possible derivations by means of beam search, and does so very efficiently: the worst case complexity of building one derivation is $O(n^2)$ and the best case complexity is $O(n)$. This approximated inference can be guided by a trained probabilistic model that can condition on larger context than standard chart-based parsers. The transitions of the parser are very similar to the transitions of bottom-up shift-reduce parsers for Context-Free Grammars, with additional transitions for online reordering of words during parsing in order to make non-projective derivations projective.

Keywords: Minimalist Grammars · Shift-reduce parsing · Transition-based parsing · Swap transition · Two-stack automata

1 Introduction

Minimalist Grammar (MG) [14] is a formalization of Chomsky's Minimalist Program (MP) [4]. MG is one of the several grammar formalisms that go beyond Context-Free Grammars (CFG) in their expressive power (both in terms of weak and strong generative capacity). The main characteristic of MG is that constituents do not only combine to make bigger constituents, but they also can move during the course of derivation.

A standard derivation in Minimalist Program (and Minimalist Grammar) roughly looks like this: first we enumerate the words that are going to be used in the sentence with operation *select*; second, we combine operations *merge* and *move* in building the derivation bottom-up. The operation *merge* (sometimes called external merge) takes two constituents and puts them together. The *move* operation (sometimes called internal merge) takes a subtree and moves it upwards to the specifier position. So even though the words enter the derivation process in one order, by the end of the derivation they might form a completely different word order. This resembles the distinction between deep structure and surface structure from the early days of Generative Grammar [3]. The distinction between deep word order and surface word order does not exist in the Minimalist approach, but we will nevertheless adopt it here because it simplifies talking about some concepts.

© Springer-Verlag GmbH Germany 2016
M. Amblard et al. (Eds.): LACL 2016, LNCS 10054, pp. 273–290, 2016.
DOI: 10.1007/978-3-662-53826-5_17

Even though intuitively it might sound simple to build a recognizer (or parser) for a formalism that contains only two simple functions such as *merge* and *move* it turned out to be quite a difficult task. Early approaches [6,16] are based on bottom-up chart parsing which does an exhaustive search trough the space of all possible derivations. Because chart parsing is based on dynamic programming, such search is formally tractable in the sense of being polynomial. However, the polynomial complexity of chart parsing is still too high–$O(n^{4m+4})$ where n is the number of words in the sentence and m is the number of unique movement licensees present in the lexicon.

Transition-based parsers are an alternative to chart-based parsers. Transition-based parsers build the derivation step by step by using a set of well defined transitions that lead from one parsing state to the next one. Because they usually do not use dynamic programming, they cannot explore the full search space and that is why only approximate inference is possible. However, this did not stop transition-based parsers from matching and outperforming their chart-based counterparts in the area of CFG, CCG and dependency parsing both in terms of accuracy and in terms of speed [2,11,20]. Part of the reason for that is that dynamic programming in chart-based parsers requires from the probabilistic scoring model to condition only on the local context, while the transition-based parsers allow conditioning on any part of the derivation that was built. So giving up on exact inference allows us not only to gain in terms of speed but also it allows replacing weak probabilistic models with a much more powerful ones.

The top-down MG parser of Stabler [9,15] can be considered as an instantiation of transition-based parsing. It builds a minimalist derivation from top clause node c by *un-merging* and *un-moving* the nodes in the derivation recursively. Adding new operations to this parser requires finding a top-down equivalent of the minimalist operations that are traditionally defined in a bottom-up manner.

This paper presents a transition-based parser that is bottom-up, thus it does not require any changes in the definition or order of application of MG operations. It is similar to shift-reduce transition-based parsers, especially those that use a *swap* transition [8,11]. Just like majority of transition-based parsers, it employs no dynamic programming and employs approximate beam search of the space of derivations.

The main idea that motivates creation of this parser is based on observing that all MG derivation trees are projective trees with respect to the "deep word order". The displacement in the surface word order is a result of applying *move* operation. So if we would reorder the words in "the right way" then parsing should be projective and almost as easy as CFG parsing.

In the next three sections we present some background material for the transition-based bottom-up MG parser which covers definition of Minimalist Grammars, description of the existing chart parser for MGs and description of transition-based shift-reduce bottom-up parser for Context-Free Grammars. After that we fully specify the deductive system of the transition-based bottom-up minimalist parser and show some formal properties of it.

2 Minimalist Grammars

Here we describe a simple version of Minimalist Grammar as presented in [18] and [7]. This simple version does not deal with adjunction and head movement, but these extensions can easily be added to our parser and they do not influence the weak generative capacity of the MG [17].

A minimalist grammar G is a tuple $(\Sigma, Sel, Lic, Types, Lex, c, \mathscr{F})$, where

$\Sigma \neq \emptyset$ is an alphabet

Sel are "selecting features"

Lic are "licensing features"

Syn are "syntactic features" defined using Sel and Lic as a union of:
$selectors = \{=f | f \in Sel\}$
$selectees = \{\ f | f \in Sel\}$
$licensors = \{+f | f \in Lic\}$
$licensees = \{-f | f \in Lic\}$

$Types = \{::, :\}$ are the lexical type and the derived/phrasal type

$C = \Sigma^* \times Types \times Syn^*$ are "chains"

$Lex \subseteq C^+$ is a finite subset of chains with form $\Sigma^* \times \{::\} \times (selectors \cup licensors)^* \times selectees \times licensees^*$

$E = C^+$ are expressions

$c \in Sel$ is the feature used to define the *complete expression* $s : c$

$\mathscr{F} = \{merge, move\}$ are partial generating functions from E^* to E

$merge : (E \times E) \to E$ is a union of the following three functions, for $s, t \in \Sigma^*, \cdot \in \{:, ::\}$, $f \in Sel$, $\gamma \in Syn^*$, $\delta \in licensees^+$, and chains $\alpha_1, \ldots, \alpha_k, \iota_1, \ldots, \iota_l (0 \leq k, l)$:

$$merge1 \quad \frac{s :: =f\, \gamma \qquad\qquad t \cdot f, \alpha_1, \ldots, \alpha_k}{st : \gamma, \alpha_1, \ldots, \alpha_k}$$

$$merge2 \quad \frac{s : =f\, \gamma, \alpha_1, \ldots, \alpha_k \qquad t \cdot f, \iota_1, \ldots, \iota_l}{ts : \gamma, \alpha_1, \ldots, \alpha_k, \iota_1, \ldots, \iota_l}$$

$$merge3 \quad \frac{s \cdot =f\, \gamma, \alpha_1, \ldots, \alpha_k \qquad t \cdot f\, \delta, \iota_1, \ldots, \iota_l}{s : \gamma, \alpha_1, \ldots, \alpha_k, t : \delta, \iota_1, \ldots, \iota_l}$$

An illustration of these functions is presented in Fig. 1. In the figure expressions are represented as tree structures and on top of them is the list of unchecked features of the first chain. Chains that are waiting to move are represented as subtrees.

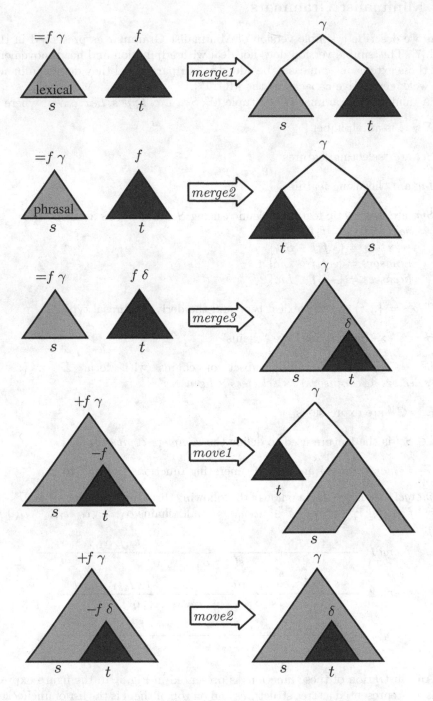

Fig. 1. Illustrations of Minimalist Grammar generating functions.

merge1 is combining the lexical head and its complement. The result is a new string in which string of the head s and the string of the complement t are concatenated and represented with st.

merge2 is combining a phrase that contains the head with the phrase that will be its specifier. Since specifier always comes on the left side, the resulting string is ts.

merge3 is combining phrases whose strings are not concatenated because the licensees δ will cause the phrase to move in the later steps of the derivation.

move: $E \rightarrow E$ is the union of the following two functions, for $s, t \in \Sigma^*$, $f \in Lic$, $\gamma \in Syn^*$, $\delta \in licensees^+$, and chains $\alpha_1, \ldots, \alpha_k, \iota_1, \ldots, \iota_l (0 \leq k, l)$ and SMC constraint defined bellow:

$$move1 \quad \frac{s : +f\, \gamma, \alpha_1, \ldots, \alpha_{i-1}, t : -f, \alpha_{i+1}, \ldots, \alpha_k}{ts : \gamma, \alpha_1, \ldots, \alpha_{i-1}, \alpha_{i+1}, \ldots, \alpha_k}$$

$$move2 \quad \frac{s : +f\, \gamma, \alpha_1, \ldots, \alpha_{i-1}, t : -f\, \delta, \alpha_{i+1}, \ldots, \alpha_k}{s : \gamma, \alpha_1, \ldots, \alpha_{i-1}, t : \delta, \alpha_{i+1}, \ldots, \alpha_k}$$

SMC is a simple version of "shortest move condition" [4]. It constrains the domain of *move* by not allowing any of $\alpha_1, \ldots, \alpha_{i-1}, \alpha_{i+1}, \ldots, \alpha_k$ to have licensee $-f$ as its first feature.

move1 is handling the movement of a subtree with yield t into a specifier of the current tree. Because *move1* moves subtree that is landing (it is not going to move any more) we can safely concatenate strings into ts.

move2 is handling the movement of a subtree that will continue moving in the later steps of the derivation because it has unchecked licensees δ, thus there is no need for concatenating strings s and t.

$CL(G)$ is a set of expressions generated by taking the closure over Lex and generating functions in $\mathcal{F}$.

$yield(e)$ is defined only over complete expressions $(s \cdot c)$ and is an alphabet component of the only chain in the expression e.

The language defined by the grammar G is $L(G) = \{s \mid \exists e \in CL(G) \wedge e \text{ is } s \cdot c\}$. In other words, a language defined by G is the set of yields of all complete expressions that are part of the closure of G.

3   Chart-Based Parser for MG

The first recognizers for MG were chart-based recognizers of Harkema [6] and Stabler [16]. The parsing strategy is presented in the form of deductive rules. These rules could be used as part of some closure computation engine, such as the ones based on "parsing as deduction" [13], in order to get efficient inference by using dynamic programming.

The general idea of "parsing as deduction" [13] is that we are trying to prove that the sentence that is parsed is part of the language defined by the grammar. We start with some claims that do not require proving i.e. axioms (for example "from position i till position $i+1$ there is a word w_i), and after that we apply deductive rules recursively until we prove the goal statement (for example "sentence with words $w_0, \ldots, w_{n-1}$ has only c as an unchecked feature") or until we exhaust all possibilities without managing to prove that the sentence is part of the language.

The statements that the deduction engine is working with are encoded in the form of "items". At any step of the parsing process, all the items can be divided in two groups: items that can trigger further deduction and items that are proved but they do not trigger future deduction. Items of the first group are stored in a queue called *agenda* and the items of the second group are stored in a data structure for efficient retrieval that is called *chart*. With this terminology we can say that the parsing process starts with putting axiomatic items in agenda and applying deduction rules on all of them in order. If the result of a deduction rule can trigger future deduction, it is added both to the chart and to the agenda, otherwise it is added only to the chart.

Items of the minimalist chart parser are essentially encodings of MG expressions which instead of using strings of alphabet use ranges of covered words in the sentence. So for example, item $(2,5) := n\ v$ can be read as "this is a phrase with features $=n$ and v and it covers continuous span of words from positions 2 until position 5 in the observed word order".

With that interpretation the following deduction rules have been proven to be sound and complete [6], where n is the length of the sentence, w_i is word at position i and i can go between 0 and n:

$$axiom \quad (i, i+1) :: \alpha \quad s.t. \quad w_i :: \alpha \in Lex$$

$$axiomEpsilon \quad (i,i) :: \alpha \qquad s.t. \quad \varepsilon :: \alpha \in Lex$$

$$goal \quad (0,n) \cdot c$$

$$merge1 \frac{(a,b) :: =f\ \gamma \qquad (b,c) \cdot f, \alpha_1, \ldots, \alpha_k}{(a,c) : \gamma, \alpha_1, \ldots, \alpha_k}$$

$$merge2 \frac{(b,c) : =f\ \gamma, \alpha_1, \ldots, \alpha_k \qquad (a,b) \cdot f, \iota_1, \ldots, \iota_l}{(a,c) : \gamma, \alpha_1, \ldots, \alpha_k, \iota_1, \ldots, \iota_l}$$

$$merge3 \frac{(a,b) \cdot =f\ \gamma, \alpha_1, \ldots, \alpha_k \qquad (c,d) \cdot f\delta, \iota_1, \ldots, \iota_l}{(a,b) : \gamma, \alpha_1, \ldots, \alpha_k, (c,d) : \delta, \iota_1, \ldots, \iota_l}$$

$$move1 \frac{(b,c) : +f\ \gamma, \alpha_1, \ldots, \alpha_{i-1}, (a,b) : -f, \alpha_{i+1}, \ldots, \alpha_k}{(a,c) : \gamma, \alpha_1, \ldots, \alpha_{i-1}, \alpha_{i+1}, \ldots, \alpha_k}$$

$$move2 \frac{(a,b) : +f\ \gamma, \alpha_1, \ldots, \alpha_{i-1}, (c,d) : -f\ \delta, \alpha_{i+1}, \ldots, \alpha_k}{(a,b) : \gamma, \alpha_1, \ldots, \alpha_{i-1}, (c,d) : \delta, \alpha_{i+1}, \ldots, \alpha_k}$$

Naturally, *move* is subject to SMC constraint.

4 Transition-Based Bottom-Up Parser for CFG

Before we move to the formal description of the transition-based Minimalist parser, we will make a small digression and informally present a type of shift-reduce parser for CFG in Chomsky Normal Form (CNF) similar to the one presented in [10,12]. This algorithm is used as a basis on which the Minimalist transition-based parser is built.

The state of the transition based parser is usually called *configuration*. A configuration consists of two data structures: stack σ and buffer (usually implemented as a queue) β. For CFG in CNF the initial configuration is an empty stack and the buffer filled with words of the sentence that is parsed. Shift action removes the first word in the current buffer and puts its POS tag on top of the stack. Reduce operation takes the two elements from top of the stack and produces a new element that goes back to the stack if there is a grammar rule that allows that.

The deduction rules are shown bellow, where [ ] represents an empty stack or an empty buffer, σ represents a stack, $\sigma|x$ represents a stack that is the result of pushing element x on top of stack σ, β represents buffer, $x|\beta$ represents a buffer (queue) with head x and tail β, G represents a CFG in CNF, w is a variable representing any word, X, Y, Z are variables representing any non-terminal, and S is the root non-terminal of the grammar G.

$$axiom \quad \langle [], [w_1, \ldots, w_n] \rangle$$

$$goal \quad \langle [S], [] \rangle$$

$$shift\{X\} \quad \frac{\langle \sigma, w|\beta \rangle}{\langle \sigma|X, \beta \rangle} \quad X \to w \in G$$

$$reduce \quad \frac{\langle \sigma|X|Y, \beta \rangle}{\langle \sigma|Z, \beta \rangle} \quad Z \to XY \in G$$

Deriving the goal configuration can be done in several ways. One of them is by using a chart-based algorithm that would compute a full closure of these deduction rules over the axiomatic configuration. Another is a transition-based approach where the algorithm would treat each deduction rule as a transition and only a predefined number of high probability sequence of transitions will be explored. The computational complexity of the transition-based algorithm is $O(n)$ because the number of shift transitions is not bigger than the number of words in the sentence, and the same holds for the number of reduce transitions. This nice property of transition-based shift-reduce parsing has caused its wide adoption in the natural language processing community which produced many extensions and implementations of the transition systems for semantic parsing [19], CCG parsing [2], non-projective constituency parsing [8] and non-projective dependency parsing [11].

5   Transition-Based Bottom-Up Parser for MG

It is striking how many of the operations and structures from transition-based shift-reduce parsing have their counterparts in Minimalist Syntax, as described in [4] and formalized in [5]. The stack plays a similar role to a minimalist *workspace*, a buffer looks similar to a *lexical array*, a configuration is like a *stage* in the minimalist derivation, shift behaves as a *select* operation and *reduce* behaves like *merge*.

A big part of Shift-Reduce parser can easily be modified to give support for Minimalist Grammars. Out of 5 rules of Minimalist Grammars, 4 are trivial to integrate: *move1* and *move2* are essentially unary feature simplification transitions, and *merge1* and *merge2* are operations that put two consecutive constituents together in almost the same way as CFG does. The only complicated cases are *merge3* and empty string terminals.

5.1 Handling Discontinuities with Online Reordering

Operation *merge3* causes complications because it merges discontinuous elements. To account for that, we introduce the possibility to reorder the elements on the stack so that the constituents that are not neighbouring can be merged. However, that is not enough because the non-head argument of *merge3* will later trigger one of the move operations that needs to satisfy neighbouring conditions. To be able to easily check if the moving constituent satisfies this constraint, the representations of the constituent that is used is the same as the representation of the constituent in the chart parser: spans and their associated chains.

5.2 Explicit Generation of Empty Strings

The problem of empty strings is mostly specific to Minimalist Grammars, since many grammar formalisms that do not have empty categories, for example CCG, do not need to account for it. Empty string terminals are introduced just like the non-empty string terminals by using an operation similar to shift, except that the buffer is not influenced by the transition. The representation of the shifted empty terminal is similar to the one in chart based parser, except that for the span we use wildcard symbols $(*, *)$ – what that means is that the constituent with this span is not a subject to the linear ordering constraints imposed by *merge1*, *merge2* and *move1*, but only to the feature matching constraints. The interpretation of the wildcard $(*, *)$ can depend on the operation that is being used in. For example, if we have a head with span $(2, 5)$ and it selects the empty constituent with span $(*, *)$ by using *merge1* then we can treat the empty constituent as if it is positioned at $(5, 5)$ (in the gap between 4th and 5th word). Note that the size of the span $(*, *)$ can never be bigger than 0 because it represents only empty elements.

Being able to explicitly generate empty strings can also cause the parser to generate empty strings *ad infinitum*, casing the parser to get stuck in the infinite loop. To prevent that we can define upper number of empty strings that can be generated for the sentence of length n which we allow to be any linear function of n. Knowing ahead of time which linear function correctly predicts the number of empty elements in a sentence is impossible because there might be no function that does that (there can be infinite number of empty strings). However, for the actual natural languages the number of empty strings is not infinite. A heuristic that can be used to determine the maximal number of empty strings is the one which assumes that: (1) empty strings appear only with function words, (2) there is a some constant of maximal number of function projections per clause (for example based on hierarchy of projections [1]) and (3) every clause contains at least one pronounced word. In that case the maximal number of empty strings is the product of the maximal number of clauses (which is the number of observed) and the maximal number of function projections per clause. Clearly this method is too conservative about the upper bound of the number of empty strings, so in practice maybe a better approach would be to estimate the number on a treebank on which the parser is trained.

5.3   Parser Description

The basic units of the minimalist transition based parser are lexical items (LI) and minimalist items (MI). Lexical items are just indices of the words in the sentence that is being parsed. Minimalist items are the same as the items in the chart-based parser that was presented in Sect. 3. For clarity we surround minimalist items with braces.

The main control structures are two stacks σ_1 and σ_2 and one buffer β (implemented as a queue). Buffer β represents the sequence of lexical items waiting to be selected for building the derivation. Stacks are sequences of already built syntactic objects i.e. minimalist items. The first stack σ_1 is the main stack that is used for actual building of syntactic objects by application of *merge, move* and

$$axiom \quad \langle\ [],\ [],\ [0,\ldots,n-1],\ 0\ \rangle$$

$$goal \quad \langle\ [\{(0,n)\cdot c\}],\ [],\ [],\ k\ \rangle$$

$$select\{\gamma\} \quad \frac{\langle\ \sigma_1,\ \sigma_2,\ i|\beta,\ k\ \rangle}{\langle\ \sigma_1|\{(i,i+1)::\gamma\},\ \sigma_2,\ \beta,\ k\ \rangle}\ w_i::\gamma \in Lex$$

$$selectEpsilon\{\gamma\} \quad \frac{\langle\ \sigma_1,\ \sigma_2,\ \beta,\ k\ \rangle}{\langle\ \sigma_1|\{(*,*)::\gamma\},\ \sigma_2,\ \beta,\ k+1\ \rangle}\ k<e \wedge \epsilon::\gamma \in Lex$$

$$tmerge \quad \frac{\langle\ \sigma_1|x|y,\ \sigma_2,\ \beta,\ k\ \rangle}{\langle\ \sigma_1|merge(x,y),\ \sigma_2,\ \beta,\ k\ \rangle}\ (x,y)\in Dom(merge)$$

$$tmove \quad \frac{\langle\ \sigma_1|x,\ \sigma_2,\ \beta,\ k\ \rangle}{\langle\ \sigma_1|move(x),\ \sigma_2,\ \beta,\ k\ \rangle}\ x\in Dom(move)$$

$$swap \quad \frac{\langle\ \sigma_1|x|y,\ \sigma_2,\ \beta,\ k\ \rangle}{\langle\ \sigma_1|y,\ x|\sigma_2,\ \beta,\ k\ \rangle}\ spanStart(x)<spanStart(y)$$

$$takeBack \quad \frac{\langle\ \sigma_1,\ x|\sigma_2,\ \beta,\ k\ \rangle}{\langle\ \sigma_1|x,\ \sigma_2,\ \beta,\ k\ \rangle}$$

Fig. 2. Deduction system for MG transition-based parser

variants of *select* operations. The second stack σ_2 is the auxiliary stack that is used for reordering minimalist items in σ_1 (it will be explained later how). The configuration (parser state) consists of σ_1, σ_2, β and an integer k that represents the count of ε transitions (transitions that generate empty strings) that led to that configuration.

The deduction system of the minimalist transition based parser is shown in Fig. 2. The starting configuration of the parser is a configuration with empty stacks (no syntactic object is built so far), buffer filled with indices of words in the sentence and the count of ε transitions set to 0. The goal configuration that the parser tries to get to is the one in which all elements of the buffer would be used, there would be no elements on hold in the auxiliary stack and the main stack has only one MI which is the complete MI (as defined for the chart parser).

The *select*$\{\gamma\}$ transition takes the first LI in the buffer and puts it on top of the main stack in the form of MI with γ chain. That happens iff there is an entry in the lexicon where word represented by LI has chain γ. That can be done only for non-ε entries in the lexicon because these are the only entries that can be directly observed in the buffer. The ε entries in the lexicon are handled by *selectEpsilon*$\{\gamma\}$ transition which does not influence the buffer, but does increase the count of the empty strings and must respect the constraint that count should not be above some prespecified number e.

Naturally, we need a transition *tmerge* that uses minimalist operation merge and transition *tmove* that uses the minimalist operation move. These transitions are applied only if the logical expressions represented by the MI on top of the main stack fall in the domain of the functions *merge* and *move* (as defined in the chart-based parser).

Discontinuity can be achieved by reordering the words in the sentence in such way that the sentence becomes contiguous. We illustrate this with the derivations of the sentence "Phong likes what Roki draws" with the following Minimalist Grammar:

$$\varepsilon :: = vc$$
$$\varepsilon :: = v + whc$$
$$\text{likes} :: = c = dv$$
$$\text{draws} :: = d = dv$$
$$\text{Phong} :: d$$
$$\text{Roki} :: d$$
$$\text{what} :: d - wh$$

The derived tree for this sentence is shown in Fig. 3a. The leaf nodes in this tree are ordered in the same way as is the surface word order of the sentence. The head for each constituent is marked with an arrow-like label, which points to the constituent which contains the head. In this derived tree it is not possible to see in which order and where the operations *merge* and *move* were applied. In order to see this we need a derivation tree like the one presented in Fig. 3b. In

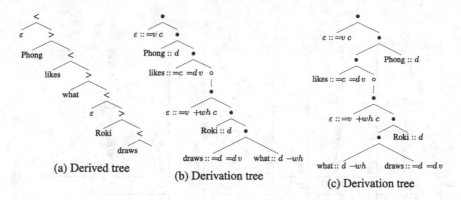

(a) Derived tree

(b) Derivation tree

(c) Derivation tree

Fig. 3. Trees for sentence "Phong likes what Roki draws"

this tree, the merge operation is marked with • and the move operation with ∘. The derivation tree is a rooted unordered binary branching tree. The ordering of nodes does not matter because merge is a commutative operation. Hence, both Fig. 3b and c represent the same derivation tree which produces the derived tree in Fig. 3a.

If the words in the sentence that is being parsed were ordered the way leaves are ordered in Fig. 3b or c then parsing would be projective and the deduction rules we defined so far would suffice. There can be exponentially many permutations of the words that would make the parsing projective and it is enough if the parser finds only one of them. We call these orderings "deep word orders".

To achieve this reordering of the elements that are participating in parsing we introduce two transitions to the parser: *swap* and *takeBack*. The transition *swap* takes the 2nd top MI from the σ_1 and puts it on top of the auxiliary σ_2. The transition *takeBack* returns these MI back to the main stack. By combining *swap* and *takeBack* we can derive any permutation of the minimalist items of the main stack. To prevent cycles of *swap* and *takeBack* there is a constraint that starting positions of the MIs that are being swapped are in the original word order.

A full transition sequence for the example sentence and example grammar is given in Fig. 4. This is only one of the possible transition sequences. We could have chosen some other sequence of *swap* and *takeBack* transitions that would produce the same derivation tree. The key part of this example are transitions *swap* and *takeBack*. These two transitions swap the order of minimalist items for the words "what" and "Roki" and in this way make parsing projective.

stage	Main stack σ_1	Auxiliary stack σ_2	Buffer β	k	transition
1			$\underset{\text{Phong}}{0}$, $\underset{\text{likes}}{1}$, $\underset{\text{what}}{2}$, $\underset{\text{Roki}}{3}$, $\underset{\text{draws}}{4}$	0	$select\{d\}$
2	$\{(0,1) :: d\}$ Phong		$\underset{\text{likes}}{1}$, $\underset{\text{what}}{2}$, $\underset{\text{Roki}}{3}$, $\underset{\text{draws}}{4}$	0	$select\{=c \,{=}d\, v\}$
3	$\{(0,1) :: d\} \cdot \{(1,2) :: {=}c \,{=}d\, v\}$ Phong likes		$\underset{\text{what}}{2}$, $\underset{\text{Roki}}{3}$, $\underset{\text{draws}}{4}$	0	$select\{d \,{-}wh\}$
4	$\{(0,1) :: d\} \cdot \{(1,2) :: {=}c \,{=}d\, v\} \cdot \{(2,3) :: d \,{-}wh\}$ Phong likes what		$\underset{\text{Roki}}{3}$, $\underset{\text{draws}}{4}$	0	$select\{d\}$
5	$\{(0,1) :: d\} \cdot \{(1,2) :: {=}c \,{=}d\, v\} \cdot \{(2,3) :: d \,{-}wh\} \cdot \{(3,4) :: d\}$ Phong likes what Roki		$\underset{\text{draws}}{4}$	0	$swap$
6	$\{(0,1) :: d\} \cdot \{(1,2) :: {=}c \,{=}d\, v\} \cdot \{(3,4) :: d\}$ Phong likes Roki	$\{(2,3) :: d \,{-}wh\}$ what	$\underset{\text{draws}}{4}$	0	$takeBack$
7	$\{(0,1) :: d\} \cdot \{(1,2) :: {=}c \,{=}d\, v\} \cdot \{(3,4) :: d\} \cdot \{(2,3) :: d \,{-}wh\}$ Phong likes Roki what		$\underset{\text{draws}}{4}$	0	$select\{=d \,{=}d\, v\}$
8	$\{(0,1) :: d\} \cdot \{(1,2) :: {=}c \,{=}d\, v\} \cdot \{(3,4) :: d\} \cdot \{(2,3) :: d \,{-}wh\} \cdot \{(4,5) :: {=}d \,{=}d\, v\}$ Phong likes Roki what draws			0	$tmerge$
9	$\{(0,1) :: d\} \cdot \{(1,2) :: {=}c \,{=}d\, v\} \cdot \{(3,4) :: d\} \cdot \{(4,5) :: {=}d\, v, (2,3) : {-}wh\}$ Phong likes Roki $\overset{\bullet}{}$ what draws			0	$tmerge$
10	$\{(0,1) :: d\} \cdot \{(1,2) :: {=}c \,{=}d\, v\} \cdot \{(3,5) : v, (2,3) : {-}wh\}$ Phong likes Roki $\overset{\bullet}{}$ what draws			0	$selectEpsilon\{=v \,{+}wh\, c\}$

Fig. 4. Part 1 of example transition sequence

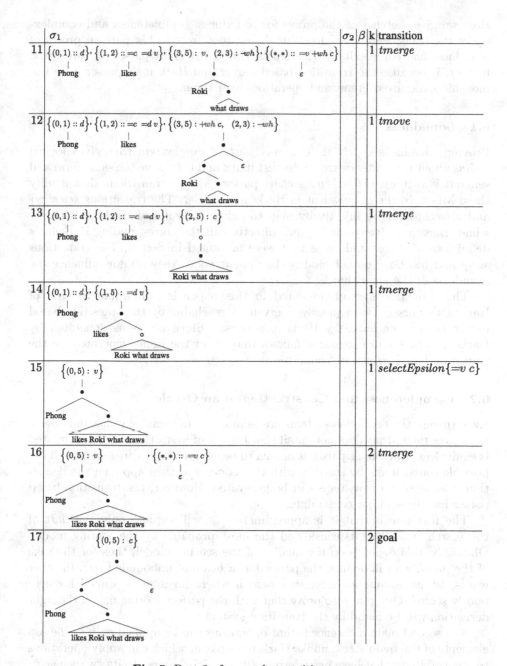

Fig. 5. Part 2 of example transition sequence

6   Soundness, Completeness and Complexity

Here we give sketches of the proofs for soundness, completeness and complexity of the transition-based algorithm. The proofs rely in big part on proofs of soundness and complexity of Harkema's Minimalist chart parser [6] (presented in Sect. 3) because the transition-based parser and Harkema's parser have isomorphic structure of items and operations over them.

6.1   Soundness

Proving soundness is trivial. The only part of our system that gives logical claims about the sentence are minimalist items and they have the same form and semantics as items in Harkema's chart parser. All the transitions that modify these MIs have their equivalent in Harkema's parser. The transitions $select\{\gamma\}$ and $selectEpsilon\{\gamma\}$ bijectivelly map to axiomatic rules of Harkema's parser while transitions $tmerge$ and $tmove$ directly call the corresponding Harkema's definitions of $merge$ and $move$ that were presented in Sect. 3. The transitions $swap$ and $takeBack$ do not modify the $mini\text{-}items$ so they do not influence the soundness of the algorithm.

The deduction system presented in this paper is isomorphic to that of Harkema's parser. Consequently, every item reachable by the transition-based parser is also reachable by Harkema's parser. Since all items generated by Harkema's parser are sound, it follows that all of the items generated by the transition-based parser are sound too.

6.2   Completeness and Construction of an Oracle

Even though the deduction systems are isomorphic in terms of items and operations over them, that does not entail that the set of items that can be generated is equivalent. Harkema's parser is proven to be complete – it can generate all the possible sound items by starting with the axiom and then applying the deduction rules until no new items can be generated. However, the transition-based parser has three major constraints.

The first one is that it is approximate – it will explore only the part of the search space that is considered the most probable by the scoring model. Obviously, this depends on the quality of the scoring model, thus for the sake of the proof, we will assume the parser has a beam of unbounded size. In other words, let us assume an exhaustive search where no item is pruned however poorly scored. Our goal is to prove that with the perfect scoring model the right derivation will be found by the transition system.

The second main difference is that operations can be applied only to the top elements of the main stack, unlike Harkema's parser which can apply operations to any two items that have been derived (it has global access to its "workspace", which is a chart). The main question is then whether this limits the set of items that can be deduced using the transition-based system. Given that Harkema's parser is complete and that all functions of Harkema's parser are present in the transition-based parser, we just need to show that for any derivation tree there is a sequence of transitions that would derive it. This conversion of a MG derivation to the sequence of transitions can be interpreted as a construction of the *oracle* sequence of transitions. The oracle is used often in transition-based parsing as a sequence of transitions on which the probabilistic parsing model is trained. There are many possible oracles for any MG tree so in the probabilistic setting all these oracles should ideally be treated as latent variables. However, experience from other grammar formalisms shows that using just one oracle seems to be good enough for most of the parsers.

The third difference is that all empty strings are explicitly generated in the transition-based parser while in the chart-based parser infinite number of empty strings can be compactly represented thanks to the dynamic programming. Since the maximal number of empty strings that can be generated is limited by some predefined constant e, any proof of completeness is limited to the trees that have less than e empty strings. Here we will assume that e is infinite. In other words, we show that for a sufficiently large e any MG derivation can be generated.

First, we cover the case in which the words of the sentence are in one of the many possible "deep word orders" (word orders in which the derivation tree is projective). In this case extracting the sequence of transitions is easy: we just need to traverse the derivation tree in the post-order traversal (the "order" of the subtrees is based on the deep order of words). Every time we encounter a leaf in the derivation tree, it will cause a *select*$\{\gamma\}$ or *selectEpsilon*$\{\gamma\}$ transition. Every time we encounter a binary branching node, it will be a *tmerge* transition and every time we encounter a unary branching node, it will be a *tmove* transition. So, if the words are processed in the projective "deep order" there is always a transition sequence that will produce any projective derivation tree.

Now we show that even if the words are in non-projective word order that they could still be processed in a projective order. Let us say that the next LI that should come on the main stack is on the mth position in the buffer. What we need to do is m *select*$\{\gamma\}$ transitions, followed by $m - 1$ *swap* transitions. The alternative situation is that the next element is on the mth positions in the auxiliary stack σ_2. The process is the same except that instead of invoking m *select*$\{\gamma\}$ transitions we invoke m *takeBack* transitions, followed by $m - 1$ *swap* transitions.

Given that we can find transitions for any derivation in projective word order, and that any non-projective derivation can be traversed in projective order, it follows that we can find a transition sequence for any non-projective derivation. This, together with Harkema's proof of the completeness of the basic functions *merge* and *move* that the transition-based parser uses, makes the transition-based parser complete.

6.3   Computational Complexity

Because the transition-based parser does not pack its derivations by using dynamic programming, its complexity with unbounded beam will be exponential in sentence length. However, since transition-based parsers are never used to search trough the full space of derivations, but always with a limited beam, we will here focus on the complexity of constructing a single derivation.

The complexity can be determined by estimating the maximal number of times each transition type will be used. The number of transitions $select\{\gamma\}$ is n because it will be used only once and for each of the observed words. By design, the transition $selectEpsilon\{\gamma\}$ will be used maximally e times which is a linear function of n. The maximal number of $tmerge$ transitions is equivalent to the maximal number of binary nodes in a binary branching tree over a $n + e$ words which is $n + e - 1$. The maximal number of $tmove$ operations is equivalent to the number of all the $licensees$ in the sentence. If the maximal number of licensees that the lexicon has per entry is some constant m, then the maximal number of licensees in the sentence is $m * (n + e)$. The number of $swap$ operations is equivalent to the number of $takeBack$ operations. In the best case there will be no swapping (all the words are in one of the possible deep word orders), and in the worst case there will be in, asymptotic notations, $O(n^2)$ $swap$s and $takeBack$s. Since m is constants we can say that for all operations the asymptotic complexity is $O(n)$, except for swapping transitions that can be between $O(n)$ and $O(n^2)$. So, the total complexity of building one derivation is dependent on the $swap$ and $takeBack$ transitions making this parser's worst case complexity $O(n^2)$ and best case complexity $O(n)$.[1]

7   Parsing of Finite-State Automaton

The minimalist transition-based parser can easily be extended to parse not only sentences, but also regular sets of sentences encoded in a finite-state automaton (FSA). All that needs to be modified are representations of buffer and the $select\{\gamma\}$ transition since it is the only transition that changes the buffer. The buffer would now instead of a queue be an FSA that is being parsed and it would additionally contain the pointer to *the current state* up until which the input was consumed. The new $select\{\gamma\}$ transition would pick one of the outgoing arcs of the current state, consume the arc's label (the same way it used to consume the word in the buffer) and change the current state to the target state of the selected arc. The initial (axiom) configuration would be with empty stacks and with buffer FSA that has its current state set to its initial state. The goal configuration would be the same as before, except for having the additional condition that the current state in an FSA is the final state of an FSA.

The simplicity of doing discontinuous parsing of FSAs can be crucial is some cases when the input is ambiguous. Take for instance morphologically rich languages: doing morphological segmentation in these languages is difficult and the selection of the right segmentation can be done only during the syntactic processing because of different forms of agreement. Now instead of parsing a potentially bad 1-best guess of the morphological analyser, we can take a full lattice that

---

[1] Interestingly, Nivre shows that the number of swap transitions in the real dataset for dependency parsing is very small which makes the transition-based parser run in expected linear time [11]. Hopefully, this will also be true for Minimalist Grammars.

would encode many hypotheses of the possible segmentation and then let the parser decide which one is the best. Another use case of FSA parsing is processing the ambiguous output of the speech recognizer which is often encoded in lattices.

8   Conclusion and Future Work

The transition-based parser presented in this paper, if supported by a good probabilistic scoring model, could handle even the longest sentences very efficiently. The very small computational complexity of building one derivation makes the transition-based parser for Minimalist Grammars as fast as its counterparts for CCG, dependency and constituency parsing.

The usual motivation for using simpler formalisms such as dependency and context-free grammars is their efficiency. However, given that the presented parser is asymptotically as fast as the approximate parsers for the simpler formalisms, the natural language processing community can start considering Minimalist Grammars as a possible more expressive alternative. In order for this transition to become a reality, a necessary next step is creation of the scoring model, as well as the creation of a Minimalist treebank on which the scoring model would be trained.

Acknowledgments. I would like to thank Raquel G. Alhama, Joachim Daiber, Phong Le, Wilker Aziz and Khalil Sima'an for useful discussions and comments on the early versions of this paper. I am also grateful to the three anonymous reviewers for their insightful comments. This work was supported by STW grant nr. 12271.

References

1. Adger, D.: Core Syntax: a Minimalist Approach, vol. 33. Oxford University Press, Oxford (2003)
2. Ambati, B.R., Deoskar, T., Johnson, M., Steedman, M.: An incremental algorithm for transition-based CCG parsing. In: Proceedings of 2015 Conference of the North American Chapter of the Association for Computational Linguistics: Human Language Technologies. Association for Computational Linguistics (2015)
3. Chomsky, N.: Syntactic Structures. Mouton & Co., The Hague (1957)
4. Chomsky, N.: The Minimalist Program, vol. 1765. Cambridge University Press, Cambridge (1995)
5. Collins, C., Stabler, E.: A formalization of minimalist syntax. Syntax **19**(1), 43–78 (2016)
6. Harkema, H.: A recognizer for minimalist languages. In: Bunt, H., Carroll, J., Satta, G. (eds.) New Developments in Parsing Technology. TSLT, vol. 23, pp. 251–268. Springer, Dordrecht (2005)
7. Hunter, T., Dyer, C.: Distributions on Minimalist Grammar derivations. In: Proceedings of 13th Meeting on the Mathematics of Language (MoL 2013), pp. 1–11. Association for Computational Linguistics, Sofia, August 2013

8. Maier, W.: Discontinuous incremental shift-reduce parsing. In: Proceedings of 53rd ACL (Long Papers), vol. 1, pp. 1202–1212. Association for Computational Linguistics, Beijing, July 2015
9. Mainguy, T.: A probabilistic top-down parser for Minimalist Grammars (2010). CoRR arXiv:abs/1010.1826
10. Mi, H., Huang, L.: Shift-reduce constituency parsing with dynamic programming and POS tag Lattice. In: Proceedings of 2015 Conference of the North American Chapter of the Association for Computational Linguistics: Human Language Technologies, pp. 1030–1035. Association for Computational Linguistics, Denver, May–June 2015
11. Nivre, J.: Non-projective dependency parsing in expected linear time. In: Proceedings of Joint Conference of the 47th Annual Meeting of the ACL and the 4th International Joint Conference on Natural Language Processing of the AFNLP, pp. 351–359 (2009)
12. Sagae, K., Lavie, A.: A classifier-based parser with linear run-time complexity. In: Proceedings of 9th International Workshop on Parsing Technology, Parsing 2005, pp. 125–132. Association for Computational Linguistics., Stroudsburg (2005)
13. Shieber, S.M., Schabes, Y., Pereira, F.C.N.: Principles and implementation of deductive parsing. J. Log. Program. **24**, 3–36 (1995)
14. Stabler, E.: Derivational minimalism. In: Retoré, C. (ed.) LACL 1996. LNCS, vol. 1328, pp. 68–95. Springer, Heidelberg (1997). doi:10.1007/BFb0052152
15. Stabler, E.: Top-down recognizers for MCFGs and MGs. In: Proceedings of 2nd Workshop on Cognitive Modeling and Computational Linguistics, pp. 39–48. Association for Computational Linguistics, Portland, June 2011
16. Stabler, E.P.: Minimalist Grammars and recognition. In: Rohrer, C., Rossdeutscher, A., Kamp, H. (eds.) Linguistic Form and Its Computation, pp. 389–440. CSLI Publications, Stanford (2001)
17. Stabler, E.P.: Recognizing head movement. In: de Groote, P., Morrill, G., Retoré, C. (eds.) LACL 2001. LNCS (LNAI), vol. 2099, pp. 245–260. Springer, Heidelberg (2001). doi:10.1007/3-540-48199-0_15
18. Stabler, E.P., Keenan, E.L.: Structural similarity within and among languages. Theoret. Comput. Sci. **29**, 345–363 (2003). Algebraic Methods in Language Processing
19. Titov, I., Henderson, J., Merlo, P., Musillo, G.: Online graph planarisation for synchronous parsing of semantic and syntactic dependencies. In: Proceedings of 21st International Joint Conference on Artifical Intelligence, IJCAI 2009 (2009)
20. Zhang, Y., Clark, S.: Syntactic processing using the generalized perceptron and beam search. Comput. Linguist. **37**(1), 105–151 (2011)

A Compositional Semantics for 'If Then' Conditionals

Mathieu Vidal[1,2](✉)

[1] Cognitions Humaine et Artificielle (CHArt), Université Paris 8, Paris, France
math.vidal@laposte.net
[2] Philosophie, Pratiques & Langages (PPL),
Université Grenoble Alpes, Grenoble, France

Abstract. This paper presents the first compositional semantics for *if then* conditionals. The semantics of each element are first examined separately. The meaning of *if* is modeled according to a possible worlds semantics. The particle *then* is analyzed as an anaphoric word that places its focused element inside the context settled by a previous element. Their meanings are subsequently combined in order to provide a formal semantics of *if A then C* conditionals, which differs from the simple *if A, C* form. This semantics has the particularity of validating contraposition for the first type but invalidating it for the second type. Finally, a detailed examination of the sentences presented in the literature opposing this schema of reasoning shows that these counterexamples do not generally concern *if then* conditionals but, rather, *even if* conditionals and that contraposition is therefore a valid means of reasoning with regard to *if then* conditionals in natural language, as this system predicts.

Keywords: Conditional logic · If · Then · Contraposition

Introduction

[16] has observed that the addition of *then* to *even if* conditionals is unsuitable:

(1) Even if John is drunk, Bill will vote for him.
(2) # Even if John is drunk, then Bill will vote for him.

This example clearly dispels the misconception whereby the addition of *then* to a conditional does not change its meaning. [1,19,30] and their followers have implicitly adopted this position in their theories, according to which no formal distinction is made between the *if A, C*, the *even if A, C*, and the *if A then C* forms. In this paper, I shall try to overcome this limitation by providing a compositional analysis of the meaning of *if then* conditionals. More precisely,

M. Vidal—I would like to thank Philippe Schlenker for suggesting to develop the ideas from [37] in a compositional way, the referees and Denis Perrin for their help to improve the paper, and the editors for the publication of these proceedings.

© Springer-Verlag GmbH Germany 2016
M. Amblard et al. (Eds.): LACL 2016, LNCS 10054, pp. 291–307, 2016.
DOI: 10.1007/978-3-662-53826-5_18

I will present a formal semantics for both *if* and *then*, ultimately showing that their combination leads to a particular semantics for the *if then* form.

This article is the first formal attempt to provide a compositional analysis of the formal semantics of the *if then* conditional in terms of the elements *if* and *then*. Indeed, previous explanations of the meaning of *then*, such as those proposed by [16, 20] were pragmatic. On the contrary, I argue for a formal analysis because the mathematical description it employs provides a more precise explanation. Furthermore, its treatment can often be automated on a computer. Finally, by having a different semantics, the various conditionals validate different forms of reasoning.

This theory concerns both indicative and subjunctive conditionals. Indeed, a similar problem occurs in counterfactuals, such as in the following examples:

(3) Even If John had been drunk, Bill would vote for him.
(4) # Even If John had been drunk, then Bill would vote for him.

As the difference between the indicative and the subjunctive has no impact on this issue, I shall refrain from examining the moods used in conditionals in this paper. I will also not be dealing with embedded conditionals (as *if A then if B then C*) and chains of conditionals (as *if A then B and if B then C*). These forms of reasoning display dynamic features, so, for the sake of simplicity, I will limit my analysis to a static approach in this initial presentation of the theory.

This article is divided into four parts. Section 1 presents the semantics for *if*. Section 2 introduces the formal semantics for *then*, for cases both inside and outside conditionals. Section 3 deals with the compositional semantics obtained by combining *if* and *then* and details the main schemas of reasoning that are valid for this approach. Section 4 focuses on contraposition in order to show that its validity, which the theory predicts, dovetails the facts observed in natural language, contrary to what is usually argued in the field.

1   If

A conditional is a linguistic expression composed of two clauses. One clause, called the *protasis* or the *antecedent*, states the conditions under which the other clause, called the *apodosis* or the *consequent*, is considered true. The marking of conditional sentences differs greatly between natural languages. However, according to [5], the general strategy is to mark the protasis using a lexical element, a particular inflectional morphology, or a purely syntactic means. If we take English as our reference language, the simplest syntactic form of a conditional is *if A, C*, with *A* being the protasis, *C* the apodosis, and *if* the marker of conditionality.

For the semantics of the *if*, I will draw upon [38]'s proposal. The semantics presented in that paper were already successfully combined with the semantics of *even* in order to provide the meaning of the *even if* conditional. The other advantage of this approach is that it can be seen as a refined version of Stalnaker's semantics, in which the problem of unconnected conditionals is resolved. For

Stalnaker, a conditional *if A, C* is true in a possible world w if the closest A-world of w is also a C-world. [7] refined this semantics in order to select not only one world, but a set of possible worlds as the closest worlds in which the antecedent is true. That is the version I shall discuss here. Since w is the closest world to itself in this semantics, as soon as the antecedent is true in the initial world, this initial world is also the world in which the consequent is evaluated.[1] The unfortunate result of this is that, if the consequent is also true in the initial world, the whole conditional is true, regardless of the links between the antecedent and the consequent. For instance, the sentence "if Mickey Mouse has four fingers per hand, Mickey Mouse has big ears" is automatically true, despite the lack of connection between its components. The associated schema of inference is called Conjuctive Sufficiency (CS): $A, C \vDash A \rightharpoonup C$.[2] This inference of a conditional from a conjunction has already been criticized from an intuitive point of view by [22] and the psychological experiments conducted in [21] and [29] show that it is also not endorsed by ordinary subjects.

To resolve this issue, Vidal proposes breaking the process of evaluation down into two parts. Prior to this evaluation, the subject can hold one of three positions concerning the antecedent: it is either believed to be true, believed to be false, or believed to be neither true nor false. During the first step of evaluation, the antecedent is inhibited. This means that it is no longer believed to be true or false. The position concerning the antecedent is now neutral (neither true nor false). The first advantage of this inhibition is that, if the antecedent was previously believed to be false, it can now be added to the stock of beliefs without leading to a contradiction. The second advantage of this inhibition is that, if the antecedent was previously believed to be true, its addition can be now accomplished with slight variations. This addition to the stock of beliefs occurs during the second step of the evaluation, the antecedent being reconstructed in several different ways. If the consequent is obtained in all of the reconstructions of the antecedent, the conditional is declared true.

These ideas can be turned into a formal semantics. In the present paper, I will only expose the main elements. The reader is referred to Appendix A of [38] for all the technical details. A stock of beliefs is represented by a set of possible worlds. In our case and to put it simply, the initial beliefs are rendered by a unique possible world. As some sentences are no longer believed to be true or false during the judgment, some of the possible worlds of this semantics are trivalent. The step of inhibition is modeled through what is called a *neutralization function*, noticed n. The step of reconstruction is modeled through what is called an *expansion function*, noticed e. There are limits on the possible reconstructions of the antecedent. Some of them are too absurd or too improbable to be considered. Hence, only a part of them is envisaged, and they are constrained by what is called a *universe of projection*. An important aspect of this

---

[1] According to Chellas's semantics, the following condition holds: if $w \in [A]$, then $f_w(A) = \{w\}$.

[2] The symbol $\rightharpoonup$ stands for the *if A, C* conditional.

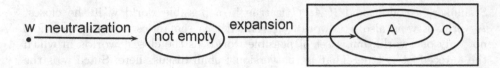

Fig. 1. Basic semantics of the conditional

universe of projection is that, at the end of the judgment, the antecedent and the consequent are either true or false in each possible reconstruction envisaged.

This semantics for the conditional "if A, C" is depicted in Fig. 1, in which w stands for the starting world of evaluation and the square for the universe of projection.

More formally, the meaning of the word *if* is the following:

Definition 1 (Meaning of If). *Let w be a possible world, λ the lambda abstractor, X and Y some sentences, $[X]$ and $[Y]$ their truth-sets, n a neutrality function and e an expansion function both governed by the universe of projection U used to evaluate the conditional at hand. Then, the **meaning of if** is*

$$[(if\ X)\ Y] = \begin{cases} 1 & if\ \lambda X \lambda Y\ n_w(X) \neq \emptyset \wedge e_{n_w(X)}(X) \subseteq [Y] \\ 0 & if\ \lambda X \lambda Y\ n_w(X) = \emptyset \vee e_{n_w(X)}(X) \nsubseteq [Y] \end{cases}$$

We obtain the following truth-conditions for the sentence *if A, C*.

Theorem 1 (Truth-Conditions of If).
$\vDash_w (if\ A)\ C$ *iff in the associated universe of projection U*

(i) $n_w(A) \neq \emptyset$
(ii) $e_{n_w(A)}(A) \subseteq [C]$

Proof (Truth-conditions of if A, C).
[if] $= \lambda X \lambda Y\ n_w(X) \neq \emptyset \wedge e_{n_w(X)}(X) \subseteq [Y]$
[if A] $= \lambda Y\ n_w(A) \neq \emptyset \wedge e_{n_w(A)}(A) \subseteq [Y]$
[if A, C] $= n_w(A) \neq \emptyset \wedge e_{n_w(A)}(A) \subseteq [C]$ $\square$

The first condition says that the inhibition of the antecedent must be successful, meaning it does not lead to an empty set of worlds. The second condition says that, by starting from this state of inhibition and by reconstructing the antecedent, all of the possible reconstructions are worlds in which the consequent is also true. In this semantics, the schema CS is no longer valid. Indeed, several reconstructions of the antecedent are considered. Hence, if the antecedent and the consequent are true in the initial world, this initial world is only one of the possibilities envisaged for the whole evaluation of the conditional. Other possibilities in which the antecedent is true but the consequent is false can also be obtained. This constraint is represented by a weaker semantic condition, compared to the one used in Chellas's system.[3]

---

[3] If $w \in [A]$ and $n_w(A) \neq \emptyset$, then $w \in e_{n_w(A)}(A)$.

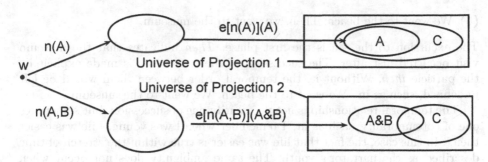

Fig. 2. Invalidity of the strengthening of the antecedent

An important consequence of this semantics is the non-monotonic behavior of the conditional. Indeed, the neutralization of two different sets of sentences can potentially lead to two different universes of projection. This feature is the main mechanism behind the invalidation of the strengthening of the antecedent: $A \rightharpoonup C \nvDash (A \wedge B) \rightharpoonup C$. Consider the oft-used example "If Tweety is a bird, Tweety flies." For the premise, the conditional can be declared true because the most well-known type of bird is envisaged, meaning a flying bird. However, in the consequence "If Tweety is a bird and Tweety is a penguin, Tweety flies," the addition of a new antecedent forces us to reconsider a few assumptions, thus calling into question the hypothesis that we were talking about a flying bird. This mechanism is illustrated in Fig. 2 in which different sets of neutralized sentences lead to different expansion sets and ultimately to different truth-values.

Notice that this feature also explains that contraposition is invalid in this system: $A \rightharpoonup C \nvDash \neg C \rightharpoonup \neg A$. Indeed, the set of neutralized sentences is $\{A\}$ for the premise and $\{\neg C\}$ for the consequence, implying that the two set relationships occur within two different universes of projection and are therefore not related. For that same reason, transitivity is also invalid: $A \rightharpoonup B, B \rightharpoonup C \nvDash A \rightharpoonup C$. Since the two premises respectively have $\{A\}$ and $\{B\}$ as inhibited sentences, their set-theoretic relationships again occur within two different universes of projection and cannot be combined to deduce the consequence. The invalidation of these three last schemas of reasoning is in line with the conditionals logics proposed by [19,30].

2 Then

I would now like to examine the semantics of *then*. It generally means *at that time*, *at that place*, or *in that case*, depending on the temporal, spatial, and logical interpretation. It embeds a phrase and relates it to a previous one. The preceding phrase can be either explicit such as the sentence that came just before or implicit. Hence, *then* is an anaphoric word that picks up the context of a prior element in order to place the subsequent element in this context. The refereed element is made redundant in order to say that the subsequent element occurs during or just after these circumstances or this place. Consider the following example.

(5) We went to the beach. Then, we went to the museum.

The excursion to the sea is the first phase. *Then* indicates that the museum visit occurred just after. The succession of the two events is made explicit by the particle *then*. Without it, the temporal order between them would be left undefined, such as in "We went to the beach. We went to the museum."

The temporal relationship is not necessarily one of succession, but sometimes one of cotemporality, such as in "I remember when I was young. Life was easier then." In this case, the fact that life was easier is true within the stretch of time described as the narrator's youth. The same ambiguity does not occur when spatial relationships are considered. The sentence "Standing beside my mother is my father, then my sister" suggests a relationship of succession, in which each person stands next to the other. But when we want to express a colocation between two elements, the particle *there* is preferred over the particle *then*, such as in the sentence "I remember this hill. A magnificent oak was standing there."

Among the researchers who are interested in the semantics of *then*, some focus on the use of *then* inside conditional sentences and do not really consider the other cases (see [10, 16, 28]). Furthermore, they wait it to be combined with the conditional before giving a precise definition of its meaning. Their solution is therefore not sufficiently general to be satisfactory. I will examine their proposals in detail later on.

I shall now give a formal definition, remaining at the most general level of abstraction. As usual for the study of natural language, this definition is contextual. The precise elements on which *then* focuses depend on the situation. Furthermore, the precise result depends on the kind of reasoning, be it temporal, spatial, or logical.

Definition 2 (Meaning of Then). *The meaning of then is a function noted* $g[then]_{C_X,Y}$ *which takes two elements* C_X *and* Y *as inputs.* C_X *is the set of circumstances linked to the first element* X. Y *is simply the second element. The value of* $g[then]_{C_X,Y}$ *is* $C_{X,Y}$. *This means that* Y *is placed in the circumstances initially linked to* X.

Hence, the meaning of *then* is a function that incorporates the focused phrase into the context stated by the preceding discourse.

I would like to examine how this semantics applies to (5). The first sentence expresses that a group to which the narrator belongs went to the beach in the past. The circumstances here are temporal. The past circumstances can be noted by P and the sentence "We go to the beach" by X. Hence, the first input element is P_X. The second input element is Y, referring to the fact that the group went to the museum. The application of *then* results in the final relationship $P_{X,Y}$. More explicitly, Y led to X in the past.

This definition involves the following consequence. The first input, of which the constituents are the circumstances, must precede *then*. Indeed, in order to be anaphoric, the refereed element must have already occurred in the discourse. Concerning conditionals, the preceding clause is the antecedent. That explains why *then* can only be used when the protasis precedes the apodosis (6), and not in the opposite order (7), contrary to an *if A, C* conditional (8–9):

(6) If it's sunny, then I'm happy.
(7) # Then I'm happy, if it's sunny.
(8) If it's sunny, I'm happy.
(9) I'm happy if it's sunny.

This constraint does not generally hold for the second input. Sometimes, it can be asserted before *then*, for instance in "Weather was milder then." But it can also appear afterward, such as in sentence (5). Another consequence of this definition is that the first input cannot be absent. Indeed, it is very difficult to randomly begin a discourse using the word *then*. Without an element to refer to, its presence would be deemed odd or cumbersome. I am not saying that it is totally impossible. One could choose to use it for rhetorical reasons. From a logical point of view, however, this missing element would render the usual function of *then* inapplicable. Similarly, the second input is generally present. If a speaker ends a sentence without providing this element, it forces the listener to guess it. For instance, it can be used to scare a disobedient child, such as in "Be careful! If you do that, THEN ... " The consequences are such a frightening prospect that they cannot be uttered.

Then is primarily used temporally and logically. The acquisition of these two different forms of reasoning must in some way be linked. [18] sums up the studies of the developmental acquisition of conditionals in the following way. Young children interpret them as conjunctions, older children as biconditionals, and adults as conditionals. Notice that the meaning of a logical conjunction does not include a relationship of succession because $A \wedge B$ is equivalent with $B \wedge A$. The biconditional is also a relationship of equivalence between $A \equiv B$ and $B \equiv A$. This symmetry does not hold for the conditional, and this part of its meaning is probably acquired through the relationship of succession in temporal reasoning. Jean Piaget's four stages of cognitive development confirm this relationship between both types of reasoning (see [23]). According to Piaget, during the *concrete operational stage*, which occurs from ages 7 to 11, the child learns to use several operations and classify elements among them. At the end of this stage, the child acquires the spatio-temporal system of common-sense reasoning. During the final stage, known as the *formal operational stage*, which occurs starting at the age of 12, the adolescent is able to reason abstractly and, in particular, to perform hypothetical reasoning. Hence, the temporal meaning of *then* is probably learned before it is applied to conditional reasoning.

This might explain why several researchers in linguistics, such as [15,27,33] favor the temporal meaning over other uses of the word. Their main discovery is that the signification of *then* depends on its position. In a clause-initial position, *then* entails a reading of succession. In a clause-final position, it is understood as having a cotemporal meaning. Thompson tries to explain this dichotomy by using the difference between the event time and the reference time used by [26] to describe the temporal discourse. She argues that, in a final position, *then* focuses on the Verb Phrase and links its event time with the event time of the preceding clause. Both event times are therefore cotemporal. On the contrary, when *then* is in an initial position, it focuses on the Inflectional Projection and links its reference time to the reference time of the preceding clause. In this case, both event times are not linked and are therefore not cotemporal, but ordered. The weakness of this argument lies in this last inference. As soon as two event times are not linked by *then*, there is no reason why any temporal ordering should occur between them and why this ordering should be a relationship of succession. The exact mechanism allowing the cotemporal or the ordered reading, depending on the position of the adverb, is at best described although not explained in this setting.

The difference in the meaning of the temporal *then* can be more easily explained in the following way. The ordering of the words in the discourse is similar to the temporal ordering of the events referred by the words. With *then* in an initial position comes the following form: A then B. Because *then* is an anaphoric word of the preceding clause, the ordering of these elements is therefore: A A B. In this case, B clearly occurs after A and succeeds to it. On the contrary, when *then* is in the final position, the pattern is as follows: A, B then. Again, as *then* has an anaphoric function, the exact ordering of the elements is: A B A. The element B is flanked by two elements A, both of which are therefore cotemporal. This iconic explanation is based on the similarity between the ordering in the discourse and the temporality of events and is simpler than Thompson's attempt.

3   If Then

We arrive now to the main objective of this paper: obtaining a compositional meaning for *if then* conditionals, based on the definitions given for the meaning of *if* and *then* in the two preceding sections. Considering the combination of *if* and *then*, it becomes clear that the association between the particle *if* and the antecedent constitutes the base for the set of circumstances that is the first input of the semantics of the particle *then*. It is also clear that the consequent is the second input of this function. However, this compositional semantics leads to the following question: what exactly are these circumstances that constitute the first input? I argue that, in this case, the set of circumstances is constituted by the possibilities representing the inhibition of the antecedent, which occurs during the first phase of evaluation of a conditional. For the record, this inhibition is also called neutralization and is represented by the function n in the formal semantics. The first reason for this choice is that, just before the assertion of the *then*, only the antecedent has been uttered. This means that only the inhibition phase can be completed at this time. Indeed, the reconstruction phase is always missing the consequent to be proceeded. The second reason for this adoption is that the neutralization can be considered as the context determining the universe of projection in which the final evaluation will be conducted. Indeed, in this semantics, a different set of inhibited sentences leads to a different universe of projection. Being the context of a conditional, the set of possibilities envisaged at the end of the inhibition phase is therefore the first input of the meaning function of the *then*.

Definition 3 (First Input of Then when applied to If). *For a conditional, the first input of the meaning function of then is the set $n_w(A)$ with n being the neutrality function, w the starting world of the evaluation, and A the antecedent of the conditional.*

With this definition, placing the second element within the context of the first element means that the consequent is also among the set of inhibited sentences. It is added to the antecedent to form the set of sentences that are explicitly neutralized. Thus, the conditional *if A then C* receives the following meaning.

Theorem 2 (Truth-conditions of if then).
$\models_w$ *if A then C iff*

(i) $n_w(A, C) \neq \emptyset$
(ii) $e_{n_w(A,C)}(A) \subseteq [C]$

Proof (Truth-conditions of if A then C).
[if] $= \lambda X \lambda Y \ n_w(X) \neq \emptyset \wedge \ e_{n_w(X)}(X) \subseteq [Y]$
[if A] $= \lambda Y \ n_w(A) \neq \emptyset \wedge \ e_{n_w(A)}(A) \subseteq [Y]$
[if A then] $= \lambda Y \ n_w(A, Y) \neq \emptyset \wedge \ e_{n_w(A,Y)}(A) \subseteq [Y]$
[if A then C] $= n_w(A, C) \neq \emptyset \wedge \ e_{n_w(A,C)}(A) \subseteq [C]$ □

The addition of *then* modifies the basic semantics by incorporating the consequent into the set of inhibited sentences. In the simple *if A, C* form, this neutralization was optional. It now becomes mandatory with the presence of *then*. This means that if the consequent reappears in each world in which the antecedent is true, both of them are linked in a stronger way. Indeed, in the *if A, C* form, the consequent was not necessarily inhibited. Its presence in the A-worlds can therefore be explained by a potential non-inhibition, which has nothing to do with the presence of the antecedent. With the *if A then C* form, the consequent is no longer believed to be true or false at the end of the inhibition phase. If the conditional is true, the reappearance of the consequent in each A-world means that its occurrence is in one way or another linked to the occurrence of the antecedent. Their copresence can no longer be fortuitous.

I would like now to review the differences between this proposal and the most well-known approaches to the meaning of *if then* conditionals. The only really formal predecessor to this theory is the one proposed by [2], who argues for a compositional analysis, but one that is restricted to counterfactuals. Indeed, Alonso-Ovalle surprisingly chooses the consequent and more precisely its inner *would* item as bearing the main meaning of conditionality. However, an extension of this analysis to the indicative case seems doomed to fail because this mood does not contain such an auxiliary modal verb. Furthermore, this choice does not dovetail with the general strategy present in most natural languages (particularly in English), which is to modify the antecedent and not the consequent by using the word *if* in order to signal a conditional sentence [5].

The biconditional approach is defended by [8], who argue that *then* locates the obtainment of the consequent in the event associated with the antecedent. By contrast, this obtainment does not hold in other alternatives. As a consequence (see p. 145), "*then* allows for the '~P,~Q' inference precisely because it is a deictic referent for a mental space. It consequently both brings up the idea that P is the unique mental space in which Q is 'located' and explicitly marks the relevant connection between P and Q as causal and/or sequential." Dancygier and Sweetser adopt both the "P, Q" and the "~P,~Q" inferences, thus defending a biconditional interpretation of the *if then* conditional. This position has some problems. First, when the biconditional is explicitly asserted in natural language, it is with a different form: *if and only if.* There is no reason why these two forms, which differ syntactically, are equivalent from a semantic point of view. This equivalence would at least need to be justified. Another problem with a biconditional interpretation is that the occurrence of the consequent entails the occurrence of the antecedent, as in the following inference: *A iff C, C. Thus A.* But this inference is not always valid for the *if then* form. For instance, speaking about tennis, I could say, "This is a match point. If she wins the next point, then she wins the match." But subsequently knowing that she won the match, I cannot conclude that she won this match point precisely because she perhaps needed some others. For both of these reasons, the biconditional interpretation is implausible.

The last important approach that I would like to examine is defended by [10, 16, 28], with some variations. For them, the *if then* conditional carries the following implicature: there are some cases in which the negation of the antecedent is present with the negation of the consequent. As Iatridou says, "Then 'carries the presupposition' that '~p is compatible with ~q', which is weaker that '~p implies ~q'." This approach has the advantage of not implying a biconditional interpretation. In Schlenker's version, *then* is a strong pronoun, and "in the case of then in if p, then q, the implicature is that some non-p worlds are non-q worlds." This is Iatridou's idea expressed in a possible world semantics.

While this idea works in a lot of cases, there are counterexamples. For instance, I can say, "All these sweets are green. If they are green, then they are peppermint candies." In that case, the speaker knows perfectly well that there are no non-green candies in the bag. The implicature is therefore implausible. However, the main problem of this approach is its direct usage of the negation in the implicature. When *then* is used outside conditionals, the negation is not a part of its meaning. Its import for the implicature is therefore completely artificial. If we again take our example (5) "We went to the beach. Then, we went to the museum," we see that the temporal interpretation only carries the idea of succession. If we also consider *then* in isolation, it is simply an anaphoric word, the meaning of which does not contain the idea of negation. To sum it up, the issue faced by this approach is that the use of negation for conditional cases cannot be derived from a more general theory of *then.*

On the contrary, the theory I am proposing in this paper makes it possible to formally derive a particular property for *if then* conditionals, without referring to any additional implicature or other pragmatic principle. As Iatridou and

Schlenker argued, along with Dancygier and Sweetser, this effect concerns the relationship entertained by the negation of the antecedent and the negation of the consequent, but this relationship is not exactly what they proposed. Indeed, the new schema of reasoning that is introduced in this semantics is contraposition: from "If A, then C", we can immediately deduce "If not C, then not A".

(10) If I have money, then I buy bread.
(11) If do not buy bread, then I have no money.

Hence, what is predicted is that all the cases in which the consequent is false are cases in which the antecedent is false. There is therefore no biconditional interpretation. Furthermore, there is no obligation for such cases to exist. Intuitively, the validity of contraposition is obtained in the following way. The truth-conditions obtained for "if A then C" implies that the inhibition of A and C was successful. Inhibiting sentence A leads to no longer believing it to be true or false. Hence, the inhibition of a sentence is equivalent to the inhibition of its negation. Therefore, the inhibition of A and C is equivalent to the inhibition of $\neg C$ and $\neg A$. They define the same context of evaluation. Inside this universe of projection, all the A-worlds are C-worlds. As all these worlds are bivalent concerning the antecedent and the consequent, all the $\neg C$-worlds are $\neg A$-worlds. This therefore makes it possible to conclude that "if $\neg C$ then $\neg A$" is true.[4]

4   A Defense of Contraposition for If Then Conditionals

The idea that contraposition is a valid schema of reasoning for *if then* conditionals goes against what is usually argued in the related literature, which is why I shall now present a detailed defense of this position. Some authors, such as [13,14,32,35,36], reject the usual counterexamples to contraposition by using pragmatic principles. They argue that these counterexamples are conversationally infelicitous and cannot be counted against their theories. They also apply the same mechanism to reject counterexamples to other patterns of inferences such as the strengthening of the antecedent or transitivity. I adopt a different position by considering the counterexamples to these different schemas of reasoning to be generally correct and that they should not be removed by a call to pragmatic principles. In particular, the invalidity of the strengthening of the antecedent shows that the conditional is a non-monotonic connective, and this invalidity should be kept in a correct theory of hypothetical reasoning. Furthermore, there is nothing to say against most of the counterexamples to contraposition, except that they do not apply to the *if then* form.

I would like to illustrate this point using a concrete case. [4] gives the following definition of a conditional early on in his study (see p. 3): "An item is

---

[4] More formally, in the settings defined in Appendix A by [38], from if A then C, we have $n_w(A, C) \neq \emptyset$ and $e_{n_w(A,C)}(A) \subseteq [C]$. From $n_w(A, C) \neq \emptyset$ and (neut), we obtain $n_w(A, C) = n_w(\neg C, \neg A)$ and $n_w(\neg C, \neg A) \neq \emptyset$. By $e_{n_w(A,C)}(A) \subseteq [C]$ and set-theoretic equivalence in the universe of projection U, we obtain $e_{n_w(\neg C, \neg A)}(\neg C) \subseteq [\neg A]$.

a conditional if it is expressed by an English sentence consisting of 'If' followed by an English sentence followed by 'then' followed by an English sentence." However, when it later comes time to give a counterexample to contraposition (see p. 172), he makes the following curious choice: "(Even) if the British and Israelis had not attacked the Suez Canal in 1956, the Soviets would (still) have invaded Hungary later in the year." Hence, Bennett considers conditionals to be part of the *if then* form but uses a concessive conditional (an *even if* form) to provide a counterexample to contraposition. This surprising choice can be explained by considering that most theories of conditionals do not provide a different semantics when the *if* is enriched by additional particles such as *even* and *then*. However, as soon as an analysis takes into account these additional particles, the assimilation between the *even if* and the *if then* forms cannot be maintained. Furthermore, if the validity of contraposition differs for these two forms in natural language, a formal theory not taking this level of analysis into account misses an essential point. It is therefore primordial to check whether the invalidity of contraposition for *if then* conditionals is a reality in everyday reasoning. To do so, I would like to give an overview of the related literature to see whether a convincing counterexample to this schema can be found.

4.1 Cases Not Using the *If Then* Form

Almost all the counterexamples to contraposition found in books or articles on the subject do not concern *if then* conditionals. As seen in [4], the majority of them concern concessive conditionals. Indeed, they can be reformulated along the "even if" pattern without their meaning being modified, as shown below by adding parentheses around the word *even* when needed.[5]

(12) (Even) if Boris had gone to the party, Olga would still have gone.
(13) (Even) if it rains tomorrow there will not be a terrific cloudburst.
(14) My car would still be white even if the maple tree in my front yard died.
(15) (Even) if Goethe hadn't died in 1832, he would still be dead now.
(16) (Even) if you open the refrigerator, it will not explode.

Notice that these sentences are taken from some of the most famous works in the field. Furthermore, some of their authors - for instance Lewis or Adams - were primarily uninterested in *even if* conditionals, and their usage precisely at this point is rather suspicious. If these authors did not find any sufficiently convincing counterexample to the *if then* form and were forced to use an *even if* form, it is surely because such a sentence is not so easily found.

A variant of this dismissal of the contraposition uses relevant conditionals (also referred to as "biscuit" conditional or "nonconditional" conditional). The following are some examples of this type sentences, taken from [3], [12] and [17] respectively.

---

[5] (10) to (14) are respectively from [1,19,24], Kratzer cited in [11] and [20].

(17) If you are hungry, there are biscuits on the table.

(18) If you don't mind, I'm trying to read.

(19) If you had needed some money, there was some in the bank.

Again, these sentences cannot constitute counterexamples to the *if then* form because the particle *then* cannot be added to them felicitously. Furthermore, it is widely agreed that relevance conditionals deserve their own particular treatment.

4.2   Simple Cases of the *If Then* Form

To my knowledge, there are only two counterexamples to contraposition that adopt the *if then* form in the literature.[6] The first one, given by [30], is well known because it was published in the first article that presented a possible world semantics for conditionals invalidating contraposition. However, this example has not been reused in later works by other authors, showing that it is certainly deficient. The sentence is as follows:

(20) If the U.S. halts the bombing, then North Vietnam will not agree to negotiate.

The sentence sounds odd and is difficult to understand. It can be explained by recalling what happened during the Vietnam War. The speaker does not mean that as long as the U.S. continues the bombing, North Vietnam will agree to negotiate and that this will change if the U.S. takes the opposite stand. Negotiations between the U.S. and North Vietnam took place in 1973, after ten years of intense bombing and only once the U.S. was considered to have lost the war, as far as public opinion was concerned. Here, the speaker wants to offer another meaning: bringing a halt to the bombing is a favorable factor for sparking negotiation, but is on its own insufficient for bringing North Vietnam to the negotiating table. A felicitous expression of this relationship would be "*Even if the U.S. halts the bombing, North Vietnam will not agree to negotiate.*"

Stalnaker thus makes a grammatical error using an *if then* form when the correct formulation requires an *even if* form. Some could consider it completely licit to use the *if then* form to carry the meaning of a concessive conditional. Indeed, language is governed not by explicit rules, but by practice. As soon as the audience is able to decipher the intended meaning, the exact syntactic form used is unimportant. In this case, this argument does not hold. Indeed, the psychological experiment presented by [9] shows that, when an antecedent is deemed as not presenting evidence for the consequent (the concessive meaning), almost all subjects do not accept the *if then* form as a correct assertion. Both expressions therefore differ in meaning and cannot be freely exchanged in a conversation without running the risk that most listeners will not understand the exact meaning or will simply reject the utterance. Stalnaker, surely knowing

---

[6] There are only two counterexamples in the major works. The literature on conditionals is so huge that nobody can claim to have read all the articles or books on this subject.

his counterexample to be deficient, adopts another one in a later book on conditionals. In his [31] study, the example becomes "If my dog were a purebred, his father would be a mutt," but as Stalnaker himself acknowledges, "One could reject the counterexample on the grounds that the conditional contraposed is an 'even if' conditional." I completely agree with Stalnaker here.

[25] presents another counterexample to contraposition, although without defining any clear context.

(21) If we take the car then it won't break down *en route*.

The first problem with this pseudo-counterexample is that it is easy to imagine a context in which the contraposition is valid. Imagine that two groups can take the car to drive on a very difficult trail and that only the group to whom the speaker belongs has a sufficiently good driver to do this safely. Learning that there was an accident, the following deduction is possible: "If the car broke down *en route*, then you didn't take it. The driver was a member of the other group." Priest perhaps had another context in mind, in which there is only one group of people. But now, if they do not take the car, it is certain that it will not also break down *en route* because the vehicle safely remains parked in the garage. The conditional relationship that the speaker wants to express is now a concessive one that would be better expressed using an *even if*: "Even if we take the car, it won't break down *en route*". The same problem as in Stalnaker's case therefor arises. The use of the *if then* form is deficient because the speaker wants to carry a concessive meaning. The counterexample therefore cannot be retained.

4.3 Cases with Modals

The last possibility for invalidating contraposition for *if then* conditionals is to incorporate modals. The following sentence is a slight modification of an example against the modus tollens initially presented in [39].

(22) If there is a break-in, then the alarm always sounds.

From the hypothesis that the alarm does not always sound, we cannot conclude that there is no break-in. Hence, the contraposition is invalid in this case. However, this invalidation is incorrect because it is based on an ambiguity. Since the consequent is the main clause of the whole sentence, the association between an adverb and its verb can result in two different scopes of application. The word "always" can be considered as either applying only to the consequent ("If there is a break-in, then always the alarm sounds") or to the whole conditional ("Always, if there is a break-in, then the alarm sounds"). In most cases, the correct interpretation is to allow the adverb to apply to the whole sentence. For sentence (22) and with a large scope of application for the adverb, the contraposition now becomes "Always, if the alarm does not sound, then there is no break-in," which is perfectly correct. Yalcin's counterexample is therefore deficient because it is based on a narrow scope of interpretation for the adverb, for which a large scope of interpretation is necessary. This issue is well known in modal logic referred to as the (conditional) Modal Fallacy by [6]. It is usually presented with explicitly modal adverb, such as in the following reformulation of (22):

(22') If there is a break-in, then the alarm necessary sounds.

The correct formal translation of (22') is $\Box(A \to C)$ and not $A \to \Box C$, with $\Box$ the symbol for necessity and $\to$ the symbol for the *if then* conditional.

4.4   Advantages

From this examination of earlier attempts, I have shown that, in order to be successful, a counterexample to contraposition for *if then* conditionals should involve two things: components that are sufficiently simple to not embed modals and an antecedent and a consequent of which the relationship is not concessive or one of relevance. I cannot prove that such a counterexample is impossible because new sentences are invented everyday in natural language. But the fact that, after fifty years of investigation in conditional logic, no convincing *if then* conditional sentence has been found to invalidate this reasoning leads me to think that this is a hopeless task. However, even if I am wrong on this point, the present semantics has its own merits. This theory is the first possible worlds semantics that manages to separate the validity of contraposition and strengthening of the antecedent. Contraposition is therefore not a non-monotonic reasoning, as argued by [11] for instance. Indeed, a non-monotonic inference involves a modification of context because a new sentence changing the initial situation is introduced. For instance, in the strengthening of the antecedent $A \to C \nvDash (A \wedge B) \to C$, the new hypothesis B makes the interpretation of the hypothesis A different between the premise and the consequence. In the transitivity $B \to C, A \to B \nvDash A \to C$, the second premise contains A, which is new and changes the context in which the sentence B must be interpreted compared to the first premise. Hence, the link between the two premises is impossible. There is no such introduction of a new sentence in contraposition. Between the premise and the conclusion, the only new element is the introduction of the negation and this cannot produce a new context. On the contrary, this only entails the consideration of the negatives cases in the initial context.

[31] argues that the validity of contraposition always entails the reintroduction of the strengthening of the antecedent. The proof is as follows:

(23) $A \to C$ (hypothesis).

(24) $\neg C \to \neg A$ (From 23 by contraposition)

(25) $\neg C \to (\neg A \vee \neg B)$ (From 24 by expansion of the consequent)

(26) $(A \wedge B) \to C$ (From 25 by contraposition)

This proof needs an additional principle: the expansion of the consequent. However, this last inference is not valid in our system: $A \to B \nvDash A \to (B \vee C)$. Since the conditional in the consequence contains more sentences than the conditional in the premise, the consequence needs the inhibition of a larger set of sentences, and its final evaluation is done in another context compared to the premise. My semantics thus introduces a principle of relevance between the antecedent and the consequent. Each part of the consequent must be related in some way to at least one part of the antecedent to make the conditional true. This choice makes it possible to explain why *if then* is the privileged form used to express causal reasoning and why the following sentence sounds false in this context:

(27) If I'm hungry, then I eat a banana or Asia is the biggest continent.

My hunger and the size of Asia are not causally linked. As a consequence, most people would consider sentence (27) to be wrong or odd for describing a causal relationship. However, notice that with the simple *if A, C* form, the reasoning is correct. The relationship being more tenuous, the schema $A \rightarrow B \vDash A \rightarrow (B \vee C)$ is valid.

The last advantage of the admission of contraposition is that it is now possible to understand why the particle *then* cannot be added in some conditionals. Indeed, when the contraposition fails, the addition of *then* is infelicitous. This is the case for relevance (biscuit) conditionals. The same rule also holds for conditionals in which the antecedent exhausts the universe, such as in [16]'s proposal "If John is dead or alive, Bill will find him." In this case, the negation of the antecedent cannot be envisaged because it corresponds to an empty set of worlds, and there is no reason to utter contraposition. Finally, contraposition is not valid for concessive conditionals which explains the initial enigma of this paper: how come an *even if* conditional cannot contain the particle *then*.

References

1. Adams, E.W.: The Logic of Conditionals. D. Reidel Publishing Co., Dordrecht (1975)
2. Alonso-Ovalle, L.: Alternatives in the disjunctive antecedents problem. In: Chang, C.B., Haynie, H.J. (eds.) Proceedings of 26th West Coast Conference on Formal Linguistics, Sommerville, MA, Cascadilla Proceedings Project (2008)
3. Austin, J.L.: Ifs and cans. In: Urmson, J.O., Warnock, G.J. (eds.) Philosophical Papers, pp. 153–180. Oxford University Press, Oxford (1961)
4. Bennett, J.: A Philosophical Guide to Conditionals. Clarendon Press, Oxford (2003)
5. Bhatt, R., Pancheva, R.: Conditionals. In: Everaert, M., Van Riemsdijk, H. (eds.) The Blackwell Companion to Syntax, pp. 638–687. Wiley, Hoboken (2006)
6. Bradley, R., Swartz, N.: Possible Worlds: An Introduction to Logic and Its Philosophy. B. Blackwell, Oxford (1979)
7. Chellas, B.: Basic conditional logic. J. Philos. Log. **4**(2), 133–154 (1975)
8. Dancygier, B., Sweetser, E.: Mental Spaces in Grammar: Conditional Constructions. Cambridge University Press, Cambridge (2005)
9. Douven, I., Verbrugge, S.: Indicatives, concessives, and evidential support. Think. Reason. **18**(4), 480–499 (2012)
10. von Fintel, K.: Restrictions on quantifier domains. Ph.D. thesis, University of Massachusetts, Amherst, MA (1994)
11. von Fintel, K.: Counterfactuals in a dynamic context. In: Kenstowicz, M. (ed.) Ken Hale: a Life in Language, pp. 123–152. The MIT Press, Cambridge (2001)

12. Geis, M.L., Lycan, W.G.: Nonconditional conditionals. Philos. Top. **21**(2), 35–56 (1993)
13. Gillies, A.S.: Epistemic conditionals and conditional epistemics. Noûs **38**(4), 585–616 (2004)
14. Gillies, A.S.: On truth-conditions for if (but not quite only if). Philos. Rev. **118**(3), 325–349 (2009)
15. Glasbey, S.R.: Distinguishing between events and times: some evidence from the semantics of then. Nat. Lang. Seman. **1**(3), 285–312 (1993)
16. Iatridou, S.: On the contribution of conditional 'then'. Nat. Lang. Seman. **2**(3), 171–199 (1994)
17. Johnson-Laird, P.N.: Conditionals and mental models. In: Traugott et al. [34], pp. 55–76
18. Johnson-Laird, P.N., Byrne, R.M.: Conditionals: a theory of meaning, pragmatics, and inference. Psychol. Rev. **109**(4), 646–678 (2002)
19. Lewis, D.K.: Counterfactuals. Harvard University Press, Cambridge (1973)
20. Lycan, W.G.: Real Conditionals. Oxford University Press, Oxford (2001)
21. Matalon, B.: Etude Génétique de l'Implication. In: Piaget, J. (ed.) Etudes d'Epistémologie Génétique XVI. PUF, Paris (1962)
22. Nute, D.: Topics in Conditional Logic. Reidel, Dordrecht (1980)
23. Piaget, J., Gruber, H., Vonèche, J.: The Essential Piaget. J. Aronson, New York (1977)
24. Pollock, J.L.: Subjunctive Reasoning. D. Reidel Publishing Co., Dordrecht (1976)
25. Priest, G.: An Introduction to Non-classical Logic: From If to Is. Cambridge University Press, Cambridge (2008)
26. Reichenbach, H.: Elements of Symbolic Logic. Dover Publications, Mineola (1947). Republished 1980
27. Schiffrin, D.: Anaphoric then: aspectual, textual, and epistemic meaning. Linguistics **20**, 753–792 (1992)
28. Schlenker, P.: Conditionals as definite descriptions (a referential analysis). Res. Lang. Comput. **2**(3), 417–462 (2004)
29. Skovgaard-Olsen, N., Singmann, H., Klauer, K.C.: The relevance effect and conditionals. Cognition **150**, 26–36 (2016)
30. Stalnaker, R.C.: A theory of conditionals. In: Rescher, N. (ed.) Studies in Logical Theory, pp. 98–112. Basil Blackwell Publishers, Oxford (1968)
31. Stalnaker, R.C.: Inquiry. The MIT Press, Cambridge (1984)
32. Starr, W.: A uniform theory of conditionals. J. Philos. Log. **43**(6), 1019–1064 (2014)
33. Thompson, E.: The temporal structure of discourse: the syntax and semantics of temporal then. Nat. Lang. Linguist. Theor. **17**, 123–160 (1998)
34. Traugott, E., ter Meulen, A., Reilly, J., Ferguson, C. (eds.): On Conditionals. Cambridge University Press, Cambridge (1986)
35. Veltman, F.: Logics for conditionals. Ph.D. dissertation, University of Amsterdam, Amsterdam (1985)
36. Veltman, F.: Data Semantics and the pragmatics of indicative conditionals. In: Traugott et al. [34], pp. 147–167
37. Vidal, M.: Conditionnels et Connexions. Ph.D. dissertation, Institut Jean Nicod, E.H.E.S.S., Paris (2012)
38. Vidal, M.: A Compositional semantics for 'even if' conditionals. Logic and Logical Philosophy (in press)
39. Yalcin, S.: A counterexample to modus tollens. J. Philos. Log. **41**, 1001–1024 (2012)

Automatic Concepts and Automata-Theoretic Semantics for the Full Lambek Calculus

Christian Wurm$^{(\boxtimes)}$

Universität Düsseldorf, Düsseldorf, Germany
cwurm@phil.uni-duesseldorf.de

Abstract. We introduce a new semantics for the (full) Lambek calculus, which is based on an automata-theoretic construction. This automata-theoretic semantics combines languages and relations via closure operators which are based on automaton transitions. We establish the strong completeness of this semantics for the full Lambek calculus via an isomorphism theorem for the syntactic concepts lattice of a language and a construction for the universal automaton recognizing the same language. Automata-theoretic semantics is interesting because it connects two important semantics of the Lambek calculus, namely the relational and the language-theoretic. At the same time, it establishes a strong relation between two canonical constructions over a given language, namely its syntactic concept lattice and its universal automaton.

Keywords: Syntactic concept lattice · Full Lambek calculus · Universal automaton · Finite automata

1 Introduction

The main contributions of this article are the following: we extend some established completeness results for the Full Lambek calculus $\mathbf{FL_\perp}$ and its fragments for syntactic concept lattices (SCL) to regular languages and finite algebras. (Throughout this article, we use completeness in the sense of *strong* completeness with respect to the *internal* consequence relation, that is, the semantics models the set of sequents which are derivable in the logical calculus.) We then present a new kind of semantics we call "automata-theoretic", where closure operators relate strings with transitions they induce in an automaton (we call the resulting structure *automatic concept lattice*). We prove its completeness for $\mathbf{FL_\perp}$ by showing that the syntactic concept lattice of a language is isomorphic to the automatic concept lattice of the universal automaton recognizing the same language (for the universal automaton, consider [14]; completeness for syntactic concept lattices has been established in [19]).

The semantics of binary relations and composition is usually associated with a "dynamic" interpretation of formulas as computations in programs (see [18]), whereas the language-semantics of stringsets and concatenation is a more "static" interpretation of formulas. Completeness (in our sense) of relational semantics has been shown by Brown & Gurr in [1], who use relational quantales and

© Springer-Verlag GmbH Germany 2016
M. Amblard et al. (Eds.): LACL 2016, LNCS 10054, pp. 308–323, 2016.
DOI: 10.1007/978-3-662-53826-5_19

prove results for a wide variety of substructural logics (or put differently, non-commutative linear logics). An even stronger result has been obtained by Pentus in [16], who also proved completeness of language (L-)models for the Lambek calculus. Automata-theoretic semantics shows how we can link language and relation models via a Galois connection.

The article is structured as follows: Sect. 2 presents established results on the full Lambek calculus and its semantics; Sect. 3 strengthens these results to regular languages, and Sect. 4 introduces the automata-theoretic semantics and proves (among other results) its completeness.

2   The Logics L, L1, FL, $FL_\perp$ and their Models

2.1   The Logics L, L1, FL and $FL_\perp$

The Lambek calculus **L** was introduced in [11]. **L1** is a proper extension of **L**, and **FL, $FL_\perp$** are each conservative extensions of **L1** and the preceding one. Let Pr be a set, the set of **primitive types**, and C be a set of **constructors**, which is, depending on the logics we use, $C_{\mathbf{L}} := \{/, \backslash, \bullet\}$, or $C_{\mathbf{FL}} := \{/, \backslash, \bullet, \vee, \wedge\}$. By $Tp_C(Pr)$ we denote the set of types over Pr, which is defined as the smallest set, such that: 1. $Pr \subseteq Tp_C(Pr)$, and if $\alpha, \beta \in Tp_C(Pr)$, $\star \in C$, then $\alpha \star \beta \in Tp_C(Pr)$. As there is usually no danger of confusion regarding the primitive types and constructors, we also simply write Tp for $Tp_C(Pr)$. We now present the inference rules corresponding to these constructors. We call an inference of the form $\Gamma \vdash \alpha$ a **sequent**, for $\Gamma \in Tp^*$, $\alpha \in Tp$, where by Tp^* we denote the set of all (possibly empty) *sequences* over Tp, which are concatenated by ','.

In general, uppercase Greek letters range as variables over sequences of types, lowercase Greek letters range over single types. In the inference rules for **L**, premises of '$\vdash$' (that is, left hand sides of sequents) must be non-empty; in **L1** they can be empty as well; besides this, the calculi are identical. In **FL** and $FL_\perp$ we also allow for empty sequents. Below, we present the standard rules of the Lambek calculus **L** (and **L1**).

(ax) $\alpha \vdash \alpha$

$$(\mathbf{I} - /) \quad \frac{\Gamma, \alpha \vdash \beta}{\Gamma \vdash \beta/\alpha} \qquad\qquad (\mathbf{I} - \backslash) \quad \frac{\alpha, \Gamma \vdash \beta}{\Gamma \vdash \alpha\backslash\beta}$$

$$(/ - \mathbf{I}) \quad \frac{\Delta, \beta, \Theta \vdash \gamma \quad \Gamma \vdash \alpha}{\Delta, \beta/\alpha, \Gamma, \Theta \vdash \gamma} \qquad (\backslash - \mathbf{I}) \quad \frac{\Delta, \beta, \Theta \vdash \gamma \quad \Gamma \vdash \alpha}{\Delta, \Gamma, \alpha\backslash\beta, \Theta \vdash \gamma}$$

$$(\bullet - \mathbf{I}) \quad \frac{\Delta, \alpha, \beta, \Gamma \vdash \gamma}{\Delta, \alpha \bullet \beta, \Gamma \vdash \gamma} \qquad (\mathbf{I} - \bullet) \quad \frac{\Delta \vdash \alpha \quad \Gamma \vdash \beta}{\Delta, \Gamma \vdash \alpha \bullet \beta}$$

These are the standard rules of **L** and **L1** (roughly as in [11]). We now add the two additional connectives $\vee$ and $\wedge$. These are not present in **L/L1**, have however been considered as extensions as early as in [12], and have been subsequently studied by [10].

$$(\wedge - \mathbf{I}\,1) \quad \frac{\Gamma, \alpha, \Delta \vdash \gamma}{\Gamma, \alpha \wedge \beta, \Delta \vdash \gamma} \qquad\qquad (\wedge - \mathbf{I}\,2) \quad \frac{\Gamma, \beta, \Delta \vdash \gamma}{\Gamma, \alpha \wedge \beta, \Delta \vdash \gamma}$$

$$(\mathbf{I} - \wedge) \quad \frac{\Gamma \vdash \alpha \quad \Gamma \vdash \beta}{\Gamma \vdash \alpha \wedge \beta} \qquad\qquad (\vee - \mathbf{I}) \quad \frac{\Gamma, \alpha, \Delta \vdash \gamma \quad \Gamma, \beta, \Delta \vdash \gamma}{\Gamma, \alpha \vee \beta, \Delta \vdash \gamma}$$

$$(\mathbf{I} - \vee\,1) \quad \frac{\Gamma \vdash \alpha}{\Gamma \vdash \alpha \vee \beta} \qquad\qquad (\mathbf{I} - \vee\,2) \quad \frac{\Gamma \vdash \beta}{\Gamma \vdash \alpha \vee \beta}$$

$$(1 - \mathbf{I}) \quad \frac{\Gamma, \Delta \vdash \alpha}{\Gamma, 1, \Delta \vdash \alpha} \qquad\qquad (\mathbf{I} - 1) \quad \vdash 1$$

This gives us the logic **FL**. This slightly deviates from standard terminology, because usually, **FL** has an additional constant 0. In our formulation, 0 and 1 coincide. In order to have logical counterparts for the bounded lattice elements $\top$ and $\bot$, we introduce two logical constants, which are denoted by the same symbol.

$$(\bot - \mathbf{I}) \quad \Gamma, \bot, \Delta \vdash \alpha \qquad\qquad (\mathbf{I} - \top) \quad \Gamma \vdash \top$$

This gives us the calculus $\mathbf{FL}_\bot$. From a logical point of view, all these extensions of **L** are quite well-behaved: they are conservative, and also allow us to preserve the important result of [11], namely admissibility of the cut-rule:

$$(cut) \quad \frac{\Delta, \beta, \Theta \vdash \alpha \quad \Gamma \vdash \beta}{\Delta, \Gamma, \Theta \vdash \alpha}$$

We say that a sequent $\Gamma \vdash \alpha$ is derivable in a calculus, if it can be derived by its rules of inference; we then write $\Vdash_\mathbf{L} \Gamma \vdash \alpha$, $\Vdash_\mathbf{L1} \Gamma \vdash \alpha$, $\Vdash_\mathbf{FL} \Gamma \vdash \alpha$ etc., depending on which calculus we use.

2.2   Interpretations of L1, FL and $\mathbf{FL}_\bot$

The standard model for **L1** is the class of residuated monoids, which are structures $(M, \cdot, \backslash, /, 1, \leq)$ such that $(M, \cdot, 1)$ is a monoid, $(M, \leq)$ is a partial order, and $\cdot, /, \backslash$ satisfy the law of residuation: for $m, n, o \in M$,

$$m \leq o/n \Leftrightarrow m \cdot n \leq o \Leftrightarrow n \leq m \backslash o.$$

This implies that $\cdot$ respects the order $\leq$. The standard model for **FL** is the class of residuated lattices, and for $\mathbf{FL}_\bot$, the class of bounded residuated lattices (for background on residuated lattices, see [9]). A residuated lattice is a structure $(M, \cdot, \vee, \wedge, \backslash, /, 1)$, where in addition to the previous requirements, $(M, \vee, \wedge)$ is a lattice; the lattice order $\leq$ need not be stated, as it can be induced by $\vee$ or $\wedge$: for $a, b \in M$, $a \leq b$ is a shorthand for $a \vee b = b$. A bounded residuated lattice is a structure $(M, \cdot, \vee, \wedge, \backslash, /, 1, \top, \bot)$, where $(M, \cdot, \vee, \wedge, \backslash, /, 1)$ is a residuated lattice, $\top$ is the maximal element of the lattice order $\leq$ and $\bot$ is its minimal element.

We call the class of residuated monoids RM, the class of residuated lattices RL, the class of bounded residuated lattices $RL_\bot$. We now give a semantics for the calculi above. We start with an interpretation $\sigma : Pr \to M$ which interprets elements in Pr as elements of the algebra, and extend σ to $\overline{\sigma}$ by defining it appropriately for $1, \top, \bot$, and extending it inductively over our type constructors $C := \{/, \backslash, \bullet, \vee, \wedge\}$ by

1. $\overline{\sigma}(\alpha) = \sigma(\alpha) \in M$, if $\alpha \in Pr$
2. $\overline{\sigma}(\top) = \top$
2' $\overline{\sigma}(\top)$ is an arbitrary $m \in M$ such that for all $\alpha \in Tp_C(Pr)$, $\overline{\sigma}(\alpha) \leq m$.
3. $\overline{\sigma}(\bot) = \bot$
3' $\overline{\sigma}(\bot)$ is an arbitrary $m \in M$ such that for all $\alpha \in Tp_C(Pr)$, $m \leq \overline{\sigma}(\alpha)$.
4. $\overline{\sigma}(1) = 1$
5. $\overline{\sigma}(\alpha \bullet \beta) := \overline{\sigma}(\alpha) \cdot \overline{\sigma}(\beta)$
6. $\overline{\sigma}(\alpha/\beta) := \overline{\sigma}(\alpha)/\overline{\sigma}(\beta)$
7. $\overline{\sigma}(\alpha\backslash\beta) := \overline{\sigma}(\alpha)\backslash\overline{\sigma}(\beta)$
8. $\overline{\sigma}(\alpha \vee \beta) := \overline{\sigma}(\alpha) \vee \overline{\sigma}(\beta)$
9. $\overline{\sigma}(\alpha \wedge \beta) := \overline{\sigma}(\alpha) \wedge \overline{\sigma}(\beta)$

Note that there are two alternative interpretations for $\top, \bot$: one which interprets them as the upper/lower bound of the lattice, which is the standard interpretation, and one which just interprets them as arbitrary elements which only have to be larger/smaller than the interpretation of any other formula. The latter will be called the **non-standard** interpretation and play some role in the sequel, but only for technical reasons.

What we interpret next is the *sequents* of the form $\Gamma \vdash \alpha$. We say that a sequent $\gamma_1, ..., \gamma_i \vdash \alpha$ is true in a model $\mathcal{M}$ under assignment σ, in symbols: $\mathcal{M}, \sigma \models \gamma_1, ..., \gamma_i \vdash \alpha$, if and only if $\overline{\sigma}(\gamma_1 \bullet ... \bullet \gamma_i) \leq \overline{\sigma}(\alpha)$ holds in $\mathcal{M}$. That is, we interpret the ',', which denotes concatenation in sequents, as $\cdot$ in the model, and $\vdash$ as $\leq$. For derivable sequents with no antecedent, we have the following convention: $\mathcal{M}, \sigma \models \vdash \alpha$, iff $1 \leq \overline{\sigma}(\alpha)$, where 1 is the unit element of $\mathcal{M}$ (this case does not arise in **L**).

More generally, for a given class of (bounded) residuated lattices (monoids, semigroups) $\mathfrak{C}$, we say that a sequent is *valid* in $\mathfrak{C}$, in symbols, $\mathfrak{C} \models \gamma_1, ..., \gamma_i \vdash \alpha$, if for all $\mathcal{M} \in \mathfrak{C}$ and all interpretations σ, $\mathcal{M}, \sigma \models \gamma_1, ... \gamma_i \vdash \alpha$ (here we have to distinguish between standard and non-standard interpretations).

2.3   Syntactic Concepts and Galois Connections

We now present a language-theoretic semantics for **FL**$_\bot$ which is based on closure operators, namely syntactic concepts. Syntactic concept lattices form a particular case of what is well-known as formal concept lattice (or formal concept analysis, FCA) in computer science (see [7]). In linguistics, they have been introduced by Sestier in [17]. They were brought back to attention and enriched with residuation by Clark in [4] (see also [6]), as they turn out to be useful representations for language learning (see [5,13]).

Let $\wp(-)$ denote the powerset. Given a language $L \subseteq \Sigma^*$, we define two maps: a map $\triangleright : \wp(\Sigma^*) \to \wp((\Sigma^*)^2)$, and $\triangleleft : \wp((\Sigma^*)^2) \to \wp(\Sigma^*)$, which are defined as follows:

(1) $\qquad$ for $M \subseteq \Sigma^*$, $M^\triangleright := \{(x,y) : \forall w \in M, xwy \in L\}$

(2) $\qquad$ for $C \subseteq (\Sigma^*)^2$, $C^\triangleleft := \{w : \forall (x,y) \in C, xwy \in L\}$

So a set of strings which is mapped to the set of contexts in which all of its elements can occur. The dual function maps a set of contexts to the set of strings which can occur in all of them. This results in a Galois connection between the two $\subseteq$-ordered structures of closed sets and contexts, see [4,19]. For extension of these maps to larger tuples, consider [20]). Importantly, all these are special applications of the general theory of Galois connections; for background, see [9]. Obviously, $[-]^\triangleleft$ and $[-]^\triangleright$ are only defined with respect to a given language L, otherwise they are meaningless. As long as it is clear about which language (if any particular language) we are speaking, we will omit however any reference to it, to keep notation perspicuous. Regardless of the underlying objects, the two compositions of the maps, $[-]^{\triangleleft\triangleright}$ and $[-]^{\triangleright\triangleleft}$, form **closure operators**. Note also that for any set of strings M and contexts C, $M^\triangleright = M^{\triangleright\triangleleft\triangleright}$ and $C^\triangleleft = C^{\triangleleft\triangleright\triangleleft}$. A set M is **closed**, if $M^{\triangleright\triangleleft} = M$ etc. The closure operator $\triangleright\triangleleft$ gives rise to a lattice $(\mathcal{B}_L, \leq)$, where the elements of $\mathcal{B}_L$ are the sets $M \subseteq \Sigma^*$ such that $M = M^{\triangleright\triangleleft}$, and $\leq$ is interpreted as $\subseteq$. The same can be done with the set of closed contexts. Given these two lattices, $[-]^\triangleright$ and $[-]^\triangleleft$ form a Galois connection between the two (see [7] for more background), that is:

(1) $M \leq N \Leftrightarrow M^\triangleright \geq N^\triangleright$, and
(2) $C \leq D \Leftrightarrow C^\triangleleft \geq D^\triangleleft$.

A **syntactic concept** is usually defined to be an ordered pair, consisting of a closed set of strings, and a closed set of contexts, so it has the form (S, C), such that $S^\triangleright = C$ and $C^\triangleleft = S$; $S^\triangleright$ is the set of all contexts in which all strings in S can occur; inversely for $C^\triangleleft$. For our purposes, we mostly need to consider only the left component, so we suppress the contexts and only consider the stringsets of the form $M^{\triangleright\triangleleft}$. An exception to this convention is Sect. 4, where we will make use of concepts as pairs (M, C) with $M = C^\triangleleft$, $C = M^\triangleright$, as it will increase readability in this case. For all operations we define below, it can be easily seen that the resulting structures are isomorphic. So when we refer to a concept, we only mean a $[-]^{\triangleright\triangleleft}$ closed set of strings (with the exception of Sect. 4), the concept in the classical sense being easily reconstructible.

Definition 1. *For $[-]^{\triangleright\triangleleft}$ defined with respect to $L \subseteq \Sigma^*$, let $\mathcal{B}_L$ denote the set of $[-]^{\triangleright\triangleleft}$-closed subsets of Σ^*. This set forms a bounded lattice $(\mathcal{B}_L, \wedge, \vee, \top, \bot)$, where $\top = \Sigma^*$, $\bot = \emptyset^{\triangleright\triangleleft}$, and for $M, N \in \mathcal{B}_L$, $M \wedge N = M \cap N$, $M \vee N = (M \cup N)^{\triangleright\triangleleft}$.*

It is also easy to verify that this forms a complete lattice, as infinite joins are defined by (closure of) infinite unions, infinite meets by infinite intersections.

2.4   Monoid Structure and Residuation for Syntactic Concepts

The set of concepts of a language forms a lattice. In addition, we can also give it the structure of a monoid: for concepts M, N, we define $M \circ N := (M \cdot N)^{\triangleright\triangleleft}$, where $M \cdot N = \{wv : w \in M, v \in N\}$. We usually write MN for $M \cdot N$, if M, N are sets of strings. '$\circ$' is associative on concepts: for $M, N, O \in \mathcal{B}_L$, $M \circ (N \circ O) = (M \circ N) \circ O$. This follows from the associativity of $\cdot$-concatenation and the fact that $[-]^{\triangleright\triangleleft}$ is a **nucleus**, that is, it is a closure operator and in addition it satisfies $M^{\triangleright\triangleleft} N^{\triangleright\triangleleft} \subseteq (MN)^{\triangleright\triangleleft}$.

It is easy to see that the neutral element of '$\circ$' is $\{\epsilon\}^{\triangleright\triangleleft}$ (which need not be $\{\epsilon\}$). The monoid operation respects the partial order of the lattice, that is, for $X, Y, Z, W \in \mathcal{B}_L$, if $X \leq Y$, then $W \circ X \circ Z \leq W \circ Y \circ Z$. A stronger property is the following: $\circ$ distributes over infinite joins, that is, we have

$$\bigvee_{Z \in \mathbf{Z}} (X \circ Z \circ Y) = X \circ \bigvee \mathbf{Z} \circ Y$$

Here $\leq$ follows algebraically ($\circ$ respects the order $\subseteq$), and $\geq$ follows from the fact that 1. $\bigcup$ distributes over $\cdot$ (infinite unions distribute over concatenation), and 2. $[-]^{\triangleright\triangleleft}$ is a nucleus. We can thus also conceive of syntactic concepts with $\bigvee, \circ, 1$ as *quantales*, and in quantales we can easily define residuals as follows:

Definition 2. *Let X, Y be concepts. We define the right residual $X/Y := \bigvee\{Z : Z \circ Y \leq X\}$, the left residual $Y \backslash X := \bigvee\{Z : Y \circ Z \leq X\}$.*

Note that this is an entirely abstract definition which does not make reference to any underlying structure. It works because of the well-known fact that for any complete lattice with a monoid operation distributing over infinite joins, residuals defined as above, we have $Y \leq X \backslash Z$ iff $X \circ Y \leq Z$ iff $X \leq Z/Y$.

Definition 3. *The **syntactic concept lattice** of a language L is defined as $SCL(L) := (\mathcal{B}_L, \circ, \wedge, \vee, /, \backslash, 1, \top, \bot)$, where $\mathcal{B}_L, \wedge, \vee, \top, \bot$ are defined as in Definition 1, $1 = \{\epsilon\}^{\triangleright\triangleleft}$, and $\circ, /, \backslash$ are as defined above.*

The syntactic concept lattice thus is a residuated lattice (see [4]). We will denote by SCL the class of all lattices of the form $SCL(L)$ for some language L, without any further requirement regarding L. We can apply the definition of interpretations to SCL, so it is clear how $\mathbf{FL}_\bot$ is interpreted in SCL.

The algebraic notion corresponding to the notion of a fragment in logic is the notion of a reduct. A reduct of an algebra is the same algebra with only a proper subset of connectives; the notion easily extends to classes. We let SCL_{FL} be the class of SCL reducts with operators $\{\circ, /, \backslash, \vee, \wedge\}$ without the constants $\top$ and $\bot$, and SCL_{L1} be the class of SCL reducts with $\{\circ, /, \backslash\}$, which all specify a unit. So it is clear how the logical fragments $\mathbf{FL}, \mathbf{L1}$ are interpreted in the reducts appropriate reducts.

2.5   Completeness: Previous Results

There are a number of completeness results for the logics we have considered here. We quickly present the ones which will be important in the sequel.

Theorem 4. *1. $RM \models \Gamma \vdash \alpha$ if and only if $\Vdash_{L1} \Gamma \vdash \alpha$*
2. $RL \models \Gamma \vdash \alpha$ if and only if $\Vdash_{FL} \Gamma \vdash \alpha$
3. $RL_\perp \models \Gamma \vdash \alpha$ if and only if $\Vdash_{FL_\perp} \Gamma \vdash \alpha$

For reference on Theorem 4, see [2,3,9]. These completeness results can actually be strengthened to the *finite model property*. A logic, equipped with a class of models and interpretations, is said to have finite model property if it is complete in the finite, that is, Theorem 4 remains valid if we restrict ourselves to finite models.

Theorem 5. *1. $L1$ has finite model property*
2. FL has finite model property
3. $FL_\perp$ has finite model property

For the first and second claim, consider [8]; the third and forth has been established in [15]. Theorem 5 is crucial to show that completeness for syntactic concept lattices and their reducts also holds if we restrict ourselves to languages over finite alphabets. The following results have been proved in [19].

Theorem 6. *1. $SCL \models \Gamma \vdash \alpha$ if and only if $\Vdash_{FL_\perp} \Gamma \vdash \alpha$*
2. $SCL_{FL} \models \Gamma \vdash \alpha$ if and only if $\Vdash_{FL} \Gamma \vdash \alpha$
3. $SCL_{L1} \models \Gamma \vdash \alpha$ if and only if $\Vdash_{L1} \Gamma \vdash \alpha$

L requires some additional considerations, as **L1** is *not* a conservative extension of it. The soundness directions follow *a fortiori* from Theorem 4. We will now strengthen the completeness result to syntactic concept lattices over regular languages, for which we have to provide a sketch of the original completeness proof.

3   Regular Languages and $SCL(REG)$

Let $\mathbf{B} = (B, \cdot, \vee, \wedge, /, \backslash, 1, \top, \perp)$ be a bounded residuated lattice. We denote the partial order of **B** by $\leq_B$, equality by $=_B$. Define $\Sigma' := \{\underline{b} : b \in B\}$, and put $\Sigma = B \cup \Sigma'$. Let $SI_B^* := \{b_1...b_i : b_1 \cdot ... \cdot b_i \leq_B 1\}$ be the set of sub-identity words over B. We define the language $L_B \subseteq \Sigma^*$ as the set of strings

$$L_B := \{b_1 b_2 ... b_n \underline{b} w : b_1 \cdot b_2 \cdot ... \cdot b_n \leq_{\mathbf{B}} b, \ w \in SI_B^*\}.$$

For a string $w = b_1...b_n \in B^*$, by $w^\bullet$ we denote the term $b_1 \cdot ... \cdot b_n$; we put $\epsilon^\bullet = 1$. By $w \sim_L v$, we mean that $xwy \in L$ iff $xvy \in L$. It is easy to prove that $w^\bullet =_B v^\bullet$ iff $w \sim_{L_B} v$ (see [19]).

Proposition 7. *For every bounded residuated lattice* **B**, *there is an faithful embedding* $\psi : \mathbf{B} \to SCL(L_B)$, *such that*

1. $\psi(\top) = B^*$, $\psi(\bot) = \{\bot\}^{\triangleright\triangleleft}$.
2. $\psi(1) = \{\epsilon\}^{\triangleright\triangleleft}$.

For proof of this fact, consider [19, 20]. Note that $\psi(\bot)$ is not the $\bot$-element of $SCL(L_B)$, as we can never substitute $\underline{b}$ with $\bot$. Note also that $\psi(\top)$ is not maximal in $SCL(L_B)$, as $\psi(\top) = B^* \subsetneq \Sigma^*$.

From here, it is easy to complete the proof of Theorem 6: just use the faithful embedding to perform the usual contraposition, where from $\nvdash_{FL_\bot} \Gamma \vdash \alpha$ and algebraic completeness then follows $SCL \not\models \Gamma \vdash \alpha$. This completes the proof of Theorem 6.1. Note however that the resulting interpretation is non-standard: we have $\psi(\top) = B^* \neq \Sigma^*$, and $\bot(\bot) = \{\bot\}^{\triangleright\triangleleft} \neq \emptyset^{\triangleright\triangleleft}$.

An important feature of our proof is that it works for all reducts of bounded residuated lattices (a reduct is the same algebra with a proper subset of connectives); and hence it allows to prove completeness for all logics $\mathcal{L}$ for which $\mathbf{FL_\bot}$ is a conservative extension (this holds for $\mathbf{L1}$ and $\mathbf{FL}$; $\mathbf{L}$ and its fragments do not satisfy this requirement). Hence we also have a proof for the other parts of Theorem 6.

By REG we denote the class of regular languages. Given an equivalence relation $\sim$ over Σ^*, we put $[w]_\sim = \{v : w \sim v\}$, and $\Sigma_\sim^* = \{[w]_\sim : w \in \Sigma^*\}$. So $\Sigma_{\sim_L}^*$ denotes the set of $\sim_L$-congruence classes over Σ^*. Recall that a language $L \subseteq \Sigma^*$ is regular if and only if $\Sigma_{\sim_L}^* = \{[w]_{\sim_L} : w \in \Sigma^*\}$ is finite. The next lemma follows easily (see also [6]):

Lemma 8. $SCL(L)$ is finite if and only if L is regular.

Let $\mathbf{B}$ be an arbitrary algebra equipped with a semigroup operation and a partial order respecting it, so we can define L_B as above (this covers all algebras we consider in this paper). Then for $w, v \in B^*$, we have $w \sim_L v$ iff $w^\bullet =_\mathbf{B} v^\bullet$. But recall that $B \subsetneq \Sigma$ in this case!

Lemma 9. $\mathbf{B}$ is a finite algebra if and only if L_B is a regular language.

Proof. $\Leftarrow$ Contraposition: if $\mathbf{B}$ is infinite, there is an infinite sequence of $=_B$-distinct objects $(w_1)^\bullet, (w_2)^\bullet, \ldots \in B$, so there are $w_1, w_2, \ldots$ which are not $\sim_L$-equivalent.

$\Rightarrow$ We construct $L_B' = \bigcup_{b \in B} \{w\underline{b} : w^\bullet \leq_B b\}$. Assume a language $\{w\underline{b} : w^\bullet \leq_B b\}$ is not regular. Then $\Sigma_{\sim_L}^*$ is infinite, and there is an infinite sequence of words $w_1, w_2, \ldots$, such that if $i \neq j$, then $w_i \not\sim_L w_j$. So there is an infinite sequence of objects $(w_1)^\bullet, (w_2)^\bullet, \ldots \in B$, such that if $i \neq j$, then $(w_i)^\bullet \neq_B (w_j)^\bullet$. Thus $\mathbf{B}$ is infinite – contradiction. Hence $\{w\underline{b} : w^\bullet \leq_B b\}$ is regular, and as B is finite, L_B' is a finite union of regular languages, which is still regular. Finally, SI_B^* is regular for the same reason as above, and so $L_B = L_B' \cdot SI_B^*$ is also regular. □

Let C be a class of languages; then by $SCL(C)$ we denote the class of structures $SCL(L) : L \in C$. So $SCL(REG)$ equals the class of finite syntactic concept lattices. As we have said, a finite algebra $\mathbf{B}$ entails a language L_B over a finite alphabet; the last lemma shows us that it also entails that L_B is regular. Moreover, as $\mathbf{L1}, \mathbf{FL}, \mathbf{FL_\bot}$ have the finite model property, for completeness it is sufficient to consider only finite algebras, and consequently we can strengthen Theorem 6 to the following:

Corollary 10. *1. $SCL(REG) \models \Gamma \vdash \alpha$ if and only if $\Vdash_{FL_\perp} \Gamma \vdash \alpha$.*
2. $SCL_{FL}(REG) \models \Gamma \vdash \alpha$ if and only if $\Vdash_{FL} \Gamma \vdash \alpha$.
3. $SCL_{L1}(REG) \models \Gamma \vdash \alpha$ if and only if $\Vdash_{L1} \Gamma \vdash \alpha$.

4   Automata-Theoretic Semantics

4.1   Automata-Theoretic Preliminaries

We now introduce a new class of bounded residuated lattices, the **automatic concept lattices**. It is very similar to SCL in that it is based on a Galois connection which, provided the certain conditions, gives rise to a nucleus. As we will learn from the main result of this section, the isomorphism theorem, if we consider structures only up to isomorphism, then automatic concept lattices form a proper generalization of syntactic concept lattices (in fact, in general they are not even residuated lattices[1]).

One can present automata in many different ways, the most standard one being probably the following: an automaton as *state-transition system* is a tuple $\mathfrak{A} = (\Sigma, Q, \delta, F, I)$, where Σ is a finite input alphabet, Q a set of states, $\delta \subseteq Q \times \Sigma \times Q$ a transition relation, $F \subseteq Q$ a set of accepting states, $I \subseteq Q$ the set of initial states. This notation of automata is somewhat clumsy in connection with the techniques we use later on, so we will choose a slightly different presentation which we call **relational**. This is a notional change we adopt for convenience. We define a **semi-automaton** as a tuple $\langle \Sigma, \phi \rangle$, where ϕ is a map $\phi : \Sigma \to \wp(Q \times Q)$, mapping letters in Σ onto relations over Q, where we use Q is an arbitrary (finite or infinite) carrier set. It is extended to strings by interpreting concatenation as **relation composition** '$,$', where $R, R' = \{(x, y) : (x, z) \in R, (z, y) \in R'\}$. So we have $\phi(aw) = \phi(a), \phi(w)$, and ϕ is a homomorphism from the free monoid Σ^* into a relation monoid over Q, and a word $w \in \Sigma^*$ then induces a relation $\phi(w) \subseteq Q \times Q$. Defining ϕ as a homomorphism, we should take care of $\phi(\epsilon)$, which we simply define by $\phi(\epsilon) = \mathrm{id}_Q := \{(q, q) : q \in Q\}$.[2] To get a full automaton, we still need an *accepting relation*. One usually specifies a set of initial and accepting states, yielding an accepting relation $I \times F$. As for us, acceptance will only play a minor role, we will take a slightly more general convention and assume that automata specify an **accepting relation** $F_R \subseteq Q \times Q$. Thus a full **automaton** is a tuple $\langle \Sigma, \phi, F_R \rangle$. We define the language recognized by an automaton $\mathcal{A} = \langle \Sigma, \phi, F_R \rangle$ by $L(\mathcal{A}) := \{w \in \Sigma^* : \phi(w) \cap F_R \neq \emptyset\}$.

---

[1] Thanks to an anonymous reviewer for pointing this out!
[2] But in principle, nothing prevents us from having $(x, y) \in \phi(\epsilon)$ with $x \neq y$ – we just have to make sure that for all $a \in \Sigma$, we have $\phi(\epsilon), \phi(a) = \phi(a) = \phi(a), \phi(\epsilon)$.

4.2   Automatic Concepts

In what is to follow, we will take the "canonical view" on formal concepts, that is: concepts are not simply $[-]^{\rhd\lhd}$-closed sets, but pairs (M, C) such that $M^\rhd = C$, $C^\lhd = M$ (this entails that both are closed). Henceforth, we will use the maps $[-]^\rhd, [-]^\lhd$ for syntactic concepts only. Given a semi-automaton $\langle \Sigma, \phi \rangle$, $M \subseteq \Sigma^*$, $R \subseteq Q \times Q$, we define the two polar maps

$$(3) \qquad\qquad\qquad M^{\blacktriangleright} = \bigcap_{w \in M} \phi(w)$$

$$(4) \qquad\qquad\qquad R^{\blacktriangleleft} = \{w : \phi(w) \supseteq R\}$$

It is easy to see that these maps establish a Galois connection and their compositions $[-]^{\blacktriangleright\blacktriangleleft}$, $[-]^{\blacktriangleleft\blacktriangleright}$ are closure operators. An **automatic concept** is then a pair (M, R) with $M^{\blacktriangleright} = R$, $R^{\blacktriangleleft} = M$ (of course, the underlying (semi-)automaton is understood as given). We denote the set of automatic concepts, given an automaton $\mathcal{A}$, by $\mathfrak{A}_{\mathcal{A}}$. Importantly, the map $[-]^{\blacktriangleright\blacktriangleleft}$ does **not** form a nucleus on Σ^*, and in general, $[-]^{\blacktriangleright\blacktriangleleft}$-closed concatenation does not distribute over infinite joins. Consequently, we cannot simply define a residuated lattice of concepts in the usual fashion. Rather, we have to restrict our attention to a certain class of automata.

Definition 11. *A (semi-)automaton* $\langle \Sigma, \phi(, F_R) \rangle$ *is* **nuclear**, *if for all* $M, N \subseteq \Sigma^*$, $(\bigcap_{w \in M} \phi(w)), (\bigcap_{v \in N} \phi(w)) = \bigcap_{wv \in MN} \phi(wv)$.

Note that $\subseteq$ always holds. The equality ensures that $[-]^{\blacktriangleright\blacktriangleleft}$ is a nucleus on Σ^*, because if $w \in M^{\blacktriangleright\blacktriangleleft}$, $v \in N^{\blacktriangleright\blacktriangleleft}$, then $\phi(w) \supseteq M^{\blacktriangleright}, \phi(v) \supseteq N^{\blacktriangleright}$. Hence $\phi(wv) = \phi(w), \phi(v) \supseteq M^{\blacktriangleright}, N^{\blacktriangleright} = (MN)^{\blacktriangleright}$, and hence $wv \in M^{\blacktriangleright\blacktriangleleft}$. So being nuclear boils down to composition distributing over (infinite) intersections of closed sets. We will later see that for every automaton there is a nuclear automaton recognizing the same language.

We define $(M, R) \wedge (N, S) = (M \cap N, (R \cup S)^{\blacktriangleleft\blacktriangleright})$, $(M, R) \vee (N, S) = ((M \cup N)^{\blacktriangleright\blacktriangleleft}, R \cap S)$, and $(M, R) \circ (N, S) = ((MN)^{\blacktriangleright\blacktriangleleft}, (MN)^{\blacktriangleright})$. It is easy to see that $\wedge, \vee$ can be extended to the infinitary operators $\bigwedge, \bigvee$ (as they are based on sets). Moreover, in case the underlying automaton is nuclear, $\circ$ distributes over infinite joins (because it is a nuclear operation), so the residuals are easily defined in the usual fashion by $M/N = \bigvee\{X : X \circ N \leq M\}$, $N \backslash M = \bigvee\{X : N \circ X \leq M\}$. We put $\top = (\Sigma^*, (\Sigma^*)^{\blacktriangleright})$, $\bot = (\emptyset^{\blacktriangleright\blacktriangleleft}, \emptyset^{\blacktriangleright})$, where by convention we put $\emptyset^{\blacktriangleright} = \bigcup_{w \in \Sigma^*} \phi(w)$. Finally, we put $1 = (\{\epsilon\}^{\blacktriangleright\blacktriangleleft}, \phi(\epsilon))$ (recall that $\phi(\epsilon) = \mathrm{id}_C$ by definition). So given a nuclear automaton $\mathcal{A}$, we have the complete bounded residuated lattice $(\mathfrak{A}_{\mathcal{A}}, \circ, \wedge, \vee, /, \backslash, 1, \top, \bot)$, which is the **automatic concept lattice** of $\mathcal{A}$, for short $ACL(\mathcal{A})$. As is easy to see, acceptance does not play a role for the automatic concept lattice, so it is sufficient to refer to semi-automata. By ACL we denote the class of all $ACL(\mathcal{A})$ for $\mathcal{A}$ an arbitrary nuclear (semi-)automaton, and we define the reducts $ACL_{\mathbf{FL}}, ACL_{\mathbf{L1}}$ in the same way we did for SCL.

We will refer to the straightforward interpretation of $\mathbf{FL}_\perp$ and its fragments into automatic concept lattices as **automata-theoretic semantics**, and write $ACL \models \Gamma \vdash \alpha$ in the usual sense that for all nuclear semi-automata $\mathcal{A}$, interpretations σ into $ACL(\mathcal{A})$, we have $\overline{\sigma}(\Gamma) \leq_{ACL(\mathcal{A})} \overline{\sigma}(\alpha)$; same for reducts $ACL_{\mathbf{FL}}$, $ACL_{\mathbf{L1}}$ etc.

For $ACL(\langle \phi, \Sigma, F_R \rangle)$, F_R is irrelevant. Still, F_R is useful because it links automata to languages, which in turn is necessary to establish the relation between ACL and SCL. For what is to follow, the phrase "automaton recognizing L" could be exchanged with "semi-automaton $\langle \phi, \Sigma \rangle$ for which there is F_R such that $L(\langle \phi, \Sigma, F_R \rangle) = L$", which however is clumsy to repeat. As automata are related to languages, there should be thus a relation between $ACL(\mathcal{A})$ and $SCL(L)$, provided that $L(\mathcal{A}) = L$. In particular, one knows that in this case, if $w \not\sim_L v$, then $\phi(w) \neq \phi(v)$ – otherwise, the automaton could not distinguish acceptance of words containing the two substrings. The inverse direction is obviously incorrect, that is, $\phi(w) \neq \phi(v)$ does not imply anything for w, v in L, as the automaton can make as many (unnecessary) distinctions as it desires (this is related to the issue of minimality of automata). From this, we can for example conclude the following: if $L(\mathcal{A}) = L$, then for $(M, C) \in \mathcal{B}_L$, there are $(M_i, R_i) \in \mathfrak{A}_\mathcal{A}$ for $i \in I$, such that $M = \bigcup_{i \in I} M_i$. However, this does **not** entail (as one might conjecture) that we have $M = (\bigcup_{i \in I} M_i)^{\blacktriangleright\blacktriangleleft}$, which by completeness of the lattice would entail that there is an automatic concept $(M, R) \in \mathfrak{A}_\mathcal{A}$.[3] In general, there is no homomorphic relation between the two structures, so there is no trivial way to extend completeness for SCL to completeness for automata-theoretic semantics via embeddings; instead, we have to recur to a peculiar automata-theoretic construction.

4.3   The Universal Automaton

There are always infinitely many distinct automata recognizing a language (even modulo a labelled-graph based notion of automaton-isomorphism). We will now consider a particular automaton type which is uniquely specified for every language and which allows us to connect syntactic concepts to automatic concepts. This is the so-called **universal automaton** (see [14]). The observation that there is some connection between syntactic concepts and the universal automaton is due to A.Clark and has been elaborated in [6]. However, the direct correlation we establish here is new to my knowledge. The universal automaton is based on the notion of a factorization of a language. (X, Y) is a **factorization** of L, iff

---

[3] Imagine the following situation: for every $w \in M$, there is $(r_w, r'_w) \in \phi(m)$, such that $\phi(x) \circ \{(r_w, r'_w)\} \circ \phi(y) \cap F \neq \emptyset$, but $(r_w, r'_w) \notin M^{\blacktriangleright}$. So every $w \in M$ has its own peculiar pair which makes sure $xwy \in L$. Obviously, for $\bigvee_{w \in M}(\{w\}^{\blacktriangleright\blacktriangleleft}, \phi(w)) = (N, R)$, we have $N \supseteq M$. Still there can be $v \in N$, $v \notin M$, because $\phi(v) \supseteq R$, but $\phi(v)$ does not contain *any* of the pairs which ensure that $xMy \subseteq L$, and in fact $xvy \notin L$.

1. $XY \subseteq L$, and
2. if $X \subseteq X', Y \subseteq Y'$ and $X'Y' \subseteq L$, then $X = X', Y = Y'$.

We denote the set of L-factorizations with $fact(L)$. So a factorization is a maximal decomposition of L into two factors. We denote the (unique) universal automaton for a language L by $U(L)$. The factorizations of L form the set of states of $U(L)$. We define I, the set of initial factorizations and F, the set of final factorizations as follows: $I = \{(X,Y) \in fact(L) : \epsilon \in X\}$, $F = \{(X,Y) \in fact(L) : \epsilon \in Y\}$. Then for $L \subseteq \Sigma^*$, one defines the **universal automaton** $U(L) := (\Sigma, fact(L), I, F, \delta)$, where for $a \in \Sigma$, $((X,Y), a, (X',Y')) \in \delta$ iff $Xa \subseteq X'$ iff $Y \supseteq aY'$. The latter bi-implication is easy to see: if $Xa \subseteq X'$, then $XaY' \subseteq L$, and so $aY' \subseteq Y$ (same for the other direction). The results of this subsection can be found in [14]; we present them as they are necessary for the proof of the isomorphism theorem, but we omit the proofs. Until now, we have given the "normal" presentation of universal automata. To proceed, we quickly need to bring the universal automaton into our "relational form" for automata: we put $U(L) = \langle \Sigma, \phi, I \times F \rangle$, where for all $a \in \Sigma$, we have $\phi(a) = \{((X,Y),(X',Y')) : (X,Y),(X',Y') \in fact(L) \text{ and } Xa \subseteq X'\}$. We define the maps $[-]^{\rightarrow}, [-]^{\leftarrow}$ by

$$(5) \qquad\qquad M^{\rightarrow} = \{w : Mw \subseteq L\}$$

$$(6) \qquad\qquad M^{\leftarrow} = \{w : wM \subseteq L\}$$

The compositions $[-]^{\rightarrow\leftarrow}, [-]^{\leftarrow\rightarrow}$ are closure operators, and $[-]^{\rightarrow}, [-]^{\leftarrow}$ establish a Galois connection between closed sets of strings (see [6] for the connection of $[-]^{\rightarrow}$ and $[-]^{\triangleright}$ etc.). A factorization is then exactly a pair of sets (M,N) such that $M^{\rightarrow} = N$, $N^{\leftarrow} = M$ (this entails $M = M^{\rightarrow\leftarrow}, N = N^{\leftarrow\rightarrow}$). Depending on L, there might be trivial factorizations $(\Sigma^*, \emptyset)$, $(\emptyset, \Sigma^*)$.

Lemma 12. *For $(X,Y),(X',Y') \in fact(L)$, $W \subseteq \Sigma^*$, the following are equivalent:*

1. $XW \subseteq X'$
2. $WY' \subseteq Y$
3. $XWY' \subseteq L$.

Lemma 13. *For every $L \subseteq \Sigma^*$, $w \in \Sigma^*$, for $U(L)$ we have $((X,Y),(X',Y')) \in \phi(w)$ iff $Xw \subseteq X'$ iff $wY' \subseteq Y$ iff $XwY' \subseteq L$.*

Lemma 14. $L(U(L)) = L$.

That is, the universal automaton of L recognizes L. It is a straightforward consequence of the Myhill-Nerode theorem that $fact(L)$ is finite if and only if L is regular. This entails the following:

Lemma 15. *$U(L)$ is a finite automaton if and only if L is regular.*

4.4   An Isomorphism Theorem for ACL and SCL

We have said that there is no homomorphic (or in fact, any simple structural) relation between $SCL(L)$ and $ACL(\mathcal{A})$ for all $\mathcal{A}$ such that $L(\mathcal{A}) = L$. This is despite the fact that $\mathcal{A}$ must make the relevant distinctions between strings distinct modulo $\sim_L$. Things change if we look at the universal automaton instead of automata in general. For two algebras $\mathbf{B}, \mathbf{B}'$, we write $\mathbf{B} \cong \mathbf{B}'$ if there is an isomorphism from one to the other, that is a bijection which preserves all results of all operations. We can establish the following, surprisingly strong connection:

Theorem 16. *(Isomorphism theorem)* $ACL(U(L)) \cong SCL(L)$

That is, the automatic concept lattice for the universal automaton over L is isomorphic to the syntactic concept lattice of L. The following generalization of Lemma 13 is quite simple, but will be very helpful in the proof of the isomorphism theorem. Let $[-]^{\blacktriangleright}, [-]^{\blacktriangleleft}$ below be defined with respect to $U(L)$.

Lemma 17. *For* $(X,Y), (X',Y') \in fact(L)$, $((X,Y),(X',Y')) \in M^{\blacktriangleright}$ *if and only if* $XMY' \subseteq L$.

Proof. *If*: Assume $XMY' \subseteq L$. Then for every $w \in M$, we have $XwY' \subseteq L$, hence $((X,Y),(X',Y')) \in \phi(w)$, hence $((X,Y),(X',Y')) \in M^{\blacktriangleright}$.

Only if: Assume $((X,Y),(X',Y')) \in M^{\blacktriangleright}$. Then for all $w \in M$, we have $((X,Y),(X',Y')) \in \phi(w)$. Hence for all $w \in M$, $XwY' \subseteq L$, so $XMY' \subseteq L$.   □

We can now show that universal automata are nuclear, so they provide a sound semantics for the full Lambek calculus.

Lemma 18. *Let* $[-]^{\blacktriangleright}, [-]^{\blacktriangleleft}$ *we defined with respect to* $U(L)$ *for some language* L. *Then* $M^{\blacktriangleright}, N^{\blacktriangleright} = (MN)^{\blacktriangleright}$. *Hence for every language* L, $U(L)$ *is nuclear.*

Proof. $\subseteq$ Holds in general, by set-theoretic properties.

$\supseteq$ Assume $((X,Y'),(X',Y)) \in (MN)^{\blacktriangleright}$. Then $XMNY \subseteq L$.

Firstly, we have $((X,Y'),((NY)^{\leftarrow},(NY)^{\leftarrow\rightarrow})) \in M^{\blacktriangleright}$: we have $(X,Y') \in fact(L)$ by assumption, $((NY)^{\leftarrow},(NY)^{\leftarrow\rightarrow}) \in fact(L)$ by definition of $[-]^{\leftarrow}$, $[-]^{\rightarrow}$, and since $XMNY \subseteq L$, we also have $XM(NY)^{\leftarrow\rightarrow} \subseteq L$. So the claim follows from Lemma 17.

Secondly, we have $(((NY)^{\leftarrow},(NY)^{\leftarrow\rightarrow}),(X',Y)) \in N^{\blacktriangleright}$: $(X',Y) \in fact(L)$ by assumption, and we have $((NY)^{\leftarrow}NY \subseteq L$ by definition of $[-]^{\leftarrow}$, hence the claim follows again from Lemma 17.

Consequently, by definition of $;$, we have $((X,Y'),(X',Y)) \in M^{\blacktriangleright}, N^{\blacktriangleright}$.   □

In the sequel, $[-]^{\rhd}, [-]^{\lhd}$ refer to SCL-closure w.r.t. to some fixed $L \subseteq \Sigma^*$, $[-]^{\blacktriangleright}, [-]^{\blacktriangleleft}$ to ACL-closure w.r.t. to $U(L)$ (referring to the same language!). Now comes the crucial lemma for the isomorphism theorem:

Lemma 19. *For all* $M \subseteq \Sigma^*$, $M^{\rhd\lhd} = M^{\blacktriangleright\blacktriangleleft}$.

Proof. $M^{\rhd\lhd} \subseteq M^{\blacktriangleright\blacktriangleleft}$. Assume $w \in M^{\rhd\lhd}$. Then whenever $xMy \subseteq L$, then $xwy \in L$. If $((X,Y),(X',Y')) \in M^{\blacktriangleright}$, then $XMY' \subseteq L$ (by Lemma 17).

However, if $XMY' \subseteq L$, then $XwY' \subseteq L$, hence (by the equivalence in Lemma 12) $Xw \subseteq X', wY' \subseteq Y$. Hence we have $((X,Y),(X',Y')) \in \phi(w)$ for all $((X,Y),(X',Y')) \in M^{\blacktriangleright}$. Hence we have $w \in M^{\blacktriangleright\blacktriangleleft}$.

$M^{\blacktriangleright\blacktriangleleft} \subseteq M^{\triangleright\triangleleft}$. Assume $w \in M^{\blacktriangleright\blacktriangleleft}$, and take an arbitrary $(x,y) \in M^{\triangleright}$. Put $X = (My)^{\leftarrow}$, $Y = (XM)^{\rightarrow}$. It is easy to see that 1. $x \in X, y \in Y$ (obvious), and 2. $((X, X^{\rightarrow}), (Y^{\leftarrow}, Y)) \in M^{\blacktriangleright}$ (by Lemma 17). Since $w \in M^{\blacktriangleright\blacktriangleleft}$, we have $M^{\blacktriangleright} \subseteq \phi(w)$, and so $((X, X^{\rightarrow}), (Y^{\leftarrow}, Y)) \in \phi(w)$, which holds iff $XwY \subseteq L$, entailing $xwy \in L$. Hence $w \in M^{\triangleright\triangleleft}$ $\qquad\square$

This already entails that the operations and constants in the respective lattices yield the same result, because they are based on the same underlying set-operations, of which we simply take the (same) closure. We denote operations in $SCL(L)$ as usual; the operation in $ACL(U(L))$ corresponding to $\star$ in $SCL(L)$ will be denoted by $\star'$. We distinguish the constants of different structures by subscripts $\top_{SCL(L)}$ etc. As concepts are tuples, we write, for tuples $(X_1, X_2), (Y_1, Y_2)$, $(X_1, X_2) =_1 (Y_1, Y_2)$ iff $X_1 = Y_1$, that is, if their first components are identical.

Corollary 20. *1. For $\star \in \{\wedge, \vee, \circ, /, \backslash\}$, $\star$ defined w.r.t. $SCL(L)$, $\star'$ defined w.r.t. $ACL(U(L))$, $(M, M^{\triangleright}) \star (N, N^{\triangleright}) =_1 (M, M^{\blacktriangleright}) \star' (N, N^{\blacktriangleright})$.*

2. $\top_{SCL(L)} =_1 \top_{ACL(U(L))}$
3. $\bot_{SCL(L)} =_1 \bot_{ACL(U(L))}$
4. $1_{SCL(L)} =_1 1_{ACL(U(L))}$

Now it is easy to construct an isomorphism $i : SCL(L) \rightarrow ACL(U(L))$: for every $(M, C) \in \mathcal{B}_L$, we put $i(M, C) = (M, M^{\blacktriangleright})$. This completes the proof of Theorem 16. The isomorphism theorem thus establishes a surprisingly strong connection between the syntactic concept lattice and the universal automaton of a language.

We now consider the consequences of the isomorphism theorem for our investigations into the semantics of substructural logics. Automata-theoretic semantics is richer than simple language-theoretic semantics, because there is a many-one relationship of recognition between automata and languages. In order to ensure soundness, we already have to restrict interpretations to nuclear automata; then it follows from more general results. To obtain completeness, the isomorphism theorem can be applied in a straightforward fashion: just compose the SCL-interpretation of $\mathbf{FL}_\perp$ (or its fragments) with the isomorphism from $SCL(L)$ into $ACL(U(L))$, and we are done.

Theorem 21. *(Completeness of automata-theoretic semantics)*

1. $ACL \models \Gamma \vdash \alpha$ iff $\Vdash_{\mathbf{FL}_\perp} \Gamma \vdash \alpha$
2. $ACL_{FL} \models \Gamma \vdash \alpha$ iff $\Vdash_{\mathbf{FL}} \Gamma \vdash \alpha$
3. $ACL_{L1} \models \Gamma \vdash \alpha$ iff $\Vdash_{\mathbf{L1}} \Gamma \vdash \alpha$

It is obvious how to further strengthen these results: let $ACL(FIN)$ denote the class of automatic concept lattices over finite nuclear automata (i.e. nuclear automata with finite state set).[4] We can depart from completeness for $SCL(REG)$: for $\not\Vdash_{\mathbf{FL}_\perp} \Gamma \vdash \alpha$ we find a countermodel $SCL(L)$ where $L \in REG$. By the isomorphism theorem, we also have a countermodel $ACL(U(L))$ which is finite. Thus we have the following:

Theorem 22. *(Completeness for finite automata)*

1. $ACL(FIN) \models \Gamma \vdash \alpha$ iff $\Vdash_{\mathbf{FL}_\perp} \Gamma \vdash \alpha$
2. $ACL_{\mathbf{FL}}(FIN) \models \Gamma \vdash \alpha$ iff $\Vdash_{\mathbf{FL}} \Gamma \vdash \alpha$
3. $ACL_{\mathbf{L1}}(FIN) \models \Gamma \vdash \alpha$ iff $\Vdash_{\mathbf{L1}} \Gamma \vdash \alpha$

5   Conclusion

We have presented a new complete semantics for the full Lambek calculus and its various fragments, the so-called automata-theoretic semantics. It is based on an automata-theoretic construction we introduced, the automatic concept lattice. What is peculiar to this semantics is that it is both language-theoretic and relational, and thus brings together two prominent types of semantics for substructural logics. Our results are based on the construction of Galois connections, closure operators and nuclei: these allow us to give rather simple proofs for completeness. This illustrates (once more) the usefulness of Galois connections in the context of substructural logics and formal language theory. Another important result concerns finiteness of models, which corresponds to regularity of languages. We showed that our completeness results can be extended to this case.

As an outlook, we hope that we can use the results established in this paper to strengthen some of the canonical completeness results regarding L-models and relational models to regular languages and/or finite relations.

Acknowledgements. I would like to thank the anonymous reviewers – one in particular – for their extremely helpful comments.

References

1. Brown, C., Gurr, D.: Relations and non-commutative linear logic. J. Pure Appl. Algebra **105**(2), 117–136 (1995)
2. Buszkowski, W.: Completeness results for Lambek syntactic calculus. Math. Logic Q. **32**(1–5), 13–28 (1986)
3. Buszkowski, W.: Algebraic structures in categorial grammar. Theor. Comput. Sci. **1998**(1–2), 5–24 (1998)
4. Clark, A.: A learnable representation for syntax using residuated lattices. In: Groote, P., Egg, M., Kallmeyer, L. (eds.) FG 2009. LNCS (LNAI), vol. 5591, pp. 183–198. Springer, Heidelberg (2011). doi:10.1007/978-3-642-20169-1_12

---

[4] This class is strictly smaller than the class of automata recognizing regular languages, as obviously there are infinite automata recognizing regular languages.

5. Clark, A.: Learning context free grammars with the syntactic concept lattice. In: Sempere, J.M., García, P. (eds.) ICGI 2010. LNCS (LNAI), vol. 6339, pp. 38–51. Springer, Heidelberg (2010). doi:10.1007/978-3-642-15488-1_5

6. Clark, A.: The syntactic concept lattice: another algebraic theory of the context-free languages? J. Log. Comput. **25**(5), 1203–1229 (2015)

7. Davey, B.A., Priestley, H.A.: Introduction to Lattices and Order, 2nd edn. Cambridge University Press, Cambridge (1991)

8. Farulewski, M.: On finite models of the Lambek calculus. Stud. Logica. **80**(1), 63–74 (2005)

9. Galatos, N., Jipsen, P., Kowalski, T., Ono, H.: Residuated Lattices: An Algebraic Glimpse at Substructural Logics. Elsevier, Amsterdam (2007)

10. Kanazawa, M.: The Lambek calculus enriched with additional connectives. J. Logic Lang. Inf. **1**, 141–171 (1992)

11. Lambek, J.: The mathematics of sentence structure. Am. Math. Mon. **65**, 154–169 (1958)

12. Lambek, J.: On the calculus of syntactic types. In: Jakobson, R. (ed.) Structure of Language and its Mathematical Aspects, pp. 166–178. Providence (1961)

13. Leiß, H.: Learning context free grammars with the finite context property: a correction of A. Clark's algorithm. In: Morrill, G., Muskens, R., Osswald, R., Richter, F. (eds.) Formal Grammar 2014. LNCS, vol. 8612, pp. 121–137. Springer, Heidelberg (2014). doi:10.1007/978-3-662-44121-3_8

14. Lombardy, S., Sakarovitch, J.: The universal automaton. In: Flum, J., Grädel, E., Wilke, T. (eds.) Logic, Automata: History and Perspectives [in Honor of Wolfgang Thomas]. Texts in Logic and Games, vol. 2, pp. 457–504. Amsterdam University Press (2008)

15. Okada, M., Terui, K.: The finite model property for various fragments of intuitionistic linear logic. J. Symb. Logic **64**(2), 790–802 (1999)

16. Pentus, M.: Models for the Lambek calculus. Ann. Pure Appl. Logic **75**, 179–213 (1995)

17. Sestier, A.: Contributions à une théorie ensembliste des classifications linguistiques. (Contributions to a set-theoretical theory of classifications). In: Actes du Ier Congrès de l'AFCAL, Grenoble, pp. 293–305 (1960)

18. van Benthem, J.: Language in Action: Categories, Lambdas and Dynamic Logic. Studies in Logic and the Foundations of Mathematics, vol. 130. North-Holland, Amsterdam (1991)

19. Wurm, C.: Completeness of full Lambek calculus for syntactic concept lattices. In: Morrill, G., Nederhof, M.-J. (eds.) FG 2012-2013. LNCS, vol. 8036, pp. 126–141. Springer, Heidelberg (2013). doi:10.1007/978-3-642-39998-5_8

20. Wurm, C.: On some extensions of syntactic concept lattices: completeness and finiteness results. In: Foret, A., Morrill, G., Muskens, R., Osswald, R., Pogodalla, S. (eds.) FG 2015-2016. LNCS, vol. 9804, pp. 164–179. Springer, Heidelberg (2016). doi:10.1007/978-3-662-53042-9_10

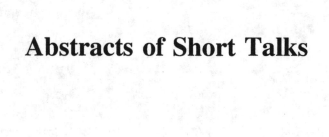

Abstracts of Short Talks

Graded Hyponymy for Compositional Distributional Semantics

Dea Bankova, Bob Coecke, Martha Lewis[✉], and Dan Marsden

University of Oxford, Oxford, UK
{coecke,marlew,daniel.marsden}@cs.ox.ac.uk

1 Introduction

The categorical compositional distributional model of natural language provides a conceptually motivated procedure to compute the meaning of a sentence, given its grammatical structure and the meanings of its words. This approach has outperformed other models in mainstream empirical language processing tasks, but needs further development towards the crucial feature of lexical entailment.

We address this challenge by exploiting the freedom in our abstract categorical framework to change our choice of semantic model. This allows us to describe hyponymy as a graded order on meanings, using models of partial information used in quantum computation. Quantum logic embeds in this graded order.

2 Results

We describe how the semantics of the categorical compositional distributional model can be lifted from vector spaces to density matrices, providing a richer environment for meaning representation.

We introduce a general setting for approximate entailment that can be built on any commutative monoid. When applied to the category of positive operators and completely positive maps, this framework generates a novel robust graded order that captures the hyponymy strength between concepts.

A procedure is given for determining the hyponymy strength between *any* pair of phrases of the same overall grammatical type. The pair of phrases can have differing lengths and even include words that are not upwardly monotonic. For example, we can determine the extent to which 'John's joyful cousin' is a hyponym of 'unhappy chaps'. This is possible because within categorical compositional semantics, phrases of each type are reduced to one common space according to their type, and can be compared within that space.

Finally, we show that, in the case of positive sentences, hyponymy strength lifts *compositionally* to the phrase level, giving a lower bound on phrase hyponymy. This means that in certain contexts, we can predict the hyponymy strength between phrases based on the hyponymy strengths of the words within each phrase. We provide numerous examples for different sentences and noun phrases, and indicate of how these can be applied within an NLP setting.

This abstracts the paper 'Graded Entailment for Compositional Distributional Semantics', available at: http://arxiv.org/abs/1601.04908.

© Springer-Verlag GmbH Germany 2016
M. Amblard et al. (Eds.): LACL 2016, LNCS 10054, p. 327, 2016.
DOI: 10.1007/978-3-662-53826-5

Minimization of Finite State Automata Through Partition Aggregation

Johanna Björklund[1](✉) and Loek Cleophas[1,2]

[1] Department of Computing Science, Umeå University, Umeå, Sweden
johanna@cs.umu.se
[2] Department of Information Science, Stellenbosch University,
Stellenbosch, South Africa
loek@fastar.org

We present a minimization algorithm for finite state automata that finds and merges bisimulation-equivalent states, identified through partition aggregation. In terms of applicability, the algorithm is a generalisation of an earlier one by Watson and Daciuk for deterministic devices. We show the algorithm to be correct and run in quadratic time in the number of states and the maximal out-degree of the transition graph, and in linear time in the size of the input alphabet. The algorithm is slower than those based on partition refinement, but has the advantage that intermediate solutions are also language equivalent to the input automaton M. Furthermore, the algorithm essentially searches for the maximal model of a characteristic formula for M, so many of the optimisation techniques used to gain efficiency in SAT solvers are likely to apply.

© Springer-Verlag GmbH Germany 2016
M. Amblard et al. (Eds.): LACL 2016, LNCS 10054, p. 328, 2016.
DOI: 10.1007/978-3-662-53826-5

Inferring Necessary Categories in CCG

Jacob Collard[✉]

Cornell University, Ithaca, NY, USA
jacob@thorsonlinguistics.com

It is possible to infer the category of unknown lexical items in categorial grammars given a two-word sentence in which one word is known. However, inferring the possible categories of words in longer sentence is a non-trivial problem that is much more difficult to solve in a general, cross-linguistic manner. I propose a probabilistic method which uses the standard combinatory rules for combinatory categorial grammars (CCGs) to learn sentences in a target language by assigning probabilities to possible categories for a given word based on the likelihood that the given category will be necessary to provide a complete grammar of the language. This algorithm is semi-supervised; it requires a seed lexicon in order to begin learning, and can then infer the categories of new words in a corpus of unannotated sentences. At no point does the algorithm require annotations such as proof nets or derivations; instead it learns primarily from strings, making it relatively naturalistic. The algorithm is sensitive to the order of sentences in the corpus, preferring shorter sentences early on before it has built up a large enough base lexicon to learn longer sentences.

As an unsupervised algorithm, the learner that I present is quite difficult to test, as it does not necessarily learn the same analysis that is presented in the test set if more than one analysis is possible. It can, however, be shown to derive reasonable analyses for small, well-understood datasets.

Because the algorithm uses generalizable algebraic structures to define its inference rules, it may be possible to extend it to other formalisms, so long as certain properties are maintained.

© Springer-Verlag GmbH Germany 2016
M. Amblard et al. (Eds.): LACL 2016, LNCS 10054, p. 329, 2016.
DOI: 10.1007/978-3-662-53826-5

Sitting and *Waiting*
An Idle Meaning of an English Posture Verb

Katherine Fraser(✉)

Universität Stuttgart, Stuttgart, Germany
fraserk4@gmail.com

Posture verbs describe spatial configuration, prototypically of humans in "at-rest" states [3]. However, it has been observed that the semantic network of these verbs is more diverse, and that the literal uses are less common than the metaphorical extensions [2]. In English, the semantic coverage of one such verb, *sitting*, includes a contingent state lacking a posture entailment and where a negative judgement of this state is particularly salient (1). In this short talk, I will provide both empirical evidence of the distribution of *sitting*'s expressive use, as well as a proposed compositional model, utilising Gutzmann's [1] use-conditional framework for the analysis. This investigation gives insight not only on posture verbs in general, but also on how multi-dimensional meaning can be formally addressed.

(1)   {A secretary had prepared a contract to hire a new employee.}
      a.   The contract **was sitting** on the CEO's desk, ready to be signed.
      b.   ?The contract **sat** on the CEO's desk, ready to be signed.
(2)   #The contract was sitting on the CEO's desk but he shouldn't sign it.

The contract in (1) is inanimate, not in a relevant posture, and judged to be in an unused state. Interestingly, the most salient part is the negative evaluation: this idle state of the object is undesired; when the unwanted aspect is negated, the sentence becomes infelicitous (2). The additional layer of meaning disappears in (1-b), where the progressive is exchanged for the simple past. The oddness of (1-b) improves with the addition of a secondary predicate like *for four days*, which eludicates the bounded interval of the idle state.

In this talk, I will describe how sitting's expressive meaning can be formalised with use-conditional theory [1]. The analysis proposed makes use of an bouletic mood operator which maps propositions onto emotional predicates, and which takes the entire descriptive proposition as its argument. As this function includes a set of use-conditional bouletic evaluator predicates, it enables the integration of a negative evaluation into the semantics, an essential aspect of the expressive meaning. By changing the modal operator, e.g., to be epistemic, the analysis can be extended to other posture verbs' expressive meaning.

References

1. Gutzmann, D.: Use-Conditional Meaning. Studies in Multidimensional Semantics. Oxford Studies in Semantics and Pragmatics, vol. 6. Oxford University Press, Oxford (2015)

© Springer-Verlag GmbH Germany 2016
M. Amblard et al. (Eds.): LACL 2016, LNCS 10054, pp. 330–331, 2016.
DOI: 10.1007/978-3-662-53826-5

2. Lemmens, M.: The semantic network of dutch posture verbs. In: Newman, J. (ed.) The Linguistics of Sitting, Standing, and Lying, pp. 103–139. John Benjamins Publishing Company (2002)
3. Newman, J.: A cross-linguistic overview of the posture verbs 'sit', 'stand' and 'lie'. In: Newman, J. (ed.) The Linguistics of Sitting, Standing, and Lying, pp. 1–24. John Benjamins Publishing Company (2002)

Types and Meaning of Relative Pronouns in Tupled Pregroup Grammars

Aleksandra Kiślak-Malinowska[✉]

University of Warmia and Mazury, Olsztyn, Poland
akis@uwm.edu.pl

Pregroup grammars were introduced by Lambek [3] as an algebraic tool for syntactic analysis of natural languages. The main focus in our study is placed on an extension of pregroup grammars—tupled pregroup grammars (TPG), proposed by Stabler [5]—and its application for a widely considered grammar issue: relative pronouns in English. We discuss the former approaches to that phenomenon proposed by several authors and compare them. Additionally, we consider expressions with relative pronouns not only from syntactical point of view (checking whether they are well formed according to grammar rules) but also go further and explore their deeper structure, creating automatically some logical forms, representing their meaning.

References

1. Buszkowski, W.: Lambek grammars based on pregroups. In: de Groote, P., Morrill, G., Retoré, C. (eds.) LACL 2001. LNCS, vol. 2099, pp. 95–109. Springer, Heidelberg (2001)
2. Kiślak-Malinowska, A.: Extended pregroup grammars applied to natural languages. Logic Log. Philos. **21**, 229–252 (2012)
3. Lambek, J.: Type grammar revisited. In: Lecomte, A., Lamarche, F., Perrier, G. (eds.) LACL 1997. LNCS, vol. 1582, pp. 1–27. Springer, Heidelberg (1999)
4. Lambek J.: From word to sentence, Polimetrica (2008). ISBN 978-88-7699-117-2
5. Stabler E.: Tupled pregroup grammars. In: Lambek, J., Casadio, C. (eds.) Computational Algebraic Approaches to Natural Language, pp. 23–52, Polimetrica (2008). ISBN 978-88-7699-125-7

© Springer-Verlag GmbH Germany 2016
M. Amblard et al. (Eds.): LACL 2016, LNCS 10054, p. 332, 2016.
DOI: 10.1007/978-3-662-53826-5

Dependent Event Types

Zhaohui Luo[1(✉)] and Sergei Soloviev[2]

[1] Royal Holloway, University of London, London, UK
zhaohui.luo@hotmail.co.uk
[2] IRIT, Toulouse, France
Sergei.Soloviev@irit.fr

Employing dependent types for a refined treatment of event types provides a nice improvement to Davidson's event semantics [3, 7]. We consider dependent event types indexed by thematic roles and show that subtyping between them plays an essential role in semantic interpretations. It is also shown that dependent event types give a natural solution to the event quantification problem in combining event semantics with the Montague semantics [1, 4, 8].

For instance, $Evt_A(a)$ is the dependent type of events whose agents are a : $Agent$. The dependent event types abide by subtyping relationships:

$$Evt_{AP}(a,p) \leq Evt_A(a) \leq Event \text{ and } Evt_{AP}(a,p) \leq Evt_P(p) \leq Event,$$

where $a : Agent, p : Patient$ and $Event$ is the type of all events. With such dependent event types, subtyping is crucial. Consider **John talked loudly**: its Davidsonian event semantics would be $\exists e : Event.\, talk(e) \,\&\, loud(e) \,\&\, agent(e,j)$, where $talk, loud : Event \rightarrow \mathbf{t}$. With dependent event types, the semantics would be $\exists e : Evt_A(j).\, talk(e) \,\&\, loud(e)$, in which the terms such as $talk(e)$ are only well-typed because $Evt_A(j) \leq Event$.

It has been argued that there is some incompatibility between (neo-)Davidsonian event semantics and the traditional compositional semantics, as the event quantification problem shows: the following two possible interpretations of **No dog barks** are both well-formed formulas, although (2) is incorrect:

(1) $\neg\exists x : \mathbf{e}.\, dog(x) \,\&\, \exists e : Event.\, bark(e) \,\&\, agent(e,x)$
(2) (#) $\exists e : Event.\, \neg\exists x : \mathbf{e}.\, dog(x) \,\&\, bark(e) \,\&\, agent(e,x)$

To exclude such incorrect interpretations, various informal solutions have been proposed [1, 8]. With dependent event types, this problem is solved naturally and formally—the incorrect semantic interpretations such as (4) below are excluded because they are ill-typed, while the correct one (3) is well-typed.

(3) $\neg\exists x : \mathbf{e}.\, (dog(x) \,\&\, \exists e : Evt_A(x).\, bark(e))$
(4) (#) $\exists e : Evt_A(x).\, \neg\exists x : \mathbf{e}.\, dog(x) \,\&\, bark(e)$

The underlying formal system C_e is the extension of Church's simple type theory [2], as used in the Montague semantics, with dependent event types and the subtyping relations. C_e can be faithfully embedded into UTT[C], i.e., the type theory UTT [5] extended with coercive subtyping in C [6], where C contains the

© Springer-Verlag GmbH Germany 2016
M. Amblard et al. (Eds.): LACL 2016, LNCS 10054, pp. 333–334, 2016.
DOI: 10.1007/978-3-662-53826-5

subtyping judgements that correspond to the above subtyping relations between dependent event types. Since UTT[C] has nice meta-theoretic properties such as normalisation and logical consistency, so does C_e.

The paper is available online at http://www.cs.rhul.ac.uk/home/zhaohui/DET.pdf.

References

1. Champollion, L.: The interaction of compositional semantics and event semantics. Linguist. Philos. **38**, 31–66 (2015)
2. Church, A.: A formulation of the simple theory of types. J. Symb. Log. **5**(1) (1940)
3. Davidson, D.: The logical form of action sentences. In: Rothstein, S. (ed.). The Logic of Decision and Action. University of Pittsburgh Press (1967)
4. de Groote, P., Winter, Y.: A type-logical account of quantification in event semantics. Logic Eng. Nat. Lang. Semant. **11** (2014)
5. Luo, Z.: Computation and Reasoning: A Type Theory for Computer Science. Oxford University Press (1994)
6. Luo, Z., Soloviev, S., Xue, T.: Coercive subtyping: theory and implementation. Inform. Comput. **223**, 18–42 (2012)
7. Parsons, T.: Events in the Semantics of English. MIT Press (1990)
8. Winter, Y., Zwarts, J.: Event Semantics and Abstract Categorial Grammar. In: Kanazawa, M., Kornai, A., Kracht, M., Seki, H. (eds.) MOL 12. LNCS, pp. 174–191. Springer, Heidelberg (2011)

Author Index

Printed in the United States
by Baker & Taylor Publisher Services

Printed in the United States
by Baker & Taylor Publisher Services